Edouard Kurstak (ed.)

Measles and Poliomyelitis

Vaccines, Immunization, and Control

Springer-Verlag Wien New York

Prof. Dr. Edouard Kurstak
Faculty of Medicine, University of Montreal, Montreal, Canada
International Comparative Virology Organization, Montreal, Canada
Medical Research Centre, King Fahad Hospital, Riyadh, Saudi Arabia
Viral Diseases Panel, World Health Organization, Geneva, Switzerland

Typesetting: Thomson Press (India) Limited, New Delhi 110 001

Printed on acid-free and chlorine-free bleached paper

With 48 Figures

ISBN-13: 978-3-211-82436-8   e-ISBN-13: 978-3-7091-9278-8
DOI: 10.1007/ 978-3-7091-9278-8

Preface

The World Health Organization's important goals are the eradication of poliomyelitis by the year 2000 and elimination of measles in 10–15 years. Despite of substantial achievements during the last decade, in global control of these two highly infectious viral diseases, an urgent and aggressive mobilization of all resources – human, technological, and financial – are needed to achieve the target. In this regard the involvement of UNESCO, Canadian Public Health Association, and the Kingdom of Saudi Arabia to promote the international immunization programmes, is an example to be followed.

Vaccines' potency and immunization programmes' efficacy, the basic elimination and eradication strategies to assure very high immunity levels and safe vaccination of populations should be discussed without restrictions, by all concerned specialists from public and private sectors as well as by experts from medical research institutions, international health organizations and medical professional associations. Only through such an approach, highly professional conclusions could be elaborated and valuable actions implemented.

The recent progress in vaccines development, establishment of new immunization programmes, molecular epidemiology of these two viral infections and their diagnosis, antigenic variations knowledge, better understanding of the post-vaccination role of individual antigens and immune responses, fully justify the publication of this very timely volume on measles and poliomyelitis.

The volume reflects the independent opinions of very well recognized experts from leading universities, medical schools, hospitals, research institutes, international health organizations, health departments, centres for disease control, and of vaccine production specialists. In thirty-one well documented chapters, the invited contributors present the latest state-of-art achievements in their field of expertise, covering practically, all subjects related to measles and poliomyelitis viruses, vaccines, and immunization. Among others, the epidemiology of infections, molecular genetic surveillance, vaccines' potency, safety, immune responses, and immunization strategies to control, and ultimately to eradicate, these deadly diseases, are discussed in depth in this volume.

It is my hope that this timely volume will benefit all professionals involved in measles and poliomyelitis control to reach their global eradication. Especially, to those responsible for the design and implementation of imunization strategies and programmes at national and international levels of health administrations, to medical practitioners, infectious diseases specialists, virologists, immunologists, and medical researchers in charge of the development of safe and highly immunogenic vaccines.

Special thanks and high appreciation are addressed to all contributors who kindly accepted to work with me and to the staff of Springer–Verlag for their part in the publication of this volume.

Prof. Dr. Edouard Kurstak

President – International Comparative Virology Organization
Member – Viral Diseases Panel, World Health Organization

Contents

Contributors

N. Ajjan, M.D., Pasteur Mérieux Serums and Vaccines, Marnes-La-Coquette, France

J. W. Almond, Ph.D., University of Reading, Whitenights, U.K.

I: Arita, M.D., Agency for Cooperation with International Health, Kunmamoto City, Japan

A. L. Atkinson, M.D., Division of Immunization, National Center for Prevention Services, Centers for Disease Control and Prevention, Atlanta, GA, U.S.A.

A. Aubert-Combiescu, M.D., Cantacuzino Institute, Bucharest, Romania

H. van der Avoort, Ph.D., Laboratory of Virology, National Institute of Public Health and Environmental Protection, Bilthoven, The Netherlands

J. Balanant M.D., Unité de Virologie Médicale, Institut Pasteur, Paris, France

J. P. Beau, Ph.D., ORSTOM, UR Population et Stanté, Dakar, Senegal

A. Candrea, Ph.D., Unité de Virologie Médicale, Institut Pasteur, Paris, France

B. Christenson, M.D., Department of Environmental Health and Infectious Diseases Control, Karolinska Hospital, Stockholm, Sweden

C. J. Clements, M.D., Expanded Immunization Programme, World Health Organization, Geneva, Switzerland

S. L. Cochi, M.D., Division of Immunization, Centers for Disease Control, and Prevention Atlanta, GA, U.S.A

M. Combiescu, M.D., Cantacuzino Instituto, Bucharest, Romania

R. Crainic, M.D., Ph.D., Unité de Virologie Médicale, Institut Pasteur, Paris, France

F. T. Cutts, M.D., The London School of Tropical Medicine and Hygiene, London, U.K.

G. Dunn, M.D., National Institute for Biological Standards and Control, Herts, U.K.

H. Faden, M.D., Ph.D., Department of Pediatrics, State University of New York, School of Medicine at Buffalo, The Children's Hospital of Buffalo, Buffalo, NY, U.S.A.

P. E. M. Fine, M.D., Department of Epidemiology and Population Sciences, London School of Hygiene and Tropical Medicine, London, U.K.

M. Furione, M.D., Unité de Virologie Médicale, Institut Pasteur, Paris, France

M. Garenne, M.D., Harvard School of Public Health, Boston, MA, U.S.A.

Y. Ghendon, Ph.D., World Health Organization, Geneva, Switzerland

G. M. Ginsberg, M.D., Department of Information and Personal and Community Preventive Health Services, Ministry of Health, Jerusalem, Israel

S. Guillot, Ph.D., Unité de Virologie Médicale, Institut Pasteur, Paris, France

A. R. Hinman, M.D., National Center for Prevention Services, Centers for Disease Control and Prevention, Atlanta, GA, U.S.A.

T. Hovi, M.D., Enterovirus Laboratory, National Public Health Institute, Helsinki, Finland

M. Kohara, Ph.D., Department of Microbiology, The Tokyo Metropolitan Institute of Science, Tokyo, Japan

F. M. LaForce, M.D., Department of Medicine, The Genesee Hospital, University of Rochester School of Medicine and Dentistry, Rochester, NY, U.S.A.

O. Leroy, M.D., Pasteur Mérieux Serums and Vaccines, Marnes-La-Coquette, France

R. W. Linkins, M.D., Division of Immunization, Centers for Disease Control and Prevention, Atlanta, GA, U.S.A

A. M. Van Loon, M.D., Laboratory of Virology, National Institute of Public Health and Environmental Protection, Bilthoven, The Netherlands

A. Macadam, Ph.D., University of Reading, Whitenights, U.K.

L. E. Markowitz, M.D., Division of Immunization, National Center for Prevention Services, Centers for Disease Control and Prevention, Atlanta, GA, U.S.A.

J. L. Melnick, M.D., Ph.D., Division of Molecular Virology, Baylor College of Medicine, Houston, TX, U.S.A.

J. Milstein, Ph.D., Biological Unit, World Health Organization, Geneva, Switzerland

P. Minor, M.D., National Institute for Biological Standards and Control, Herts, U.K.

M. Mulders, Ph.D., Laboratory of Virology, National Institute of Public Health and Environmental Protection, Bilthoven, The Netherlands

A. Nomoto, M.D., Department of Microbiology, The Institute of Medical Science, The University of Tokyo, Tokyo, Japan

W. A. Orenstein, M.D., Ph.D., Division of Immunization, Centers for Disease Control and Prevention, Atlanta, GA, U.S.A.

D. Otelea, Ph.D., Unité de Virologie Médicale, Institut Pasteur, Paris, France

P. A. Patriarca, M.D., Division of Immunization, National Center for Prevention Services, Centers for Disease Control and Prevention, Atlanta, GA, U.S.A.

L. Piirainen, M.D., Enterovirus Laboratory, National Public Health Institute, Helsinki, Finland

P. Poelstra, M.D., Laboratory of Virology, National Institute of Public Health and Environmental Protection, Bilthoven, The Netherlands

A. Ras, Ph.D., Laboratory of Virology, National Institute of Public Health and Environmental Protection, Bilthoven, The Netherlands

B. K. Rima, Ph.D., Division of Genetic Engineering, School of Biology and Biochemistry, The Queen's University of Belfast, Belfast, Northern Ireland

M. Roivainen, M.D., Enterovirus Laboratory, National Public Health Institute, Helsinki, Finland

P. Saliou, M.D., Pasteur Mérieux Serums and Vaccines, Marnes-La-Coquette, France

A. A. Salmi, M.D., Ph.D., Department of Virology, University of Turku, Turku, Finland

B. D. Schoub, M.D., Ph.D., National Institute for Virology and Department of Virology, University of the Withwatersrand, Johannesburg, South Africa

I. Sene, M.D., Centre Médical de Fatick, Fatick, Senegal

M. Strassburg, M.D., Data Collection and Analysis, Los Angeles County Department of Health Services, Los Angeles, CA, U.S.A.

R. W. Sutter, M.D., Ph.D., Division of Immunization, National Center for Prevention Services, Centers for Disease Control and Prevention, Atlanta, GA, U.S.A.

J. Thipphawong, M.D., Clinical and Medical Affairs Department, Connaught Laboratories, Willowdale, Ontario, Canada

C. Torel, Ph.D., Expanded Programme on Immunization, World Health Organization. Geneva, Switzerland

T. H. Tulchinsky, M.D., Personal and Community Preventive Health Services, Ministry of Health, Jerusalem, Israel

O. Van-Ham, D.Sc., Connaught Laboratories Limited, Wollowdale, Ontario, Canada

N. A. Ward, M.D., Expanded Programme on Immunization, World Health Organization. Geneva, Switzerland

T. F. Wild, M.D., Ph.D., Unité d'Immunologie et Stratégie Vaccinale, Institut Pasteur, Lyon, France

R. Wittes, M.D., Ph.D., Clinical and Medical Affairs Department, Connaught Laboratories Limited, Willowdale, Ontario, Canada

B. Yang, M.D., Department of Epidemic Prevention, Ministry of Public Health, Beijing, China

Editor's introduction

Measles and poliomyelitis: vaccines, immunization, and control

Introduction

Measles and poliomyelitis happen to be two well known viral diseases which inconceivably remain yet to be fully eradicated worldwide. Similarities exist in their both, being preventable with highly effective and safe vaccines, and in causing acute self limiting infections exclusively in humans. Despite their successful control in the industrialized countries of the world, their threat as two major diseases of the developing world still remains. Immunization programmes proved to be perhaps more a successful experience with poliomyelitis than measles in controlling and eliminating infection.

Measles elimination – vaccines and immunization problems

Three decades ago, measles was envisioned as an inevitable childhood disease, one of the greatest killers of children in the history of the world, but today in industrialized countries, it is rather a sporadic viral infection prevented by effective vaccination. However, this highly infectious disease with potentially fatal complications, remain an important killer of infants and children in developing world. Measles virus produces explosive outbreaks in non-immune populations.

Clinically, the disease is charaterized differently in patients with typical measles than in those with modified or atypical measles. Modified measles occurs in partially immune individuals, whereas atypical measles is known as a clinical syndrome following exposure to natural measles in some subjects previously immunized with inactivated vaccine. Modified measles, on the other hand, has been characterized as a mild illness, with essentially the sequence of events identical to typical measles. The mild characteristic condition is attributable to the presence of specific measles antibodies in the affected child. A less common modified measles, usually a manifestation of vaccine failure, may also be encountered. In such cases, the prodromal period is generally shortened with minimal cough, coryza and low-grade fever. Koplik spots are usually known to be absent and few in number and transient when they do occur. The exanathem has been noted to be consistent in distribution to typical measles but with rare confluence.

Measles is a serious disease which has been noted to occur worldwide. The transmission occurs by direct contact with infectious droplets or in limited cases by airborne spread. In the pre-vaccine era, measles was experienced as an epidemic disease with biennial cycles in urban areas occurring mostly in pre-school and young school-age children. Epidemic measles is a winter and spring disease with no noticeable gender difference. It is spread more commonly under crowed situations as in day-care centers, with higher morbidity and mortality in immunocompromised hosts, children under 2 years of age, and malnourished children.

Because of the morbidity and mortality associated with measles, its control is a high public health priority for all countries [1]. Despite improved vaccine coverage in recent years the transmission of measles virus goes unabated claiming the lives of about 1 million children in a year, almost all of which die in developing countries. Before the introduction by the World Health Organization of Expanded Programme on Immunization (EPI), 130 million measles cases and 13 to 16 million deaths annually were estimated. Coverage with measles vaccine has risen rapidly in the last eight years so that roughly about 80% of the world's children are now receiving one dose of the vaccine [2]. The twenty per cent unreached would need special efforts to be protected. In fact, the Measles Immunization Programme has prevented an estimated 2.1 million deaths in 1990.

Newer approaches need to be devised to ensure that groups at high risk receive the vaccine. One of the high risks is children in poor urban environment, who live under ideal conditions for transmission of the disease and in general, severely underprevileged in terms of vaccination services. Idea of concentrating efforts on those, most in need, is relatively new for measles control. Such a development would undoubtedly maximize chance of interrupting transmission of virus with scarce resources available. In addition to urban poor, those at high risk to contract measles include children in hospitals, those visiting health facilities, those living in low vaccination coverage areas, and those in refugee camps. Schlenker et al [3] have recently examined the association between incidence of measles and immunization coverage among pre-school children. These authors conclude that modest improvements in low levels of immunization coverage among 2 years old confer substantial protection against measles outbreaks. In fact, coverage of 80% or less may turn out to be sufficient so as to prevent sustained measles outbreaks in an urban community.

Standard dose attenuated vaccines in usage worldwide since the past twenty years rank alongside the safest vaccines. There is some evidence that measles vaccines may in rare cases be associated with encephalitis or encephalopathy and conditions as subacute sclerosing panencephalitis (SSPE).

Risk of SSPE associated with conventional attenuated measles vaccine is at most a tenth associated with wild type infection, in the order of 0.7 per million doses [4].

Age at vaccination has been shown to be a major determinant of vaccine efficacy. Children vaccinated too early do not seroconvert and as such do not appear to be protected by the vaccine. While it is commonly expected that as immunization coverage increases, the age of patients with measles will rise, and at the same time the risk of exposure before age 9 months will diminish considerably, but these predictions unfortunately have not always materialized [5].

Aaby et al [6] have carried out a prospective study in two districts in Guinea Bissau, Africa, where vaccine coverage for children was 81% and 61%. Their data are suggestive of the necessity to vaccinate children before the age of 9 months to control measles in hyperendemic urban African areas. This situation is in sharp contrast to that in most western countries where the recommended age at vaccination with standard vaccine ranges from 12 to 15 months.

The recommended titer of measles vaccine is about 1000, 50% tissue-culture infective dose ($TCID_{50}$) as defined by the World Health Organization (WHO) and standard vaccines often contain 5000 to 10 000 $TCID_{50}$. Studies that shown that increasing the titer of the vaccines grown on human diploid cells by 100 fold increased the immunogenicity and that the Edmonston-Zagreb (EZ) strain produced a higher rate of seroconversion than the Schwarz strain [7]. These results prompted the suggestion of children receiving measles vaccination at an early age (4–6 months). The use of the EZ-high titer vaccines at 6 months of age, in developing countries, when measles poses a significant mortality factor, was recommended [8]. Questions, however, remain on the safety of the high titer EZ measles vaccine, especially when used before 9 months of age. Due to recent data associating the high titer measles vaccines with increased long term mortality, the WHO/EPI recommendation to give such vaccine to infants before 9 months of age has been suspended recently, and additional study recommended. Garenne et al [9] have evaluated the immunogenicity and clinical efficacy of high titer measles vaccines in a randomized controlled trial in Senegal. Two high titer vaccines were studied – a higher titer Edmonston-Zagreb vaccine and a high titer Schwarz vaccine. Both high titer vaccines were administered at 10 months of age and compared to a Schwarz standard vaccine administered at 10 months of age. The results of this study clearly indicate that vaccination with Schwarz measles vaccine at 9 months of age remains as yet the safest and the most effective strategy for controlling measles.

The strategy of one dose of measles for every child has been known to be successful in raising immunization coverage and reducing significantly the incidence of measles. However, this traditional single dose approach has failed to fully control and eliminate measles, even in the most industrialized countries of the world [10]. It has been reported that a usage of one dose of MMR (measles, mumps, rubella triple) vaccine at 12–24 months in a industrialized western country enabled a measles immunization coverage at 80 per cent. In order to reach fully protective herd immunity coverage in the range of 94%–97% would be necessary. Obviously, this level of coverage would be an exremely formidable task to reach, with the single dose in immunization strategy of infants, even in the most industrialized countries. However, it appears to be within sight, should a second compulsing measles vaccination at school age be adopted. In fact, despite disappointing experiences with this two-dose vaccine strategy a decade ago, there has been recently renewed interest in it. Several experts consider that only with two-dose schedule the measles could be eliminated. A rationale of importance for a two-dose schedule strategy is the problem of failure to seroconvert, particularly, when the vaccine is administered prior to optimal time. Evidences indicate that schools play a vital role in the transmission of measles epidemics. Furthermore, secondary spread of measles from epidemics originating in school has been found to be greater than when the primary case is a child under 5 years of age [3]. Prompted by school child epidemic of measles, a two-dose measles vaccination policy was recommended in the United States by Mast et al [11] in 1990. These authors analyzed data from a large measles outbreaks in 1986, in a highly vaccinated, predominantly school-aged population. Since a few children were revaccinated in controlling this outbreak, an opportunity was also provided to conduct a cost effective analysis of various revaccination strategies during measles outbreaks in school. Based on their results, it was suggested [11] that revaccination in all schools, assessed to be at risk of measles, may be necessary to prevent large outbreaks, until the full implementation of a two-dose vaccination policy. The phenomena of deficient coverage rates, school child epidemics, primary and secondary vaccination failures, clearly seem to reinforce the need for a strategy for measles control by adoption of a second booster dose [12]. The two-dose policy has already proved to be successful in many countries and a cost beneficial strategy. Controversy in some countries, however, prevails as to whether or not adopt this two-dose policy.

Although it might be impossible to avoid some infants contracting measles, in the remainder of the decade, many lives could, however, be possibly saved simply by a reduction in case fatality rates. Treatment of complication appears to be reasonably simpler, for the most part moderately priced, and could become widely available. Chest infections could be treated with antibiotics, and diarrhea with oral rehydration. Vitamin D administration can reduce complications dramatically, and especially can prevent children from going blind.

Research into aspects of natural history of measles are providing a vital insight into ways of controlling the disease. Animal models will allow evaluation candidate vaccines, prior to human field trials. In future, new technologies, as rapid field test for diagnosis of the virus, will make significant contribution for furthering control initiatives.

Elimination of measles virus from the globe now appears to be a real possibility. Considerable problems, associated with such a strategy, however, remain. It is obvious that spread of measles does indeed depend on the number of susceptible not immune individuals and not only on the coverage level, as envisioned earlier. Even high coverage with present measles virus vaccines may not be adequate so as to interrupt transmission. It, thus, appears that a strategy must be developed to permit earlier immunization, one that would overcome maternal antibody blocks. It is apparent that the necessary improved vaccine efficacy may only be achieved with a two-dose measles vaccination schedule. The two-dose strategy should be applied to obtain high level immunization coverage, necessary to measles elimination.

The ideal measles vaccine would appear to be one which is highly antigenic, totally effective in a single injection with no side effects and attainment of solid life-long immunity in 100% of vaccinated individuals. Undoubtedly, some have reason to believe that we are indeed still a long way from such a vaccine. However, research is progressing to provide in near future a highly immunogenic recombinant measles virus vaccine. The possible use of canarypox and fowlpox viruses as carriers of measles virus fusion and hemagglutinin antigens is considered promising [13,14,15].

Ajjan and Saliou [16] have recently reported simplified immunization schedule by simultaneous administration of measles, mumps and rubella (MMR) vaccine with a booster dose of quadruple absorbed vaccine containing diptheria, tetanus, pertussis and inactivated trivalent poliovirus vaccine, injected in the same site of infants, 16 to 26 months of age. No serious adverse reactions were observed and a seroconversion or a booster effect in 95% to 100% vaccinees was dramatically achieved. This study appears to be very interesting. However, menengoencephalitis associated with MMR vaccines containing the Urabe AM/9 strain, as reported recently [17] should carefully be investigated.

Several experts reached a consensus that with improved vaccines, better immunization coverage and adequate two-dose schedule strategy, the elimination of measles worldwide appears as a feasible goal.

Poliomyelitis eradication-vaccines and immunization strategies

The initiative for the global eradication of poliomyelitis is currently being addressed by the Expanded Programme on Immunization (EPI), established in 1974 by the World Health Assembly. The initial objective of EPI was – providing immunization by 1990 to all the children worldwide during their first year of life against diptheria, tetanus, pertussis, measles, tuberculosis and poliomyelitis. As of 1991, 83% of children around the world were reportedly receiving three doses of diptheria-pertussis-tetanus (DPT) vaccine and 85%, the three doses of trivalent oral poliovirus vaccine (TOPV) in their first year of life. Having attained these levels of coverage, the priorities of the EPI for the 1990's have been – achieving and sustaining full immunization coverage in all countries using all the vaccines targeted by EPI and controlling the targeted diseases by utilization of measures as eradicating poliomyelitis globally by year 2000.

The poliovirus appears to meet the criteria necessary to allow for eventual eradication. Poliomyelitis, in fact, happens to be the first disease whose eradication is the goal of EPI. It is inherently, however, more difficult to eradicate than smallpox. Among the epidemiologic characteristics which offer differentiation between them are the asymptomatic illness that characterizes most poliovirus infections and the enteric transmission which presents potential difficulties towards identification and containment of cases of poliomyelitis. Contrastingly, smallpox cases were rather clinically clear-cut and easily confirmable. In terms of vaccines, important differences also seem to surface. The smallpox vaccine has been characterized as heat stable with a single dose attributing to protection over a period of several years, whereas trivalent oral poliovirus vaccine (TOPV) reportedly loses substantial potential after a day or so at 37°C and the necessity of multiple doses for complete protection. Futhermore, properly administered smallpox vaccine has been noted to be a highly effective immunogen, while seroconversion rates after 1–4 doses of TOPV have been suboptimal in developing countries [18]. Despite these and perhaps other differences, the eradication of smallpox has indeed provided the EPI with a model of success worth considering [19].

Experiences in the industrialized countries have shown that geographic areas could become poliovirus free over a period of time through the usage of poliovirus vaccines. In fact, the more promising evidence that poliomyelitis could be possibly eradicated has come from the Americas, where the WHO TOPV policy was applied in mass vaccination campaigns [20,21]. Field experience has indeed demonstrated that within a short period of time, continental areas could invariably be freed of the wild virus with the thorough application of technically correct immunization policies. Based on this experience, the World Health Assembly in 1988 drafted and endorsed the Plan of Action for the Global Eradication of Poliomyelitis by the year 2000. This plan identified needs for immunization coverage, surveillance, investigation of outbreaks and their control, and quality control for vaccines [22]. The plan identified four stages of eradication:

Stage A – Countries are considered to be virtually free of poliovirus. They have a reliable reporting system, have reported no indigenous cases of poliomyelitis for at least the previous three years and have rate of immunization coverage of at least 80 per cent with a full course of vaccine among children reaching their first complete year of life.

Stage B – Countries/areas having immunization coverage exceeding 50 per cent and reporting fewer than 10 cases of poliomyelitis annually.

Stage C – Countries/areas with immunization coverage exceeding 50 per cent and reporting 10 or more case of poliomyelitis per year.

Stage D – Countries/areas with immunization coverage of 50 per cent or less or perhaps an unknown rate of coverage or reporting 10 or more cases of poliomyelitis annually.

As of 1990, most of the world's population were assessed as living in areas that fit into the criteria of Stage A or B and 8 per cent in Stage D. About 442 000 cases of paralytic poliomyelitis were, in fact, averted by adoption of EPI policy but nevertheless 116 000 cases went through. During the next half decade, efforts towards eradication would be centered on establishing and extending poliomyelitis-free geographic zones. In a number of areas, interruption of transmission of wild-type poliovirus might, however, require higher levels of coverage than with routine immunization. Use of national or local immunization days, outbreak control and identification of high-risk areas may prove beneficial. As in the Americas, these strategies could possibly include immunization with all vaccines used by EPI of persons who have not received them previously, and ensuring the further strengthening of the health care infrastructure.

The importance of developing active surveillance for early detection of possible poliomyelitis cases, institution of energetic control, plus universal sustained immunization as early in life as possible, needs to be emphasized. Current strategy includes efforts to respond immediately to suspected cases in countries where effective surveillance indentifies fewer than 50 cases of poliomyelitis annually. Part of such efforts include full characterization of the illness and its causative agent. Furthermore, case definition need to be specific, sensitive and appropriate to the national scope of poliomyelitis eradication. It seems that even industrialized countries need to develop appropriate accute flacid paralysis (AFT) surveillance, targeted at the age group likely to suffer from poliomyelitis with details expert investigation of suspect cases. Development of poliovirus laboratories network would greatly enhance molecular epidemiology surveillance. There now is available a test for poliovirus virulence that probably will be used to screen vaccines for safety. It may, in time, replace the present neurovirulence tests [23].

The final push towards total eradication of the remnants of poliomyelitis will depend heavily on the application of vaccines and immunization strategies already in existence. A re-evaluation of various vaccination strategies for the developing world would be useful. In particular, two aspects require to be critically assessed – the type of vaccine used and the dosage scheduling. Any design of immunization strategies must address two issues, namely, maximization of community immunity and reduction of reservoir of poliovirus (the human gastrointestinal tract). Risk factors of continual wild virus spread including high riks, even in immunized populations, after natural disasters and grosily contaminated water supplies. Inactivated poliovirus vaccine singly does not appear to be completely suitable for eradication of poliomyelitis in developing countries, where the reservoir problem would obviously remain unaddressed due to high fecal–oral transmission. However, the continued use of enhanced potency IPV and OPV could be useful [24,25]. The combined IPV/OPV immunization programme was resulted in elimination of poliomyelitis disease from Denmark, Israel, West Bank and Gaza [26]. For mass immunization campaigns, incorporating both IPV and OPV into the programme for the first year would be beneficial. Subsequently, OPV strategy alone would be sufficient.

Although much of the incidence of poliomyelitis is attributable to cases occurring in unvaccinated or partially vaccinated children, vaccine failure remains an important problem, particularly, with respect to poliovirus types 1 and 3. Factors contributing to oral poliovirus vaccine (OPV) failure tend to be complex and appear to include high levels of secretory antibody present in colostrum and breast milk, malnutrition, a high risk of concurrent infection with non-polio enteroviruses and other enteric pathogens as well as other obscure host factors' [18]. Furthermore, another factor which may also influence the immune response to vaccination is the formulation of the vaccine itself. In order to achieve desired results towards poliomyelitis eradication, a need exists to maximize the efficiency of vaccine delivery by achieving higher rates of seroconversion early in life through development of enhanced immunogenic formulations of TOPV as recommended by WHO.

Despite of evidences providing support for the currently WHO recommended policies of OPV vaccination at birth, 6, 10 and 14 weeks of age, the vaccinees may still be susceptible to poliovirus 1 and 3 after the fourth dose. Expansion of the routine schedule of OPV to five or more doses, and especially, the combined use of both oral and inactivated vaccines are worth considering. In Africa, where access to patients appears to be limited and health infrastructures weak, present OPV schedules could be optimally supplemented with doses of new enhanced potency inactivated poliovirus vaccine (EP-IPV). Experience indicates that while IPV promotes predictable high levels of humoral immunity, there is also evidence of some in concommitant development of gut immunity (secondary IgA antibodies). For IPV, there is certainly a place in immunization programmes, especially in countries with no known wild virus transmission and good surveillance laboratories network. Further developing IPV and possibility of future clinical trials of trypsincleared IPV would help provide better levels of intestinal protection than the existing vaccine. Combined use of IPV and OPV based on clinical study and observation of its routine use in several countries appears to be a useful strategy. The use of both vaccines, OPV and IPV, in India, is also considered as a valuable option [27]. The strategy of combined enhanced potency IPV with diptheria and tetanus and pertussis (DPT) vaccine was suggested in the efforts of poliomyelitis eradication [25].

One of the lessons from the past poliovirus immunization strategies is that IPV or OPV given alone has been insufficient to interrupt the circulation of wild poliovirus, as testified by outbreaks, and that OPV alone, administered in infancy, is not completely protective for the life and could induce a paralytic disease [26,28,29]. It is estimated that the combined IPV/OPV schedule vaccination strategy could reduce the risk of paralytic poliomyelitis from approximately 1 in 560 000 for the first dose of OPV to 1 in 10.5 million, among vaccinees and their contacts [25]. Basic research to develop a safe poliovirus recombinant vaccine could be an appropriate answer to current TOPV safety problems [19,30,31]. Several of recombinants between polioviruses were constructed. Data suggest that safer vaccine strains Sabin type 2 and type 3 polioviruses can be constructed by replacing the sequence of viral capsid proteins of the Sabin 1 poliovirus by the corresponding genome sequences of the Sabin type 2 and type 3 polioviruses. The obtained poliovirus recombinants had antigenicity and immunogenicity of type 2 and type 3 polioviruses. The monkey neurovirulence tests and in vitro phenotypic marker tests qualified these poliovirus constructs as candidates for new strain of type 2 and type 3 recombinant oral poliovirus vaccine (Dr. M. Kohara, 1993, personal communication).

In order to avoid problems with cold chain (system of storage and transport of a vaccine at safe tempeature) failure, the WHO is interested in making poliovirus vaccines less exposed to higher tempers. OPV is noted to be the most thermally labile of all vaccines used. Recently, research progress toward the development of more thermostable OPV was noted. The experiments have shown that by introducing or deleting the appropriate nucleotides, it is possible to make viruses that are less temperature sensitive than the original Sabin polioviruses. Such thermostable viruses may also induce better "take rates" although this remains as yet to be proven.

Even though eradication of poliomyelitis would for the major part depend on application of improved vaccines and immunization programmes, operational research to improve immunization strategies, epidemiologic research towards improvement of surveillance techniques would further enhance the eradication effort. The eradication of poliomyelitis by the year 2000, at an estimated cost of one and half billion dollars, may indeed will be the best gift to the children of the future [13].

Edouard Kurstak

References

1. Henderson RH, Keja J, Hayden G (1988) Immunizing the children of the world: Progress and prospects. Bull WHO 66: 535–543
2. Clements CJ, Strassburg M, Cutts FT, Milstein J, Torel C (1992) Global control of measles. In: Kurstak E (ed) Control of Virus Diseases, Sec. Edition, 179–211, Dekker: New York, p 179–211
3. Schlenker TL, Bain C, Boughman AL, Hadler SC (1992) Measles herd immunity. The association of attack rates with immunization rates in pre-school children. JAMA 267(6): 823–826
4. Preblud SR, Katz SL (1988) Measles vaccine. In: Plotkin SA, Mortimer EA (Eds) Vaccines, Saunders, Philadelphia, p 182–222
5. Taylor WR, Mambu RK, Ma-Disu M, Weinman JM (1988) Measles control efforts in urban Africa complicated by high incidence of measles in the first year of life. Am J Epidemiol 127: 788–794
6. Aaby P, Samb B, Simondon F, Whittle H, Sec AM, Knudsen K, Bennett J, Markowitz L, Rhoses P (1991) Child mortality after high titre measles vaccines in Senegal, the complete data set. Lancet 338: 1518
7. Whittle PF, Mann G, Eckles M (1988) Effect of dose and strain of vaccine on success of measles vaccination of infants aged 4–5 months. Lancet i: 963–966
8. WHO/EPI (1990) Measles immunization before 9 months of age. Wkly Epidem Rec 2: 8–9

9. Garenne M, Leroy O, Beau JP, Sene I (1991) Child mortality after high titre measles vaccine: Prospective study in Senegal. Lancet 338: 903–907

10. Orenstein W (1992) World conference on poliomyelitis and measles. New Delhi, January 7–12, 1992

11. Mast EE, Berg JL, Hanrahan LP, Wassel JT, Davis JP (1990) Risk factors for measles in a previously vaccinated population and cost-effectiveness of revaccination strategies. JAMA 264(19): 2529–2533

12. Centers for Disease Control (1990) Reported cases of measles by state, United States, weeks 14–17. MMWR 39: 300

13. Kurstak E (1993) World conference on poliomyelitis and measles: Vaccines and immunization. Vaccine 11(1): 93–95

14. Plotkin SA (1992) World conference on poliomyelitis and measles. New Delhi, January 7–12, 1992

15. Taylor J, Weinberg R, Tartaglia J, Richardson C, Alkhatib G, Briedis D, Appel M, Norton E, Paoletti E (1992) Nonreplicating viral vectors as potential vaccines: Recombinant canarypox virus expressing measles virus fusion (F) and hemagglutinin (HA) glycoproteins. Virology 187: 321–328

16. Ajjan N, Saliou P (1992) World conference on poliomyelitis and measles. New Delhi, January 7–12, 1992

17. WHO (1992) Meningitis associated with measles-mumps-rubella vaccines. Wkly Epidem Rec 67(41): 301–302

18. Patriarca PA, Wright PF, John TJ (1991) Factors affecting the immunogenicity of oral poliovirus vaccine in developing countries. Rev Infect Dis 13: 926–939

19. Wright PF, Robert J, Kim-Farley RJ, De Quadros CA, Robertson SE, Scott RMcN, Ward NA, Henderson RH (1991) Strategies for the global eradication of poliomyelitis by the year 2000. N Eng J Med 325(25): 1774–1779

20. de Quadros CA, Andrus SK, Olive SM, de Macedo CG, Henderson DA (1992) Polio eradication from the Western Hemisphere. Ann Rev Public Health 13: 239–252

21. Ward NA, Kim-Farley RJ, Milstein JB, Hull HF, Tarel C, de Quadros C (1992) Poliomyelitis eradication. The Expanded Programme on Immunization of the World Health Organization. In: Kurstak E (ed) Control of Virus Diseases, Sec. Edition, Dekker, New York, p 247–266

22. World Health Assembly (1988) Global eradication of poliomyelitis by the year 2000. Resolution WHA 41.28, Geneva

23. Kew O, De L, Yong CF, Nottay B, Pallansch M, da Silvia E (1992) The role of virologic surveillance in the global initiative to eradicate poliomyelitis. In: Kurstak E (ed) Control of Virus Diseases, Sec. Edition, 215–246, Dekker, New York, p 215–246

24. Faden H, Modlin SF, Thomas ML, McBean AM, Ferdon MB, Ogra PL (1989) Comparative evaluation of immunization with live attenuated and enhanced-potency inactivated trivalent poliovirus vaccines in childhood: Systemic and local immune responses. J Infect Dis 162: 1291–1297

25. Marcuse EK (1989) Why Wait for DTP-E-IPV? Am J Dis Children 143: 1006–1007

26. Melnick JL (1992) Poliomyelitis: Eradication in sight. Epidemiol Infect 108: 1–18

27. John TJ, Moore R, Gibson JJ (1992) Choice between inactivated and oral poliovirus vaccine in India. Lancet 339: 504

28. Hovi T, Huovilainen A, Kuronen T (1986) Outbreak of paralytic poliomelitis in Finland: Widespread circulation of antigenically altered poliovirus type 3 in a vaccinated population. Lancet i: 1427–1435

29. Sutter RW, Patriarca PA, Brogan S (1991) Outbreak of paralytic poliomyelitis in Oman: Evidence for widespread transmission among fully vaccinated children. Lancet 338: 715–720

30. Kohara M, Abe S, Komatsu T, Tago K, Arita M, Nomoto A (1988) A recombinant virus between the Sabin 1 and Sabin 3 vaccine strains of poliovirus as a possible candidate for a new type 3 poliovirus live vaccine strain. J Virol 62: 2828–2835

31. WHO (1990) Potential use of new poliovirus vaccine. Bull World Health Organ 68: 545–548

Part I
Measles Global Control

1

Challenges for the global control of measles in the 1990's

C. J. Clements[1]**, M. Strassburg**[2]**, F. T. Cutts**[3]**,
J. Milstein**[4]**, and C. Torel**[1]

[1] Expanded Programme on Immunization, World Health Organization, Geneva, Switzerland
[2] Data Collection and Analysis, Los Angeles County Department of Health Services, Los Angeles, California, USA
[3] The London School of Tropical Medicine and Hygiene, London, U.K.
[4] Biologicals Unit, World Health Organization, Geneva, Switzerland

Summary

Measles is a highly infectious disease which has a major impact on child survival, particularly in developing countries. The importance of understanding the epidemiology of this disease is underlined by its ability to change rapidly in the face of increasing immunization coverage. Much is still to be learned about its epidemiology and the best strategies for administering measles vaccines. However, it is clear that tremendous progress can be made in preventing death and disease from measles with existing knowledge about the disease, and by using the presently available vaccines and applying well tried methods of treating cases. Research in the coming decade may provide improved strategies and more effective vaccines for use in immunization programmes.

Introduction

Not without reason has measles been called the greatest killer of children in history. Despite having an excellent vaccine against measles for twenty years, the world is still struggling to control this disease. While certain island populations may go many years virtually free from measles, western nations such as the United States have experienced enormous difficulties in controlling measles. Many African countries are continuing to experience epidemic and endemic disease which decimates the child populations.

The eradication of smallpox from the world in 1977 proved the feasibility of infectious disease eradication. Public health officials, researchers, politicians and others have been systematically evaluating other candidates for potential eradication. In May 1988, the World Health Assembly selected poliomyelitis as another disease to be targeted for eradication. In 1990, international consensus was reached at the World Summit for Children to embark on a well coordinated global control programme as a major step towards the global eradication of measles in the longer term.

To date, measles has not been successfully controlled globally. The disease has proved to be a formidable adversary. Measles infection is caused by a virus which is extraordinarily uniform wherever it occurs around the globe [1]. Unlike viruses such as poliovirus and influenza, no major strain variations appear, and the genotype is stable with little evidence of geographical variation. Nonetheless, the virus continues to cause unparalleled morbidity and mortality in many parts of the world despite the availability of a vaccine for over 20 years.

The present article looks at key issues in the global control of measles and identifies reasons why increased attention to the subject is warranted. It is anticipated that successful control programmes will result in great reductions in measles during the 1990's, which will pave the way for its eventual eradication.

Global situation

Before measles vaccine became available, virtually all children contracted measles, an estimated 130 million cases and about 16 million deaths globally each year. The activities of the Expanded Programme on Immunization (EPI) have resulted in a dramatic increase in coverage which has contributed significantly to reducing both measles morbidity and mortality. In 1991, approximately 80% of the world's children under one year of age were reported

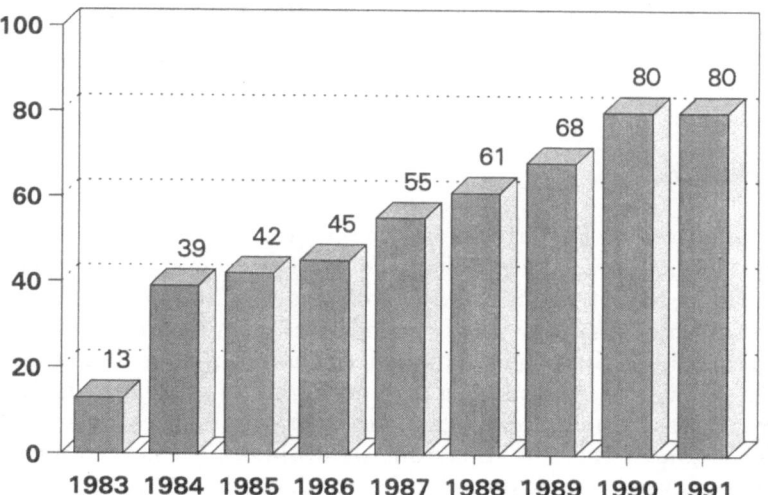

Fig. 1. Global coverage percentage with measles vaccine 1983–1991
Source: information provided to WHO/EPI

to have received measles vaccine (Fig. 1), and it was estimated that around 1.6 million deaths were prevented.

Despite this important progress to date, it is estimated that 45 million cases and around 1 million deaths occur annually. Thus measles is still responsible for more deaths than any other EPI target diseases. The true number dying as a result of measles may be twice the estimated 1 million if the recently documented delayed effect of the disease is taken into account. Measles ranks as one of the leading causes of childhood mortality in the world. In one community study in Kenya [2], measles accounted for 35% of reported deaths in infants one to 12 months of age and for 40% of deaths in children one to four years old. The impact of measles on the lives of millions of young children every year, particularly in developing countries, indicates the urgent need for further reduction in measles incidence.

Changing epidemiology

The epidemiology of measles changes rapidly in the face of increasing coverage. Even with countries placing additional efforts behind measles control, outbreaks of measles are expected to continue to occur and do not necessarily indicate programme failure [3]. After reaching high coverge levels, a temporary period of low measles incidence usually follows which may last a number of years. This low incidence or "honeymoon" period may be followed by a resurgence of outbreaks of considerable size. Pockets of susceptibles are

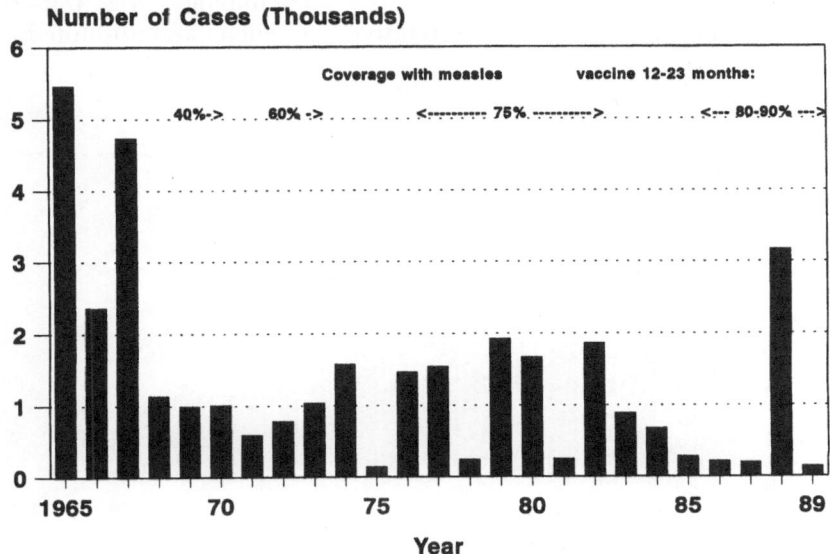

Source: Officially reported to WHO

Fig. 2. Monthly measles case notifications, Hong Kong, 1965–1988. Weekly Epidemiological Record 101 (1990) 49:379–381.

capable of sustaining outbreaks (Fig. 2) due to the highly infectious nature of measles.

Even so, high coverage is likely to lengthen the period between epidemics. With control programmes and effective surveillance systems in place, the size of the expected outbreaks can be minimized, and it is important to note that measles incidence has been greatly reduced from that of the pre-immunization era.

Reported measles cases have declined since the introduction of the first measles vaccines in the 1960's (Fig. 3). Some countries (industrialized and developing) have experienced more than 99 percent reduction in incidence of reported cases. However, in recent years, the number of reported cases in some of those countries has temporarily increased, although still far below pre-vaccine era levels. Outbreaks have occurred among school-aged children, for a variety of reasons. In developing countries, measles remains endemic in most areas, with the largest proportion of cases reported to occur in children under 5 years, and over a quarter of all reported cases being below 9 months of age [4]. Transmission in some urban settings is due to a combination of factors including crowding, low coverage, an uneven vaccine coverage, and migration of unimmunized children into the cities. Transmission also occurs in both urban and rural areas in age groups which are below the recommended age of immunization. It was previously thought that the achievement of high coverage in older children would protect young children by herd immunity, but this has not always been observed due to the highly infectious nature of measles virus.

When assessing changes in measles epidemiology, it is important to distinguish between proportions, relative frequencies and absolute incidence

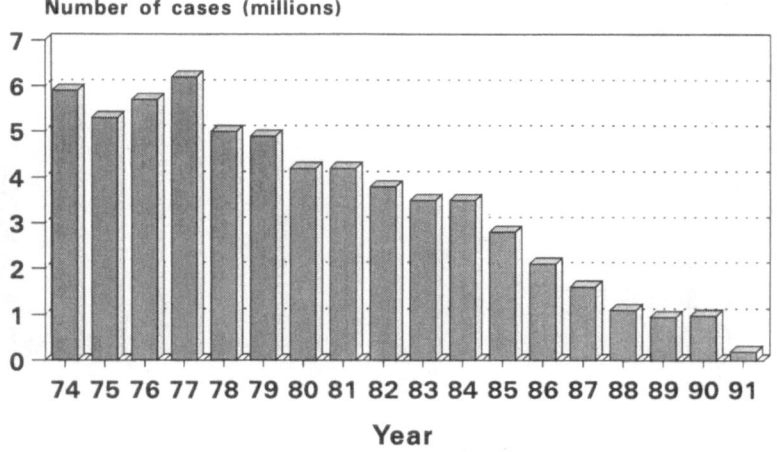

Fig. 3. Number of cases of measles reported globally 1974-91*
*1990 data 97% of countries reported
*1991 data 74% of countries reported

or incidence rates. Changes in proportions do not always represent the situation accurately. Following a period of successful immunization in a community, two events may commonly occur – an increase in proportion of measles cases which occurs is immunized children, and an increase in the proportion of children who acquire measles before the scheduled age of immunization.

The age at which children contract measles depends on a variety of local circumstances. Because maternal antibodies decay at variable rates in different parts of the world, different proportions of infants are protected against measles in the first few months of life. As maternal antibodies wane, the attack rates for measles increase each month to a maximum between one and two years of age in developing countries. A hospital-based study in Afghanistan reported a typical picture of the age distribution of cases, with 74 per cent of all cases occurring between 4 months and 3 years of age [5].

Compared with the age distribution of cases in the pre-immunization era, there is now a greater proportion of cases occurring before 9 months of age in some developing countries. While this does not represent an absolute increase in numbers, these data have served as a critical impetus for researchers to find a measles vaccine which could successfully immunize infants earlier in life. Over the next decade, it is hoped that a vaccine will be developed which can be administered early in life, thus protecting infants from early attacks of measles.

Protection by immunization

Since the introduction of measles vaccines in the 1960s, there have been a number of opportunities to study their field efficacy at nine months of age. Vaccine efficacy is typically determined by comparing the rate of measles in immunized children with that of measles in un-immunized children. Under most field conditions, the vaccine efficacy is thought to be about 85% for children receiving the vaccine at 9 months of age or older. At present it is not clear what the efficacy would be for those measles vaccines which might be used at younger ages.

The issue of duration of protection following measles immunization has been raised. If measles could be effectively controlled in a country in a relatively short period of time, this issue might not play a critical role. Previously it was thought that if children were protected during the first few years of their life, the usual time of peak transmission, the likelihood of further exposure would be low. It is generally assumed that an infection with the wild measles strain confers life-long immunity. The measles vaccine virus is an attenuated strain of the live virus, and therefore there is the theoretical possibility that the antigenic properties of the vaccine strain are not equal to that of the wild virus. Recent studies, however, now show that protection from the standard measles vaccine (administered at an age when maternal antibody has been lost) lasts at least 21 years [6]. Some countries are now experiencing a second peak of transmission in older children and young adults. This does not indicate waning immunity from vaccine. Rather, the cases occur mainly among primary vaccine failures (children who were vaccinated but failed to seroconvert) and children who were never vaccinated. Young people exposed during this peak in their second decade of life will only be protected using a vaccine with ability to confer immunity lasting at least 20 years.

When an immune individual is exposed to the wild virus, an amnestic response occurs, boosting antibody levels. As outbreaks of measles become rare, exposure to the wild virus will diminish, and therefore this boosting effect by the wild virus will diminish also. The net result will be that an individual's protection against measles will be solely dependent on the duration of immunity conferred by the vaccine.

Various levels of herd immunity have been suggested as being necessary to interrupt transmission of measles. Mathematical models of infectious diseases have shown that once a population reaches a certain level of immunity, the remaining un-immunized individuals will be protected by the immunity of the surrounding "herd". But due to the highly infectious nature of the measles virus, the required level is thought to be very high in urban areas. It is also recognized that the non-random distribution of unimmunized in a population will also confound predictions of herd immunity effects.

It has been demonstrated that even in communities with high coverage, outbreaks can occur. In 1988, Harare, Zimbabwe experienced an outbreak when coverage was reported to be 83% [7]. Another study from Texas, United States of America, reported a school outbreak which occurred with a reported 99% coverage rate. This same Texas school population had a 95% level of demonstrable antibodies to measles [8]. There are other examples of outbreaks occurring among populations with reported high coverage, such as the outbreak which occurred in Hungary in 1988 at a time when reported coverage was 98% [9]. It is difficult to ascertain with a high degree of certainty what the precise level of herd immunity was, based on such reports, since a large number of factors may have contributed to these outbreaks. Reported coverage may not have reflected the true levels of immunity in the community, and there was always the possibility of measles virus being introduced from nearby areas with lower immunity.

It appears that the level of coverage needed to interrupt transmission will vary with the epidemiological situation. A rural or island population may need much lower levels of coverage to achieve herd immunity, whereas it has been demonstrated that coverage of 83% and a vaccine efficacy of around 85% were not sufficient for a city like Harare. In crowded conditions, it is likely that even 95% vaccine coverage using a vaccine whose efficacy is less than 90% may not be sufficient to interrupt transmission completely. Vaccine efficacy of measles vaccines in current use at nine months of age is in the order of 85–95%. Thus, even a coverage level of 100% may only protect as little as 85% of immunized infants in some situations in developing countries. It is not yet clear whether coverage in excess of 90% throughout a population will be sufficient to prevent outbreaks, and investigation of the impact of very high coverage in all sections of the population would seem to be a prerequisite before exploring additional mechanisms to boost vaccine efficacy (such as two doses of vaccine) as a way of meeting the 1995 targets.

High-risk areas and groups

As measles vaccine coverage increases, it is expected that the majority of future outbreaks will probably originate from pockets of low coverage. Such pockets are not randomly distributed, but are known to occur in certain settings. Densely populated urban areas are particularly at risk, where the virus may be kept circulating by numerous pockets of susceptibles, despite a relatively high overall level of coverage. The urban poor, who usually have less access to immunization services, are usually the most at risk. In addition, there may be ethnic or cultural reasons which these groups face which might hinder their accepting immunization [10].

Additionally, high-risk groups may include specific age groups, e.g. school children (who represent cohorts from previous years when coverage was lower and who may not have been exposed to measles infection); ethnic minorities (who may have been under-served or may have rejected immunization for cultural reasons); hospitalized children (who are at high risk of nosocomial transmission); and children in refuge camps (where crowding facilitates the spread of the virus) [11].

Medical settings have been the foci of a number of measles outbreaks. It is paradoxical that the immunization clinic, the very place parents bring their children to be protected against disease, may become a place a child acquires a life-threatening illness.

A study in the United States of America confirmed that in 1984 nearly 3% of measles cases acquired their infection in a medical setting [12]. A study in Côte d'Ivoire revealed that two thirds of measles cases treated in a health centre were nosocomial in origin [13]. Prevention of measles must include routine immunization of children brought to clinics for any reason, and improvements in waiting areas so that patients with measles can be easily isolated.

Children admitted to hospital and siblings who visit them must be immunized at the earliest opportunity within the national schedule. In some settings, medical personnel have been the source of measles spread. Policies requiring immunization or proof of immunization among health workers may need to be implemented in such settings.

The highly communicable nature of measles

The highly communicable nature of measles and the explosives nature of outbreaks has been well documented. An outbreak in the Faroe Islands in 1846 documented almost every susceptible individual contracted measles [14]. More recently, a study among an unimmunized population in India

[15] reported that 80% of children had been infected with measles by the time they reached 10 years of age.

Among highly susceptible populations in crowded settings, the opportunity for explosive point source outbreaks is great. In one study in the United States, one infected person was believed to have transmitted measles to dozens of others at a sports meeting [16], again emphasizing the highly infectious nature of measles.

In households, secondary attack rates of 80% have frequently been reported. In contrast to densely populated areas, nomads and other groups in sparsely populated areas experience only sporadic disease transmission [17]. However, when such peoples come together for festivals or markets, explosive outbreaks may occur, hence such occasions and gatherings should be targeted for immunization activities.

There appears to be an inter-relationship between continuous urban transmission and rural outbreaks. Barlet [18] and Black [19] suggest that in an isolated situation, a minimum population of 400 000 to 500 000 susceptible individuals is required to sustain measles transmission. If this is correct, rural transmission is likely to be maintained by a continuous process of re-contamination by cases from nearby cities. This observation underlines the importance of improving coverage not only at the country level but even at the village and community levels as well. Improved surveillance and special studies are needed to obtain better documentation of the foci of measles transmission in countries with high coverage.

Measles is a serious disease

Contrary to popular belief, measles is not a benign childhood illness, but affected children are at risk of various complications with varying degrees of seriousness. Even in countries with adequate health care and healthy child populations, the complication rate can reach 10% [20]. The commonest complications in industrialized countries are ear infection (around 5%), pneumonia (around 2.5%), diarrhoea (around 2.5%) and encephalitis (around 0.1%).

The picture is significantly different in developing countries where measles affects children at a younger age and many children are already experiencing a background of viral, bacterial and parasitic infections, and may in addition be malnourished. One community study from India showed 82% of children with measles had complications [15]. A study of hospitalized children in Afghanistan [21] showed 84% of children with measles had pneumonia, 30% had diarrhoea and 22% laryngotracheobronchitis (some had multiple complications). In a hospital study from Mexico, half the cases with complications were less than one year of age, and 70% of these were malnourished [22].

Due to measles complications, a large number of children become handicapped due to deafness, impaired vision or blindness. In areas where vitamin

A deficiency is a problem, children with measles may be precipitated into acute vitamin A deficiency with resulting impairment or loss of vision from corneal ucleration and xerophthalmia. A less common but devastating complication is subacute sclerosing panencephalitis (SSPE) which develops in between 1 and 5 per 100 000 cases and occurs several years after the initial measles infection.

Mortality

The degree to which the measles virus inflicts death on a population varies, as there is a wide range in the reported measles case-fatality rates (CFRs). There is inter-country, inter-regional and even inter-epidemic variation in CFRs.

Crowded conditions have been found to be associated with higher case-fatality rates [23]. Typically, large urban agglomerations present ideal conditions for measles transmission, especially among dense populations of poor who are not only under-immunized, but also experience high rates of other infectious conditions as well as reduced access to medical care. In such conditions, the CFR can be as high as 20%.

The CFRs in industrialized countries are currently estimated to be less than 0.1%. This is partly due to the difference in age-specific attack rates, underlying good nutrition, supportive treatment for bacterial chest infections, and other environmental and sociological factors.

Quantifying CFRs is difficult and may be subject to bias, in that most information is usually obtained from sources which have an unrepresentative case load. For example, hospitals tend to admit and report cases who are the most serious. Such hospitalized cases are at increased risk of dying compared to non-hospitalized cases. Recognizing such difficulties, WHO/EPI reviewed community studies of CFRs, and in the absence of country-specific information, now uses a value of 3% as an estimate of the CFRs within one month of onset of acute measles infection among persons of all ages in the developing world. Such estimates must be viewed with caution in light of some community based studies which suggest this may be a significant underestimate [24], and further work is needed to refine how CFRs are calculated. For instance an outbreak in Afghanistan resulted in a CFR of 28% for all ages and a CFR of 42% for children 0–4 years. The results from this study underline the added danger of acquiring measles at a young age [25].

A number of studies suggest that measles is more severe and may result in death in situations where intense exposure occurs, such as "secondary" transmission (transmission from an index case to other children in the same household), crowding and outbreaks. A study in Machakos District, Kenya, [26] showed that in families with several cases of measles, secondary cases had a higher risk of dying than those who were infected from outside. Older children with their increased mobility appear to have a greater opportunity for infection outside the home with resulting lower case fatality rates due to this and more mature immune systems. Another study from the Gambia showed measles mortality was strongly associated with more than five measles cases in a household [27].

In contrast, in an industrialized country, Sutter found no difference in severity between primary and secondary cases in a very different sociological setting where the overall severity of measles was much less and mortality was minimal [28].

Delayed mortality

In addition to well documented complications and mortality associated with the acute illness, the long term impact of the infection is now being reported [29, 30]. In several African studies it has now been documented that for many months after the acute attack, there is a reduced survival rate in the affected infants. Thus it appears that those infected with measles virus who survive the acute disease have a greatly increased risk of dying after the period of the acute illness when compared with those not infected with measles (Table 1). Death may result from a variety of apparently unrelated conditions in the months following measles infection. One possible explanation of these observations may be that there is a reduced cell-mediated immune response as a result of measles infection. Some authors postulate the persistence of measles virus to account for increased delayed mortality.

The younger the age of infection, the more dramatic appears the effect of this delayed mortality phenomenon, emphasizing the urgent need to protect children at the youngest possible age.

Based on such studies, it is now thought that the true impact of measles virus infection is much greater than is actually recorded in the number of deaths generally attributed to measles, making the control of this disease an even greater priority. To date, studies describing delayed mortality have all been conducted in Africa. It is important to confirm these observations in Asia and the Americas.

Treatment

Vitamin A administered to children acutely ill with measles has been shown to reduce mortality. Results from a trial in South Africa showed children

Table 1. Mortality during nine months of follow-up for measles cases and community controls, The Gambia

Age at infection	Mortality of measles cases		Mortality of community controls 0–9 months later
	Acute	Delayed 1–9 months later	
3–11 months	18%	56%	3%
1–2 years	9%	13%	2%

Adapted from Hull (1983) [30]

treated with vitamin A had a reduced risk of dying, recovered more quickly from pneumonia and diarrhoea and had less croup [31]. In addition, symptomatic treatment for cases requires antibiotics to combat bacterial complications, and oral rehydration salts for dehydration following diarrhoea. Case-fatality rates can be lowered if cases reach health care facilities early where appropriate care is offered. For uncomplicated cases, supportive fluids, antipyretics and nutritional therapy may be required. Many children need increased food intake for four to eight weeks to recover their pre-measles nutritional status.

References

1. Black FL (1989) Measles active and passive immunity in a worldwide perspective. In: Melnick JL (ed) Prog Med Virol, Vol 36. Karger, Basel, pp 1–33
2. Spencer HC et al (1987) Impact on mortality and fatality of a community based malaria control programme in Saradidi, Kenya. Ann Trop Parasitol 81 [Suppl 1]: 36–45
3. Cutis FT, Henderson RH, Clements CJ, Chen RT, Patriarca PA (1991) Principles of measles control. Bull WHO 69: 1–7.
4. Taylor WR, Kalis R, ma-Disu M, Weinman JM (1988) Measles control efforts in urban Africa complicated by high incidence of measles in the first year of life. Am J Epidemiol 127: 788–794
5. Arya LS, et al (1987) Spectrum of complications of measles in Afghanistan: a study of 784 cases. J Trop Med Hyg 90: 117–122
6. Miller C (1987) Live measles vaccine: a 21 year follow-up. B Med J (Clin Res) 295: 22–24
7. Kambarami RA, Nathoo KJ, Nkruma FF, Pirie DJ (1991) Measles epidemic in spite of high measles immunization coverage rates in Harare. Zimbabwe. WHO Bull 69: 213–219
8. Gustafson TL, Lievens AW, Bourell PA (1989) Measles outbreak in a fully immunized secondary school population. N Engl J Med 316: 717–724
9. Measles outbreak. Wkly Epidemiol Rec (64): 137–138 (1989)
10. Coetzee N, Yach D, Blignaut R, Fisher SA (1990) Measles vaccination coverage and its determinants in a rapidly growing peri-urban area. S Afr Med J 78: 733–737
11. Toole MJ, Stecketee RW, Waldman RJ, Nieburg P (1989) Measles prevention and control in emergency settings. Bull WHO 67: 381–388
12. Davis RM, Whitman ED, Orenstein WA, Preblud S, Markowitz LE, Hinman AR et al (1987) A persistent outbreak of measles despite appropriate preventive and control measures. Am J Epidemiol 126:438–449
13. Klein-Zabban ML et al (1987) Frequency of nosocomial measles in a maternal and child health center in Abidjan. Bull WHO 65: 197–201
14. Panum PL (1939) Observations made during the epidemic of measles on the Faroe Islands in the year 1846. Med Classics 3: 839–886
15. Nairn JP, Khare S, Rana SRS, Banergee KB (1989) Epidemic measles in an isolated unvaccinated population, India. Int J Epidemiol 18: 952–957
16. Chen RT, Goldbaum GM, Wassiklak SGF, Markowitz LE, Orenstein WA (1988) An explosive point-source outbreak in a highly vaccinated population: mode of transmission and risk factors for disease. Am J Epidemiol 129: 173–182
17. Loutan L, Paillard S (1993) The epidemiology of measles in a nomadic community. In press.
18. Bartlet MS (1957) Measles periodicity and community size. J R Statist Soc A 120: 48–70
19. Black FL (1966) Measles endemicity in insular populations: critical community size and its implication. J Theor Biol 11: 207–211
20. Measles – United States, first 26 weeks, 1989. Morbid Mortal Rep 38: 863–871 (1990)

21. Srivastava RN et al (1979) Morbidity and mortality of children hospitalized with medical disorders in Afghanistan. J Trop Pediator 168–172
22. Avila-Figuera C, Navarette-Mavarro S, Martinez-Aguilar M, Ruiz-Gutierrez E, Santos JI (1990) Complication en ninos con serampion. Boletin medico del Hospital Infantil de Mexico 47: 520–524
23. Aaby P (1988) Introduction to community studies of severe measles: comparative test of crowding/exposure hypothesi. Rev Inf Dis 10: 451
24. Aaby P, Clements CJ, Orinda V (1993) Mortality from measles: assessing the impact. Bull WHO (in press)
25. Wakeham PC (1978) Severe measles in Afghanistan. J Trop Peditar 24: 87–88
26. Aaby P, Leeuwenburg J (1990) Patterns of transmission and severity of measles infection: a reanalysis of data from the Machakos Area, Kenya. J Inf Dis 161: 171–174
27. Hull H (1988) Increased measles mortality in households with multiple cases in the Gambia, 1981. Rev Inf Dis 10: 463–467
28. Sutter RW, Markowitz LE, Bennetch JM, Morris W et al (1991) Measles among the Amish: a comparative study of measles severity in primary and secondary cases in households. J Inf Dis 163: 12–16
29. Aaby P, Clements CJ (1989) Measles immunization research: a review. Bull WHO 67: 443–448
30. Hull HF, Williams PJ, Oldfield F (1983) Measles mortality and vaccine efficacy in rural West Africa. Lancet i: 972–975
31. Hussey GD, Klein M (1990) A randomized controlled trial of vitamin A in children with severe measles. N Engl J Med 323: 160–164

2

The experience with measles in the United States

W. A. Orenstein[1], **L. E. Markowitz**[1],
W. L. Atkinson[1], and **A. R. Hinman**[2]

[1] Division of Immunization, Centers for Disease Control and Prevention, Atlanta,
Georgia, USA
[2] Centers for Disease Control and Prevention, Atlanta, Georgia, USA

The experience with measles in the United States

Prior to measles vaccine licensure, virtually every child in the United States suffered from measles. While the reported number of cases averaged approximately 525,000 annually, an incidence rate of 315 per 100,000 total population, the actual yearly number of cases approached 4 million [1–4]. Measles vaccine offered the opportunity to prevent this substantial health burden. This report will describe the results achieved using a single dose measles vaccine schedule, recent recommendations for a two-dose schedule, and the efforts to control a resurgence of measles in the United States.

Strategy

When measles vaccine was licensed in 1963, a single dose was recommended routinely at 9 months of age [5]. In 1965, because of substantial numbers of vaccine failures in persons vaccinated prior to the first birthday, the age for routine measles vaccination was raised to 12 months. Persons vaccinated before the first birthday were to be revaccinated. The high failure rate associated with vaccination before the first birthday was presumably due to interference with vaccine virus multiplication caused by the persistence of transplacentally acquired maternal antibody. In 1976, the age for routine vaccination was increased from 12 months to 15 months when data from a variety of studies showed that protection rates in those vaccinated at 12 months were slightly lower than in those vaccinated at 15 months and that there was little risk of acquiring measles in the 12–15 month period. Maternal antibody apparently persisted beyond the first birthday in some children. Because protection at 12 months was still relatively high, $\geq 80\%$, it was not recommended that persons vaccinated from 12 to 14 months of age be revaccinated. Thus, through 1989, except for small numbers of persons vaccinated prior to the first birthday and other selected small groups who had been revaccinated, the United States had a single dose policy for measles vaccination.

By 1968, the reported occurrence of measles had dropped to 22 000 cases, a $> 95\%$ decline from prevaccine levels (Fig. 1). Between 1969 and 1977, reported measles occurrence varied from a high of 75 000 cases in 1971 to a low of 22 000 cases in 1974. In the spring of 1977 a childhood immunization initiative was begun with the goal of achieving 90% immunization levels in all of the nation's children. Because of the success of the childhood immunization initiative, in October 1978 a goal was announced to eliminate indigenous measles in the United States by October 1, 1982 [2].

The strategy adopted had three components: (1) high immunization levels; (2) careful surveillance; and (3) aggressive outbreak control. Of these, high immunization levels was deemed the most important. Implementation was helped by the enactment and enforcement by states of school laws requiring evidence of measles immunity of all students in kindergarten through 12th grade. Exclusion from school of students who lacked evidence of measles immunity was vigorously pursued [6,7].

Measles immunity was defined as one of the following: (1) documentation of receipt of live measles vaccine on or after the first birthday; (2) documentation of physician diagnosed measles; or (3) laboratory evidence of measles immunity [8].

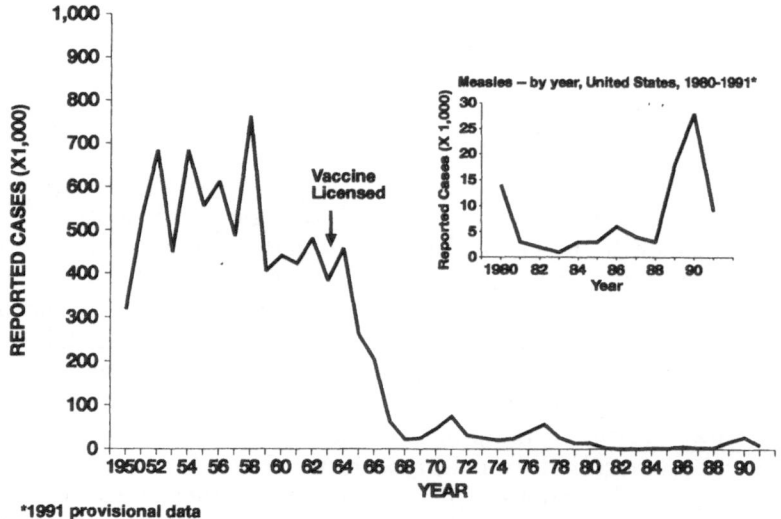

*1991 provisional data

Fig. 1. Measles (rubeola) – By year, United States, 1950–1991

High immunity levels, particularly in the schoolage population, have been documented. In 1979, a survey of more than 28 million students, kindergarten (age 5–6 years) through 12th grade (age 18 years), revealed evidence of measles immunity in 91% [1]. Since 1980, ≥ 96% of entering schoolchildren have had evidence of immunity to measles.

These high immunity levels led to marked reductions in measles morbidity (Fig. 1). Reported measles decreased from a high of 57 000 cases in 1977 to a low of approximately 1 500 cases in 1983, an incidence rate of < 1/100 000 total population, and a > 99% decrease from the prevaccine era. During the 1980's, measles incidence averaged approximately 3000 cases annually ranging from 1500 cases in 1983 to 6300 cases in 1986. While remarkable progress had been made in the control of measles, the goal of measles elimination had not been achieved. During the late 1980's a reevaluation of the single dose strategy was undertaken. Data from 1985 and 1986 were analyzed in detail to help understand the reasons for the failure to eliminate measles.

Outbreaks were defined as 5 or more epidemiologically linked cases [9]. Three outbreak patterns were seen with very different epidemiologic characteristics based on the predominant age group involved. The preschool and school-age outbreaks are presented in Table 1. Forty preschool outbreaks were identified. These outbreaks were generally small with a median of only 13 cases although the largest outbreak, 945 cases in New York City, was classified in this category. A median of 45 percent of the cases were considered preventable, that is, occurring in children who were unvaccinated although vaccine was indicated [10]. The most common reason for nonpreventability was being younger than the routine age for measles vaccination, i.e., < 16 months, accounting for a median of 32 percent of the cases in these outbreaks.

In contrast, school age outbreaks were more numerous (101) and were generally larger, with a median of 25 cases. There were fewer cases in unvaccinated children, a median of only 27 percent. Further, the reasons for nonpreventability were substantially different. A median of 60 percent of the

Table 1. Measles outbreaks in the United States from 1985 through 1986

Varible	Outbreak classification (Age)	
	Preschool (N = 40)	School (N = 101)
	Median (range)	
Number of cases	13(5–945)	25(5–363)
Preventable cases (%)	45(0–100)	27(0–81)
Patients characteristics (%)		
> 16 Mo. of age	32(0–80)	2(0–29)
Vaccinated at ≥ 12 mo.	14(0–42)	60(0–100)

NEJM 1989; 320: 75–81

Table 2. Characteristics of school-age measles outbreaks of $\geq$ 100 cases, 1985–1986

	Median	Range
No. cases	138	101–363
% Appropriately vaccinated	55%	1%–69%
% Appropriately vaccinated age 5–19 years	72%	2%–90%
Interval since last report of $\geq$ 5 cases	7 years	1–15 years

NEJM 1989; 320: 75–81

cases were in appropriately vaccinated persons with a range as high as 100 percent.

In some instances, measles transmission occurred in highly vaccinated populations with vaccine failures playing a major role in transmission. A more detailed analysis is available for the 15 school-age outbreaks with 100 or more cases reported in 1985 and 1986 (Table 2). The outbreaks ranged in size from 101 cases to 363 cases with a median of 138 cases. A median of 55 percent of the cases occurred in persons who had a history of vaccination at 12 or more months of age with a range of 1 percent to 69 percent. Transmission in most of these outbreaks began in the school-age population and then sometimes involved preschool and post-school individuals. When evaluating school-age children only during these outbreaks, the proportion with appropriate vaccination increases dramatically. A median of 72 percent of 5–19 year old persons with measles had a history of appropriate vaccination with a range of 2 percent to 90 percent. Also of interest is that these large outbreaks occurred primarily in areas that had been relatively measles free in prior years. It had been a median of 7 years with a range of 1 to 15 years since the last report of 5 or more measles cases in these communities.

Further evidence that measles could be sustained among vaccine failures comes from an outbreak investigation in Corpus Christi, Texas, by Gustafson et al. in which they documented transmission for almost two months in one of the schools with vaccination levels of 99 percent and immunity levels determined by serology to be 96 percent [11] (Fig. 2).

Are these vaccine failures due to an initial failure to seroconvert or do they represent secondary vaccine failure, a waning of vaccine-induced immunity? Our thinking traditionally has been that primary vaccine failure accounts for the majority of cases with histories of vaccination on or after the first birthday [12]. Measles vaccine-induced immunity is thought to be long-term, probably lifelong. This is based, in part, on reasoning by analogy since measles disease-induced immunity appears to be lifelong. Also, reported measles morbidity is at near record lows among vaccinated populations more than 29 years since measles vaccines were first licensed. If waning immunity were a substantial problem, attack rates should be much higher. Further evidence against waning immunity is that primary vaccine failure is known to occur in 2–5 percent of vaccinees. Attack rates in most measles outbreaks

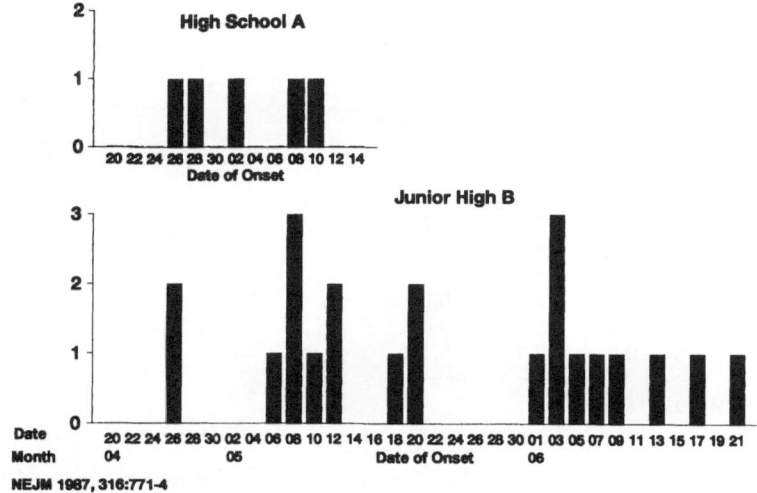

Fig. 2. Measles outbreaks at a Junior High and High School in Corpus Christi, Texas, 1985 (Reprinted by permission of NEJM 1987, 316: 771–774)

are consistent with this rate of primary vaccine failure with most incidence rates in schools less than 5 percent [9].

Nevertheless, some persons appear to lose immunity over time. An investigation from Canada in 1989 suggested that as many as 5 percent of persons who initially seroconverted against measles developed disease within the 10 years following vaccination [13] (Table 3). In contrast, follow-up of two persons who did not seroconvert (primary vaccine failures) indicated that both reported having had measles. Additional data raise the possibility that small numbers of people may lose immunity over time. Table 4 shows data from outbreak investigations examining attack rates by increasing time since vaccination [12]. In no outbreak was there a significantly increased risk with time since vaccination. However, in all there was a trend in this direction.

In summary, the preponderance of the data suggests that primary vaccine failure is the major cause of vaccine failure. However, because waning immunity may be the cause of some vaccine failures, further efforts are needed to determine better the magnitude of the problem.

The high communicability of measles and the demonstration that measles transmission can be sustained among highly vaccinated populations in some

Table 3. Reported measles cases by seroconversion status*

	Total no.	No. cases	%
Initial seroconversion	175	9	5
Nonconversion No revaccination	2	2	100

*AJPH 1989; 79: 475

Table 4. Epidemiologic studies in the U.S.
attack rates by years since measles vaccination

Study	Years since vaccination			
	0–4	5–9	10–14	> 15
Shasby	9.4%	6.9%	15.4%	
Marks	4.0	4.2	5.4	11.7
CDC	0.0%	3.2%	9.9%	8.8%
Nkowane	0.0%	1.1%	1.4%	
Davis	1.1%	1.7%	2.6%	0.0%
Robertson	0.0%	1.4%	3.4%	0.0%
Hutchins	0.0%	0.0	2.2%	3.1%

PIDJ 1990; 9: 101–110

instances suggested that more than one dose would be necessary to eliminate measles.

Beginning in 1989, the United States adopted a two dose schedule [14]. The first dose is usually administered at 15 months of age except in areas at high risk for preschool measles such as in the cities where there are large numbers of unvaccinated preschool children. In these areas, MMR should be given at 12 months. In outbreak settings measles vaccine can be given as young as 6 months, although doses given prior to the first birthday do not count toward the two dose schedule. The second dose is administered at either entry to primary school, approximately 4–6 years of age or at entry to middle or junior high school, approximately 12 years of age.

Since those recommendations were made, the epidemiology of measles has changed dramatically. From 1981–1988, the average incidence was approximately 3000 cases, and the median incidence rate was 1.4 cases/100 000 (Fig. 1). In 1989, more than 18 000 cases were reported, rising to more than 27 000 cases in 1990, an incidence rate of 11.2 cases/100 000 [15]. The increase was associated with a major change in the characteristics of cases. In contrast to the 1970's and most of the 1980's when measles was predominantly a disease of school-age children, ages 5–19 years, measles in 1989, 1990, and 1991 has been a disease primarily of preschool children (Fig. 3). For the first time in 1990, more cases were reported in preschoolers age 0–4 years, a 5 year age group, than in the school-age group, 5–19 years which comprises 15 birth cohorts of children.

From 1980–1988, children < 12 months of age accounted for a median of 8 percent of all reported cases (Fig. 4). In 1989, children < 12 months of age increased to 11 percent of reported cases, and in 1990, increased further to 17 percent of all reported cases, the highest proportion of cases in this age group ever reported.

This trend has continued in 1991, with children < 12 months of age accounting for 19 percent of all reported cases, and 39 percent of cases occurring in children < 5 years of age. This trend is most likely due to

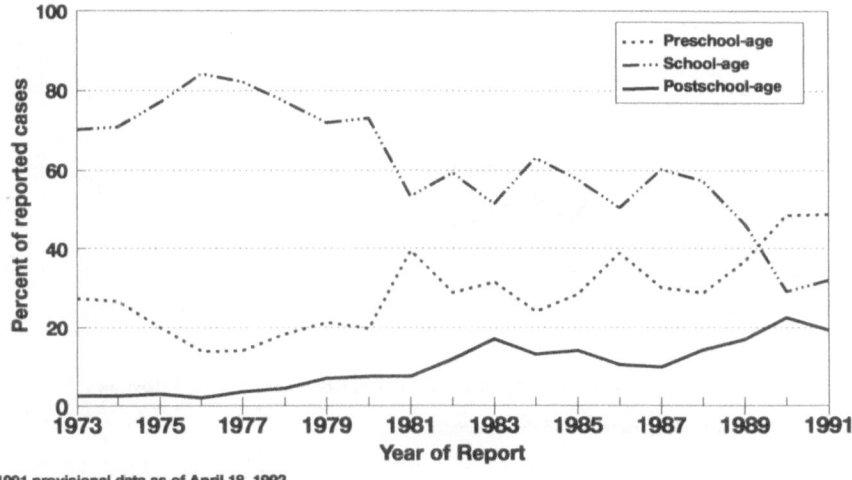

Fig. 3. Measles – United States, 1973–1991. Proportion of total cases by age group

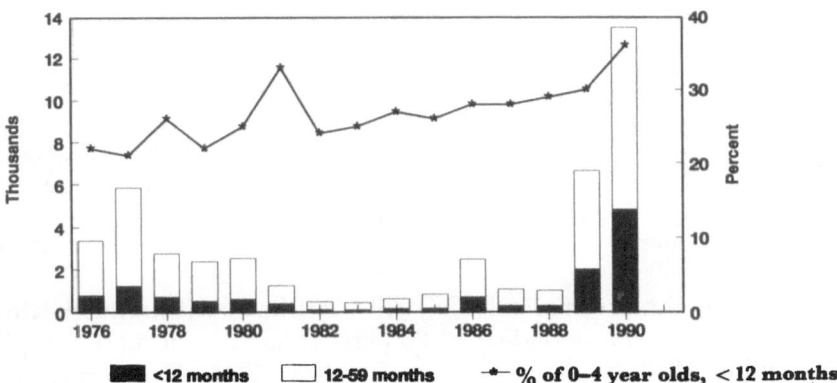

Fig. 4. Measles – United States, 1976–1990. Age distribution < 5 years

crowding that occurs in the inner cities in the United States allowing more opportunity for exposure and the fact that children of vaccinated mothers are more likely to become susceptible at an earlier age than those of mothers who had measles disease [18,17]. Vaccinated mothers may have lower levels of antibody to transfer across the placenta.

The measles epidemic disproportionately affected persons of racial and ethnic minorities living in crowded inner city populations (Fig. 5). Among children < 5 years of age, black and hispanic children were at 4–9 fold greater risk of measles than non-hispanic whites. While the outbreak was widespread, only 13 of the more than 3000 countries in the U.S. accounted for 50 percent of the total cases in the two year period.

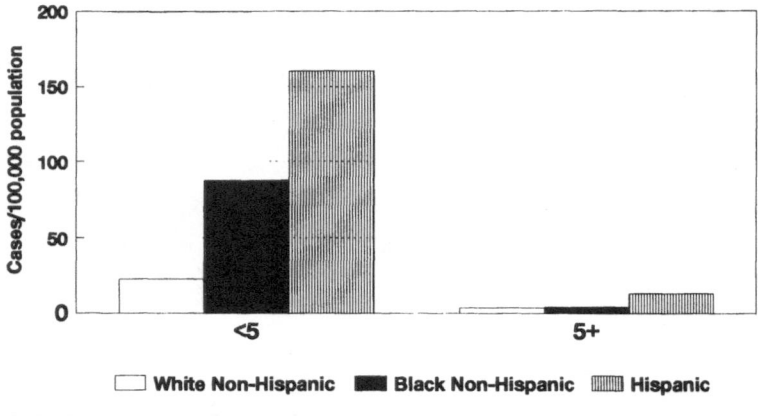

□ White Non-Hispanic   ■ Black Non-Hispanic   ▥ Hispanic

*States reporting race for >67% of cases

Fig. 5. Measles incidence by age and race in 35 states, 1990

Table 5. Complications from measles, 1989–1990 United States

Complications	Number	Percent
Diarrhea	4942	9
Otitis media	4244	8
Pneumonia	3489	6
Encephalitis	80	0.1
Hospitalization	11 251	20
Death*	166	0.3

*Provisional data

The resurgence of measles in 1989–1991 had a substantial health impact (Table 5). Of all reported cases, 20 percent had one or more complications, including 4942 (9 percent) with diarrhea, 4244 (8 percent) with otitis media, 3489 (6 percent) with pneumonia, and 80 (0.1 percent) cases of encephalitis. Twenty percent of reported cases required hospitalization, for more than 44 000 hospital days.

The resurgence was associated with the greatest number of measles associated deaths in almost 2 decades. Fifty-five percent of the deaths were in preschool children including 22 percent in infants less than 1 year of age. The reported death-to-case ratio increased from a median 0.64 deaths/1000 cases during 1980–1988 to 2.3 deaths/1000 cases in 1989, 3.2 deaths per 1000 cases in 1990 and 3.7 deaths per 1000 cases in 1991. While some of this increase was due to more cases in preschoolers who have a higher death-to-case ratio than school age children, this cannot account for the total difference [18].

In addition to the shift in age, there was a marked change in the vaccination status of cases (Fig. 6). From 1985–1988, a median of 42 percent of the

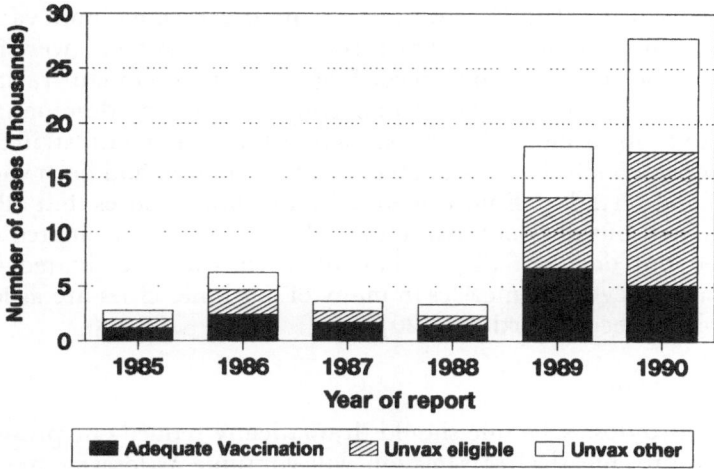

Fig. 6. Measles – United States, 1985–1990. Vaccination status by year

cases were in vaccinated children. In contrast, only 20 percent of the measles patients in 1990 and 1991 had a history of vaccination. More than 80 percent of the cases occurred in unvaccinated persons, most of whom should have been vaccinated. In addition, a substantial portion, 24 percent, occurred in persons for whom vaccine was not routinely indicated generally because they were younger than the routine age for vaccination (15 months at the time).

The measles epidemic was caused by a failure to vaccinate. Data from Chicago, one of the cities with the largest outbreaks, indicate that virtually all children are vaccinated against measles by the time they enter school,

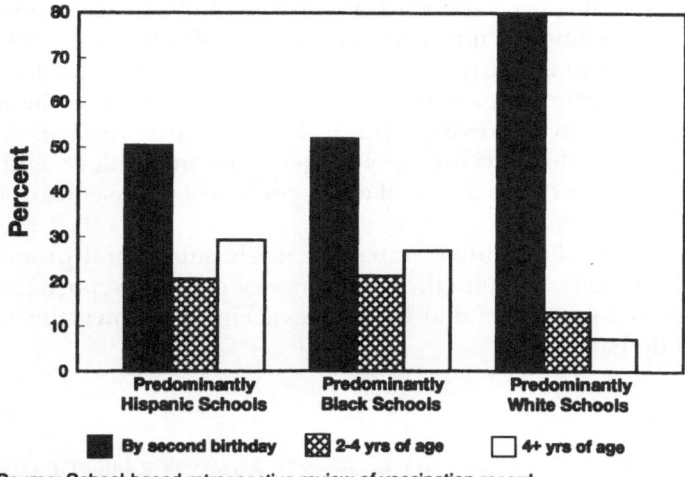

Source: School-based retrospective review of vaccination record.

Fig. 7. Age vaccinated against measles by race-city of Chicago, March 1990

about 5 years of age [19]. However, many children do not obtain vaccination at the recommended ages. In 1990, records of school-enterers were reviewed and the schools were classified according to the predominant racial/ethnic group of students. Almost 80 percent of students in the predominantly white schools had been vaccinated by the second birthday. In contrast, in the black and Hispanic schools only about 50 percent of children had been vaccinated by the second birthday. Enforcement of school laws ensures that all persons are vaccinated but does not assure that all persons are vaccinated on time. vaccinated but does not assure that all persons are vaccinated on time. Immunization levels for measles in many of our inner cities are as low as 50 percent by the second birthday [20].

What is the future?

While the two dose schedule should dramatically reduce our problem with vaccine failure, measles transmission will continue as long as reservoirs of unvaccinated preschool children exist. During 1991, there has been a 65 percent reduction in reported measles but the characteristics of cases are similar to those in 1990 and measles epidemics are likely to recur in the future unless preschool immunization levels improve.

Many of the issues addressed by developing countries in terms of improving their immunization levels are applicable to the United States. Substantial numbers of opportunities are missed and parents often face significant obstacles that deter immunization [19,21,22]. The resurgence of measles has generated unprecedented interest in improving immunization levels of preschool children. In June 1991, the President held a special ceremony in the White House Rose Garden to stress the importance of immunization and to charge the Secretary of Health and Human Services to visit various cities around the country to determine how coverage could be improved. An Infant Immunization Initiative has been launched consisting of efforts to improve service delivery, Federal funds to improve the delivery infrastructure, development of better methods to measure coverage, increased emphasis on motivation of parents through education, and operational research to find the most cost-effective strategies to improve coverage. The United States will also pursue efforts to use measles vaccine at younger ages since more and more of our cases are younger than a year of age, ages in which the efficacy of present vaccines is not optimal.

In conclusion, the United States has made substantial progress in the fight against measles. Despite the resurgence of measles, reported morbidity is still less than 5 percent of that in the prevaccine era. Nevertheless, we can and should do better.

References

1. Hinman AR, Brandling-Bennett AD, Bernier RH, Kirby CD, Eddins DL (1980) Current features of measles in the United States: Feasibility of measles elimination. Epidemiol Rev 2: 153–170

2. Hinman AR, Brandling-Bennett AD, Nieburg PI (1979) The opportunity and obligation to eliminate measles from the United States. AMA 242: 1157–1162
3. Bloch AB, Orenstein WA, Stetler HC et al (1985) Health Impact of Measles Vaccination in the United States. Pediatrics 76: 524–532
4. Langmuir AD (1962) Medical importance of measles. Am J Dis Child 103: 224–226
5. Orenstein WA, Markowitz LE, Preblud SR, Hinman AR, Tomasi A, Bart KJ (1986) Appropriate age for measles vaccination in the United States. Dev Biol Stand 65: 13–21
6. Robbins KB, Brandling-Bennett AD, Hinman AR (1981) Low measles incidence. Association with enforcement of school immunization laws. Am J Public Health 71: 270–274
7. Orenstein WA, Hinman AR, Williams WW (1993) The impact of legislation on immunization in the United States. Proceedings of the 1991 Australian Second National Immunization Conference, Immunization: the old and the new. Aust J Public Health (in press)
8. Centers for Disease Control (1987) Recommendations of the Immunization Practices Advisory Committee. Measles Prevention. Morbid Mortal Weekly Rep 36: 409–418, 423–425
9. Markowitz LE, Preblud SR, Orenstein WA, Rovira EZ, Adams NC, Hawkins CE, Hinman AR (1988) Patterns of transmission in measles outbreaks in the United States, 1985–1986. N Engl J Med 320: 75–81
10. Centers for Disease Control (1982) Classification of measles cases and categorization of measles elimination programs. Morbid Mortal Weekly Rep 31: 707–711
11. Gustafson TL, Lievens AW, Brunell PA, Moellenberg RG, Buttery CMG, Schulster LM (1987) Measles outbreak in a fully immunized secondary-school population. N Engl J Med 316: 771–774
12. Markowitz LE, Preblud SR, Fine PEM, Orenstein WA (1990) Duration of live measles vaccine-induced immunity. Pediatr Infec Dis J 9: 101–110
13. Mathias R, Meekson J, Arcand T, Schechter M (1989) The role of secondary vaccine failures in measles outbreaks. Am J Public Health 79: 475–478
14. Centers for Disease Control (1989) Measles Prevention: recommendations of the Immunization Practices Advisory Committee (ACIP) Morb Mortal Weekly Rep 38: 1–18
15. Atkinson WL, Orenstein WA, Krugman S (1992) The resurgence of measles in the United States, 1989–1990. Annu Rev Med 43: 451–463
16. Lennon JL, Black FL (1986) Maternally derived measles immunity in era of vaccine-protected mothers. J Pediatr 108: 671–676
17. Pabst HF, Spady DW, Marusyk RG et al (1992) Reduced measles immunity in infants in a well vaccinated population. Pediatr Infect Dis J 11: 525–529
18. Englehandt SF, Halsey NA, Eddins DL, Hinman AR (1980) Measles mortality in the United States 1971–1975. Am J Public Health 70: 1166–1169
19. Orenstein WA, Atkinson WL, Mason D, Bernier RH (1990) Barriers to vaccinating preschool children. J Health Care Poor Underserved 1: 315–320
20. Centers for Disease Control (1992) Retrospective assessment of vaccination coverage among school-aged children-selected U.S. cities. Morbid Mortal Weekly Rep 41: 103–107
21. National Vaccine Advisory Committee (1991) The Measles Epidemic. The problems, barriers, and recommendations. JAMA 226: 1547–1552
22. Cutts FT, Orenstein WA, Bernier RH (1992) Causes of low preschool immunization coverage in the United States. Annu Rev Publ Health 13: 385–398

3

Mass vaccination programme aimed at eradication of measles in Sweden

B. Christenson

Department of Environmental Health and Infectious Diseases Control, Karolinska Hospital, Stockholm, Sweden

Summary

In 1982, a two-dose vaccination programme with the combined vaccine against measles, mumps and rubella (MMR) was introduced in Sweden [20]. The vaccination coverage was sufficiently high to reduce transmission of these viral infections [1, 2, 4, 5, 7]. The reported incidence of measles decreased from 76 per 10 000 individuals in 1982 to 1.2 per 100 000 in 1988[7]. Vaccination against measles had been introduced in Sweden in 1971 and was offered to children from the age of 18 months. In 1981, reports from child-health centres showed that 56% of all the pre-school children had been vaccinated against measles. In 1982, 88% of the 12-year-olds born in 1970 received the trivalent MMR vaccine. The coverage continued to increase and during 1989 an average of 95% of the children received the trivalent vaccine at both occasions. The vaccination coverage has been sufficiently high to reduce transmission of measles in Sweden. Evaluation of the immune status before and after vaccination is essential for the prevention of infections. At the same time as the vaccination schedule was introduced a vaccination study was started. Each year between 400 and 800 12-year-old school-children were tested on serum samples obtained prior to and after vaccination.

Only minor variations of the prevaccination immunity to measles were seen during the period 3–7 years after introduction of the programme. The age groups studied had partly been vaccinated against measles earlier. Between 12 and 16% lacked prevaccination or natural immunity. The seroconversion rate of children seronegative for measles was high, i.e. 100% in 1985 and later varied between 96 and 97%. During the follow-up period there was a declining incidence of measles, mumps and rubella. The relationship between the vaccination and reduction of disease and natural immunity strongly suggests that the association is causal and that this vaccination policy reduced the transmission of infection.

Introduction

The two-dose programme of vaccination with the combined vaccine against measles, mumps and rubella (MMR-vaccine) was introduced in Sweden with the ultimate aim of the elimination of these three virus diseases. The vaccination is optional and is offered free of charge to all children at two ages – at 18 months at Child-Health Centres and at 12 years at School Health Centres. The second dose, which is given irrespective of earlier diseases or vaccinations, is given in order to prevent the build-up of a susceptible population in older age-groups.

The epidemiological considerations and the first experiences with the two-dose vaccination schedule have already been published [1, 2, 4–7]. Vaccination against measles had been possible in Sweden since 1971. Between 1972 and 1977 the acceptance of the measles vaccination was low, only 10–15% of the child population were vaccinated. It increased thereafter much due to that many pediatrics who had been negative to vaccination against measles in the beginning changed and accepted the immunization. In 1981, reports from child-health centres showed that 56% of all the pre-school children had been vaccinated against measles and between 1988 and 1990 it was found that 70% of the 12-year-old children already had received a measles immunization prior to the MMR vaccination. Even before the vaccination was introduced in Sweden few children < 1 year were infected with measles. The peak beeing in the ages 3–9 and 7–10 years. After that the vaccination against measles was introduced the peak of measles cases has been displaced to higher ages > 20 years.

Subjects and methods

Data collection

The Swedish national register of infectious diseases constituted the source of basic information. General practitioners have monthly between 1968–1988 reported all cases of clinical measles.

All laboratories in Sweden report cases of verified measles together with all other microbiological diagnosis to the Department of Epidemiology at the National Bacteriological Laboratory in Stockholm. Hospitalized cases of measles have been reported since 1981.

Study design

Since the introduction of the two-dose regimen blood samples have each year been collected from vaccinees immediately before vaccination and two months after. A first study was performed in 1982 on 18-month-old children from well-baby clinics [4]. In 1983, the first study of the 12-year-olds was carried out on 247 children from four regions [5] and in 1985 a similar vaccination study was performed on 496 schoolchildren [1]. The latest studies were performed between 1986 and 1990 on 12-year-old children from five different regions in Sweden, constituting a representative selection of this population [7]. The vaccination was carried out as part of the routine health work in schools but with extra assistance to the blood sampling of the children.

Serological testing

For measure of antibodies against measles the haemolysis-in-gel (HIG) method, [21] complemented by neutralization titre (NT), was used in 1985–7. In 1988 and 1990, the enzyme-linked immunsorbent assay (ELISA) technique replaced the HIG method for measles after the assay had been tested in parallel with HIG and NT [6,8].

Vaccines

The vaccine used was manufactured by Merck Sharp and Dohme. It contained the Edmonston measles strain (strength $> 4–8000\,TCID_{50}$), the Jeryl Lynn mumps strain ($> 100\,000\,TCID_{50}$) and the RA 27/3 rubella strain ($> 3–6000\,TCID_{50}$).

Results

It is of great importance when vaccination schedules aiming at the elimination of virus diseases are introduced to simultaneously commence a sero-epidemiological programme to check the efficacy and effect of the immunization efforts. Defining the serum antibody response after vaccine-induced immunity is therefore of particular interest. For the determination of measles immunity, several techniques have been used. It has earlier been found that haemagglutination inhibition (HI), which has been the most commonly used method of measuring measles antibody, is less ideal, in that it is cumbersome to perform and not sensitive enough [6].

The ELISA was found to be a very sensitive method of monitoring changes in the immunity pattern of vaccine-induced immunity. It was also simple to perform and thus suitable for large-scale sero-epidemiological screening. The ELISA was sensitive and reliable in the testing of low-titred serum which is necessary when measuring vaccine-induced immunity. The specificity was very high and in no instance yielded a discrepant positive or negative result compared with NT. In epidemiological screening it was possible to discriminate between late post-vaccination and early post-vaccination titres [6,7].

In the first survey (1983) of 247 12-year-old children it was found that 20% of the children coming from urban areas and 6% of the children from rural areas were susceptible to measles [5]. These differences were dependent on the prior rate of vaccination against measles. At that time, information on the immunization status of the individuals had not routinely been collected. The ante-vaccination statuses of the vaccinees were therefore recorded from 1987 onwards. In 1989 64% had had a prior vaccination against measles and in 1990 69% of the children. In spite of that these children, borne in 1977, had not been reached with the general vaccination schedule which was introduced in 1982, a prior measles vaccination was very common. There was no increase

Table 1. Percentage of 12-year-old children seronegative to measles before vaccination with the combined measles, mumps and rubella vaccine (MMR). Between 56% and 69% had received a measles vaccination earlier

Year	Children born in	Prior to vaccination (%)
1985	1972	64/496 (13)
1986	1973	83/542 (15)
1987	1974	93/756 (12)
1988	1975	68/513 (13)
1989	1976	66/420 (16)
1990	1977	68/476 (14)

Table 2. Seroconversion against measles in 12-year-old children

Year	Seroconversion (%)
1985	100%
1986	99%
1987	98%
1988	96%
1989	97%
1990	97%

in the proportion of seronegative children prior to MMR vaccination in 1985 compared with 1990 (Table 1). The seroconversion rate was 100% in 1985, when conversion was seen in all seronegative children (64/64). In 1988 the conversion rate was 96 and in 1989 and 1990 to 97% (Table 2).

Since 1972, the reported number of clinical cases of measles has decreased from 228 to 1.2 cases per 100 000 individuals in 1988. In 1982 and 1983, these figures were 76 and 56 cases per 100 000 individuals (Fig. 1). The figures are based on the clinical diagnoses reported monthly by the general practitioners.

The age distribution of the measles cases in the beginning of the 70-ties (1970–1974) just at the introduction of the measles immunization compared with the measles cases 1986–1990 is seen in Fig. 2. The age distribution has changed to older ages the last years. The peak before being in the young ages and lately in the ages of 20–29 years. Measles in children under the age of 12 months has been rare in Sweden. As can be seen in Fig. 2 the peak was in the age 7–10 years before the immunization was introduced.

Cases hospitalized for measles since 1981 is shown in Table 3. In 1981, 372 cases were hospitalized and among these 15 were encephalitis cases.

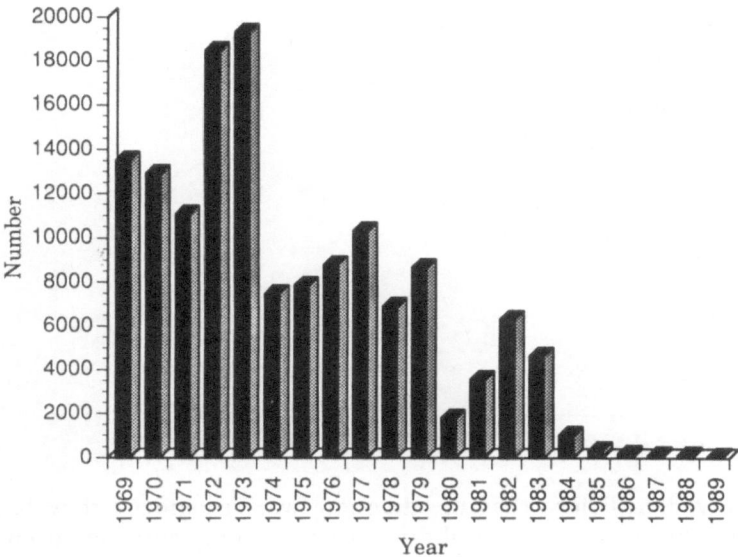

Fig. 1. Measles: number of cases annually reported by GPs in Sweden

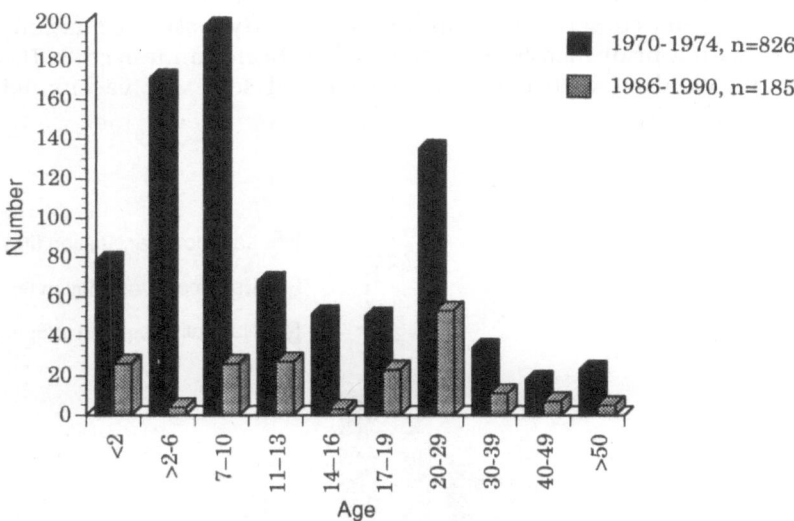

Fig. 2. Age distribution of laboratory verified measles cases in Sweden, 1970–1974 and 1986–1990

Since 1985 no encephalitis case has been reported. The 16 measles cases observed in 1989 was due to an outbreak which occurred among immigrants in Sweden. It also came to involve some hospital staff who had had contact with these cases. This outbreak involved about 30 persons, 13 of whom were children to refugees who were hospitalized in Gothenburg, the second biggest

Table 3. Cases hospitalized for measles in
Sweden 1981–1989

Year	Total number of cases	Encephalitis cases
1981	372	15
1982	388	15
1983	248	8
1984	81	1
1985	9	–
1986	11	–
1987	10	–
1988	1	–
1989	16	–

town in Sweden. Before that the diagnose was recognized they had been
waiting in the pediatrics consultation room among other children in the
hospital. Five hospital staff borne in the 1960ties became infected and 10
other children 16–18 months old who had been exposed in the hospital.

The Fig. 3 shows NT-antibody distribution of titre values in 160 12-year-
old children who had a natural immunity compared with children who had
late post-vaccinated sera and children with early post-vaccination sera.
Children with a prior measles vaccination had been immunized 8–10 years
earlier and children with early post-vaccinated sera was vaccinated two
months previously.

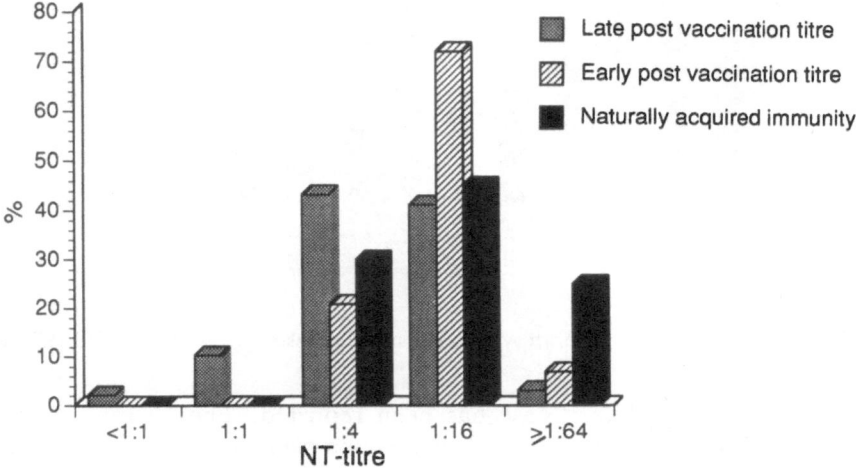

Fig. 3. NT-antibody distribution against measles in connection with MMR vaccination seen
in 160 12-year-old children.

The figure demonstrates that there was a continuous decrease of antibody levels after vaccination. This development strengthens the motivation for the two-dose vaccination schedule carried out in Sweden and also to have regular follow-up studies of the immunity status of the population in relation to the vaccinations.

Discussion

Measles virus infection is about to be eliminated in Sweden. The incidence of the disease become so low that the sentinel surveillance system was abandoned in 1989 and only cases verified by laboratory diagnose is now reported. Cases associated with death and disability no longer exist. Infection with measles have became so rare that the disease is hardly recognized any longer. Younger physicians and medical students have never seen a measles case. It has therefore become important that all cases where measles could be suspected is laboratory verified.

Infection by natural measles virus has been believed to induce lifelong immunity. These are based on observations that reports of second attacks of measles are rare. From an outbreak on the isolated Faroe Island [18] it was observed that the only residents who became infected were those who had not had measles before in an outbreak that had occurred 65 years earlier. Other reports have however indicated that measles can occur in persons who prior have been exposed to measles if antibody titres fall under a certain level [3,19]. But this has not been proved. It has also been beleived that immunization with live-attenuated measle vaccine will give life long protection [9]. It is beleived that secondary vaccine failures do not paly an important role in measles outbreaks [10]. Even if one dose vaccination would give life long protection this schedule would build up a susceptible population in older age groups. Our data together with other observations suggest however that the vaccine induced immunity could wane with time which further strengthens the need for a two-dose programme [6,12].

The main aims of the two-dose vaccination programme against measles, mumps and rubella were to avoid building up a large number of young adults being susceptible to these diseases and to ensure a high coverage and long lasting immunity. The purpose of the second vaccination was to immunise those who were not vaccinated the first time and those who failed to respond to the first inoculation, but it would also achieve a booster effect in children with low titres. If 90% of the children are vaccinated with 90% efficiency on both occasions, fewer than 5% of the adolescents will lack immunity. Since the two-dose programme began in 1982, the morbidity from measles, mumps and rubella decreased considerably (Fig. 1). The vaccination efficacy and coverage thus appeared to be high enough to interfere with the transmission of the viruses.

When a vaccination aiming at the elimination of virus diseases is introduced the effect of the immunization efforts should also be followed. A sero-epidemiological study should be commenced simultaneously to ensure that the vaccination programme gives a satisfactory immunity. The seroconversion to measles was high, between 96 and 100%. The antibody titres were lower after vaccination than after naturally occurring disease in all three diseases [6,12]. It was found that there was a continuous decrease of antibody levels after vaccination. This suggests that the percentage of susceptible individuals in the grown-up population may increase with time. There have lately also been reports of immunization failures and that infection occurred among persons who had seroconverted 10 years earlier [12]. Outbreaks of measles and mumps occurred in vaccinated school populations in the USA [14,16]. This problem has recently attracted the attention of the American Academy of Pediatrics [15]. A two-dose measles vaccination schedule has recently been recommended in USA [13,15]. This development strengthens the motivation for the two-dose vaccination schedule carried out in Sweden since 1982 and regular follow-up studies of the immunity status of the population in relation to the vaccinations.

Booster effects after vaccination with the MMR vaccine were seen in sera from children who had previously received a measles vaccination. A significant decrease in antibody levels in late post-vaccination sera, compared with early post-vaccination sera was discovered. Children with naturally acquired measles immunity had significantly higher ELISA titres than those with a prior measles vaccination.

The significant difference of antibody levels in late post-vaccination sera compared with early post-vaccination sera suggested that the vaccine-induced immunity may be lost in time [6]. The response to the live vaccine in the presence of low levels of antibodies in pre-vaccination sera may also indicate that low levels of antibody do not prevent infection. It is uncertain what the secondary vaccine failures will mean in time and its role in measles outbreaks [17]. A second dose might have a significant impact on the long lasting immunity status in thus prevent measles. The role of waning immunity is poorly understood [11]. This stresses the necessity of continuous, sero-epidemiological, follow-up studies to monitor changes, both quantitatively and qualitatively, of the immunity status in the population when one has the goal to eradicate infectious diseases.

References

1. Böttiger M, Christenson B, Taranger J and Bergman M (1985) Mass vaccination programme aimed at eradicating measles, mumps and rubella in Sweden: Vaccination of schoolchildren. Vaccine 3: 113–116
2. Böttiger M, Christenson B, Romanus V, Taranger J and Strandell A (1987) Swedish experience of two-dose vaccination programme aiming at eliminating measles, mumps and rubella. Br Med J 295: 1264–1267
3. Chen RT, Markowitz LE, Albrecht P, Stewart JA, Mofenson LM, Preblud SR and Orenstein WA (1990) Measles antibody: reevaluation of protective titers. J Infect Dis 162: 1036–1042

4. Christenson B, Böttiger M and Heller L (1983) Mass vaccination programme aimed at eradicating measles, mumps and rubella in Sweden: First experience. Br Med J 287: 389–391
5. Christenson B and Böttiger M (1985) Vaccination against measles, mumps and rubella (MMR): A comparison between the antibody responses at the ages of 18 months and 12 years and between different methods of antibody titration. J Biol Stand 13: 167–172
6. Christenson B and Böttiger M (1990) Comparison of methods of screening for naturally acquired and vaccine-induced immunity to measles. Biologicals 18: 207–211
7. Christenson B and Böttiger M (1991) Changes of the immunological patterns against measles, mumps and rubella. A vaccination programme studied 3 to 7 years after the introduction of a two-dose schedule. Vaccine 9: 326–329
8. Kirsten A, Weigle M, Murphy D, Brunell PA (1984) Enzyme-linked immunosorbent assay for evaluation of immunity to measles virus. J Clin Microbiol 19: 376–379
9. Krugman S (1983) Further-attenuated measles vaccine: characteristics and use. Rev Infect Dis 5: 477–481
10. Markowitz LE, Preblud SR, Orenstein WA, Rovira EZ, Adams NC, Hawkins GE, Hinman AR (1989) Patterns of transmission in measles outbreaks in the United States 1985–1986. N Engl J Med 320: 75–81
11. Markowitz LE, Preblud SR, Fine PEM, Orenstein WA (1990) Duration of live measles vaccine-induced immunity. Pediatr Infect Dis J 9: 101–110
12. Mathias RC, Meekison WG, Arcand TA, Schecter MT (1989) The role of secondary vaccine failures in measles outbreaks. Am J Public Health 79: 475–478
13. Centers for Disease Control. Measles prevention: recommendation of the Immunization Practices Advisory Committee (ACIP) (1989) M M Wly R 38 [Suppl 9]: 1–18
14. Measles: Hungary (1989) M M Wkly Rep 38: 665–668
15. Committee on Infectious Diseases, American Academy of Pediatrics. Measles: reassessment of the current immunization policy (1989) Pediatrics 84: 1 110–1113
16. Mumps outbreaks on university campuses Illinois, Wisconsin, South Dakota (1987) M M Wly R 36: 496–505
17. Nokes DJ and Anderson RM (1988) Measles, mumps and rubella vaccine: what coverage to block transmission? Lancet ii: 1374
18. Panum PL (1940) Observations made during the epidemic of measles on the Faroe Islands in the year 1846. New York: American Public Health Association, p52
19. Pelner L (1941) Repeated attacks of measles. Am J Dis Child 62: 358–361
20. Rabo E and Taranger J (1984) Scandinavian model for eliminating measles, mumps and rubella. Br Med J 289: 1402–1404
21. Strannegård Ö, Grillner L and Lindberg IM (1975) Hemolysis-in-gel-test for demonstration of antibodies to rubella virus. J Clin Microbiol 6: 491–494

4

Identification of high risk groups for measles immunization

F. M. LaForce

Department of Medicine, The Genesee Hospital and the University of Rochester School of Medicine and Dentistry, Rochester, N.Y., USA

Summary

Global immunization programmes can take justifiable pride in the dramatic increase in the fraction of infants under one year who have received measles vaccine which has been linked to a corresponding drop in measles-related mortality. These successes have dramatized conditions and situations associated with continued circulation of wild measles virus. Enhanced immunization efforts should be focused on these high risk groups at the same time that measles surveillance systems are improved.

Introduction

Because of the morbidity and mortality associated with measles, its control is a high public health priority for all countries. National immunization programs have responded to this challenge and global measles vaccination coverage by 12 months of age has increased from 20 percent in 1974 to 70 percent in 1990 (Expanded Programme on Immunization, 1991). Measles immunization programs have prevented an estimated 2.1 million deaths in 1990 (Expanded Programme on Immunization, 1991). Despite these dramatic successes there remain foci where transmission of measles virus continues unabated. This paper will highlight what is known about these high risk groups and situations and suggest techniques by which they can be identified so that they may be targeted for special emphasis.

Well recognized high risk groups

Table 1 enumerates five high risk groups or situations that have been associated with endemic and epidemic measles.

Table 1. Groups at high risk for measles

1. Urban poor
2. Refugees
3. Hospitalized children
4. Infants visiting health care facilities
5. Infants living in areas with
   low measles vaccine coverage

Urban poor

Infants of the urban poor are particularly vulnerable to measles because case fatality ratios are inversely related to age of acquisition of measles. Disease transmission in West African cities is intense because of crowding and the common custom of mothers carrying their infants to market. In these settings measles is a disease of the first two years of life.

In addition, the global trend towards urbanization shows no sign of slowing. Demographers predict that urban populations in developing countries will increase by almost 50 percent from 1985 to 2025 (Table 2). The number of cities in Africa of over 500,000 inhabitants has grown from 3 in 1969 to 28 in 1980. This trend will continue to place major stresses on urban health facilities and city-based immunization programmes. Lastly "chains of transmission" of measles often begin in cities and spread to rural areas.

Studies from the EPI in Zaire have shown that achieving immunization rates between 50 and 60 percent in Kinshasa using a strategy of a single dose of measles vaccine at 9 months, did not result in a decrease in reported measles cases [5]. Control of measles in this setting was further complicated in that over one-quarter of all cases were reported in children under 9 months of age. These data and other studies, have suggested that epidemic cycles of measles in West African urban areas will not be broken unless there is either a change in vaccination strategy or the attainment of significantly higher vaccination rates.

Table 2. Percentage change in urban and rural populations in developing countries 1985–2025* (Base year 1970)

Year	Cumulative percent change	
	Urban	Rural
1985	7	3
2005	24	6
2025	48	4

*From Universal Child Immunization, UNICEF, 1990

Coetzee and associates have studied measles immunizations in a rapidly growing peri-urban township in South Africa [1]. Most of the residents originated from a rural area over 1000 km away. Only 15 percent of the population lived in formal housing with piped water; the majority of residents were squatters living in shacks. Sixty three percent of children from 12 to 23 months of age had received measles vaccine. This coverage rate over-estimated protection since the immunization strategy is a two dose schedule at 6 and 9 months and the survey noted that only 16 percent of measles doses were given after 9 months. Lower immunization rates were noted in children who lived in the township less than six months and those born in rural areas. These findings have been noted by others who have stressed the concept of the constant inflow of susceptibles in growing urban areas as a barrier to measles control.

Refugees

Measles is a well known health hazard in all refugee populations. Toole and Waldman have recently summarized mortality trends among refugee populations in Somalia, Sudan and Thailand [6]. Their data showed that during the early "emergency phase" mortality rates in refugees were up to 40 times higher than those for non-refugee populations. Deaths were skewed to a younger population as noted in Table 3.

Table 3. Age-specific mortality in refugees (emergency phase)*

	Year	Mort. rate (deaths/10 000/day)			
		Age group (years)			Crude
		< 1	1–4	15–44	
Thailand	1979	10.7	7.6	2.2	3.8
Somalia	1980	27.0	14.0	2.6	4.7

*From Toole and Waldman, 1989

Table 4. Cause-specific mortality – three Sudan camps in 1985*

Month	Deaths/1000/month		
	Measles	Other causes	Total
February	16.0	13.4	29.4
March	5.0	9.2	14.2
April	5.2	14.8	20.0
May	2.8	15.1	17.9
June	0.8	9.2	10.0
July	0.2	4.9	5.1

*From Toole and Waldman, 1989

Measles was an important contributor to mortality [6]. An analysis of cause-specific mortality in 1985 refugee camps in the Sudan showed that measles accounted for a major share of mortality (Table 4). Fortunately, measles outbreaks can be brought under control with aggressive immunization programs.

Infants visiting health facilities

The infectivity of measles is legendary. This phenomenon has been best studied when outbreaks of measles have occurred in well immunized populations. Over the last decade the importance of measles transmission in the health care setting has been increasingly recognized. Infants with measles who are brought to physicians' offices or to health care clinics place all unprotected infants who are visiting that site at risk from measles. This risk may occur in the absence of any direct contact between the index and the secondary case. Of 443 children exposed to measles in an Abidjan, Côte d'Ivoire clinic over a four day period, 39 returned to the clinic 10–20 days later [3]. Fifteen of those exposed children had acquired measles for an attack rate of 3.4 percent. Health facilities may inadvertently serve as foci of.amplification of measles transmission. The end result is the exposure of unprotected infants brought for routine well baby care or, even worse, exposure of susceptible infants who are debilitated from a disease requiring medical attention.

Hospitalized infants

Nosocomial measles can be devastating; mortality in malnourished children with measles in a hospital setting is frequently above 50 percent. In a 1972 randomized study from Zimbabwe hospitalized children who had not had measles were randomized to receive immunoglobulin, no treatment or measles vaccine [4]. Children given vaccine had a mortality of 7 percent compared to 17 percent in the no treatment group. Immunoglobulin recipients had a mortality of 11 percent. The difference in mortality was due to deaths among children who acquired measles at the hospital.

Areas with low measles vaccine coverage

The axiom equating lack of measles immunization with the certainty of acquiring clinical measles remains valid. Hence, low measles vaccine coverage under one year, for whatever reason, defines a group of infants at high risk.

Suggested strategies for the identification and management of groups at high risk of measles

The best data for the determination of high risk groups are those coming from a comprehensive surveillance system which follows measles incidence over time. Some countries, like Morocco, which have high coverage rates throughout the country, successfully use data from routine reporting to identify areas which need special emphasis. Unfortunately, countries with well developed surveillance systems are in the minority, since disease surveillance is the weakest component of most immunization programs. Furthermore, when measles data are available, there is often insufficient analysis at the district or provincial level.

The control of measles requires an enumeration of measles cases. The best source of case data will vary from country to country and from area to area within a country. Identifying the best source of measles cases is an important task for EPI personnel at the district and urban level.

Urban areas

There are two important sources of information to assess the adequacy of urban immunizations. The first is the fraction of infants under one year who have been given measles vaccine. Coverage rates based on results from cluster sample surveys or vaccine use in a defined population are standard techniques to evaluate vaccination coverage. Because of the intensity of exposure in urban settings, measles immunization rates should be targeted to be above 70 percent.

A second source of informtion, and one which may be more helpful in identifying high-risk areas, involves the analysis of surveillance data. If these data are not available, a focused review of hospital or health center data on measles cases may be necessary. Admitting and out-patient registers can be used as a source of descriptive information on measles, and the following information, either whole or in part, may be obtained: (a) age, (b) outcome (hospital cases), (c) geographic origin of cases, (d) immunization status (often not available).

These data can be used to estimate the fraction of measles cases occurring in infants under nine months of age. For example, if more than 25 percent of reported measles cases occur in infants under 9 months it may be appropriate to consider, if the resources are available, introducing a two dose schedule. In addition, the geographic origin of cases can be used to identify high-risk groups with presumably low immunization rates.

Refugee areas

There is little need to conduct detailed epidemiologic studies of measles in refugee camps; rather, emphasis should be placed on the delivery of measles vaccine to all eligibles and the establishment of some type of a surveillance system. All refugee areas should be enumerated and a policy should be in place which insures that measles immunization has a high priority in relief activities. The following questions or points may offer useful landmarks to evaluate the completeness of refugee-related measles immunization activities.

Measles immunizations in hospitals

Efforts should be concentrated on developing a policy whereby all pediatric admissions are given measles vaccine on admission. The likelihood is so high that these infants will come into contact with children with clinical measles that a two dose measles vaccine strategy is recommended for infants between 6 and 9 months of age.

Infants visiting health facilities

As previously mentioned unimmunized infants are at risk of being exposed to measles during visits to health centers. Measles vaccine should be offerred to all health facilities for all encounters.

Low coverage areas

Cluster sample surveys will continue to serve EPI programme managers well in helping to define general levels of measles vaccine coverage in the under one population. At the district or provincial level measles vaccine utilization data should be reviewed to identify areas where measles vaccine coverage in infants under 1 year is less than 70 percent. Many mature immunization programmes have realized that surveillance data can be more useful in identifying nests of low measles immunization coverage and helping to prioritize areas where special efforts are necessary.

References

1. Coetzee N, Yach D, Blignaut R, Fisher SA (1988) Measles vaccination coverage and its determinants in a rapidly growing peri-urban area. SAMJ 78: 733–737
2. Expanded Programme on Immunization, Global Advisory Group (1990) WHO Weely Epidem Rec 1991 (1/2): 3–7; (3): 9–12
3. Foulon G, Klein-Zabban ML, Gnasnov-Nezzi L, Martin-Bouyer G (1983) Preventing the spread of measles in children's clinics. Lancet ii: 1498–1499
4. Glyn-Jones R (1972) Measles vaccine and gamma globulin in the prevention of cross infection with measles in an acute pediatrics ward. Central Afr J Med 18: 4–9
5. Taylor WR, Ruti-Kalisa, Mambu Ma-Disu, Weinman JM (1988) Measles control efforts in urban Africa complicated by high incidence of measles in the first year of life. A J Epidemiol 127: 788–794
6. Toole MJ, Waldman RJ (1989) An analysis of mortality among refugee populations in Thailand, Somalia, and Sudan. Bull WHO 66: 237–247
7. Universal Child Immunization (1990) UNICEF

5

Is measles eradicable?

A. R. Hinman and **W. A. Orenstein**

Centers for Disease Control and Prevention, Atlanta, Georgia, USA

Introduction

Measles remains one of the major killers of children. In the absence of vaccination, the World Health Organization estimates that more than 2.8 million would die each year as a result of measles. Even though current measles vaccination coverage worldwide is estimated to be 80%, more than 800 thousand infants and children died in 1991 as a result of measles [1]. Consequently, measles is certainly a worthy target for eradication.

In this chapter we will consider definitions of eradication (drawing heavily on a 1984 paper entitled "Prospects for disease eradication or elimination" [2]), factors favoring or not favoring eradication of measles, describe the findings of mathematical models of measles transmission as they relate to eradicability, and conclude with a description of the practical experience with measles eradication.

Definitions

The *Oxford English Dictionary* defines eradication as "the action of pulling out by the roots, total destruction, extirpation [3]". Cockburn defined eradication as the "extinction of the pathogen that causes the infectious disease in question; so long as a single member of the species survives, then eradication has not been accomplished. The definition implies action on a worldwide scale [4]".

Soper said that the objective of eradication was the complete elimination of "the possibility of occurrence of a given disease, even in the absence of all preventive measures [5]". Finally, Andrews and Langmuir defined eradication as the "purposeful reduction of specific disease prevalence ... to the point of continued absence of transmission with a specified area [6]".

There are some differences in these definitions, but they all share the concepts that eradication is a purposeful act and that the term should not be applied to the disappearance of a particular pathogen or disease in the absence of specific eradication efforts. Although not necessarily stated overtly, they also share the concept that after an initial period of effort, measures against the disease can be terminated without reappearance of the disease. This carries with it a global implication, which was made explicit in Cockburn's definition. Although Cockburn agrees that "regional" eradication could be achieved, he states that this "implies a basically unstable situation because at any time the infection may be reintroduced by carriers or vectors from outside [4]." Soper states that "eradication can, therefore, be regarded as that state in which the infection does not return from infected areas after control measures have been abandoned. If procedures have to be continued to prevent return of the infection, then the state is one of control and not of eradication [5]".

Attempting to synthesize the various definitions, we believe it is useful to consider a spectrum of disease incidence with five levels that depend on the degree of success of the efforts taken against the disease (Fig. 1). In the natural state, disease incidence is affected by the traditional triad of host, agent, and environment, and although evolutionary changes might lead to disappearance of the disease, there is no specific control action.

The application of specific measures to reduce the occurrence of a disease may lead to a situation of control, in which continuing efforts leave the occurrence of the disease at a reduced, but presumably acceptable, level. This is the situation with many infectious diseases in the United States at present. From a state of control, elimination of a disease from a particular area can be achieved. In this situation, the disease no longer occurs on a continuing basis in the area, but the threat of reintroduction of disease from outside this area (with subsequent reestablishment of transmission) is so great that continuing control efforts are required. This situation describes our objectives for measles and rubella in United States [7,8]. As the risk of reintroduction of disease is lowered through measures applied in other countries, regional eradication may be attained. In this case it would not be necessary to pursue actively the control measures; surveillance and prompt response to importation are capable of maintaining the area free of disease. This represents the present status of malaria in many countries and islands. The final leap, from regional eradication to global (or "true") eradication,

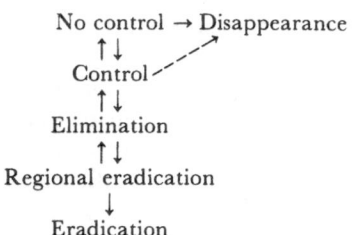

Fig. 1. Spectrum of disease incidence

would only occur when all areas had achieved regional eradication. To date, smallpox is the only example of a disease which has been truly eradicated.

Strategies proposed for elimination or eradication contain three major elements: achieving and maintaining high levels of immunity, effective surveillance to detect all suspected cases, and aggressive response to the occurrence of disease. This paper will focus on the first element – achieving and maintaining the necessary high levels of immunity.

Factors favoring and not favoring eradication of measles

The factors which make measles a candidate for eradication are also those which make it a favorite of mathematical modelers:

1) short and predictable incubation period (about 10 days),
2) short communicable period (about 8 days) and absence of a long-term carrier state for infectious virus,
3) lack of a natural non-human reservoir,
4) life-long immunity following infection,
5) high clinical expression of disease (few inapparent infections),
6) relative antigenic stability of the measles virus,
7) highly effective vaccine producing long-term (probably life-long) immunity.

There are also several factors which do not necessarily favor eradication:

1) high infectivity, making it necessary to achieve very high levels of immunity (often in very young populations) in order to interrupt transmission,
2) airborne transmission, which means that face-to-face contact may not be necessary to become infected and complicates investigation of outbreaks,
3) presence and persistence of maternally-derived passive immunity in infants for varying periods, making it impossible to successfully vaccinate newborns with current vaccines,
4) the fact that current vaccines are not perfectly effective,
5) lack of visible evidence of immunity,
6) lack of a pathognomonic clinical syndrome, complicating surveillance and case investigations, and making close laboratory support essential,
7) lack of popular belief in many countries that measles is a serious enough problem to warrant an eradication effort.

Models of measles transmission

Hedrich studied the patterns of measles occurrence in Baltimore in the period 1900–1931 and reported that when the level of immunity was higher than 53%, epidemics did not develop [9]. This suggested a threshold of herd immunity that would protect against a measles epidemic in that city. Subsequent experience has demonstrated that such a low proportion of immunes will not interrupt transmission.

Sencer et al announced in 1966 that the epidemiologic basis existed for the eradication of measles from the United States [10]. Without making a specific estimate of the proportion of children who would have to be immunized they said that "it is difficult to estimate whether the threshold of herd immunity for an average American city now would be higher or lower than Hedrich's estimate for Baltimore 30 to 70 years ago. There is not reason, however, to question the validity of the basic assumption that the occurrence of measles epidemics depends upon the balance of immunes and susceptibles, and that for all areas and special groups in this country the immune threshold is considerably less than 100%."

Yorke and Nathanson also reviewed data on measles in Baltimore as well as other U.S. cities and estimated that if 94% or more of the population is immune to measles, transmission would not be sustained and measles would die out [11]. In the United Kingdom, Anderson and May estimated that more than 95% of children in successive cohorts would have to be successfully immunized in order to achieve eradication [12]. Hethcote estimated that herd immunity would be achieved and measles would disappear from the United States if 93.5% or more of the population was immune [13]. Choi estimated there would still be a 10–30% chance of secondary cases even when immunity levels in a population were 96% [14]. Anderson and May and Hethcote also point out that, in order to be able to eradicate a disease, it is essential to vaccinate children at an age earlier than the median age of infection. In most developed countries, the median age of infection is on the order of 5–6 years; by contrast, in some developing countries (and in selected areas elsewhere) the median age of infection may be between 1–2 years, offering a much narrower period in which to carry out vaccination.

Given the fact that measles vaccine is not perfectly effective,' achieving the levels of immunity required for eradication necessitates exceedingly high level of vaccine coverage with a single dose or administration of more than one dose of vaccine to a high proportion of susceptibles. At the same time, it must be remembered that the actual number of susceptibles and their degree of contact are as important, or more important, than the proportion of susceptibles in terms of the liklihood of transmission following introduction of measles virus. In a 1971 paper, Fox et al stated "... the potential for contact spread of an agent depends entirely on the number of susceptibles and their opportunities for contact with each other ... No matter how large the proportion of immunes in the total population, if some pockets of the community, such as low economic neighborhoods, contain a large enough number of susceptibles among whom contacts are frequent, the epidemic potential in these neighborhoods will remain high ... Success of a systematic immunization program requires knowledge of the age and subgroup distribution of the susceptibles and maximum effort to reduce their concentration throughout the community, rather than aiming to reach any specified overall proportion of the population [15]."

Using measles surveillance data from West Africa, the late Professor George Macdonald produced mathematical models simulating the transmission of measles in the area in an attempt to identify the prerequisites for interrupting measles transmission [16,17]. Unlike the situation in industrialized countries, where measles transmission primarily takes place in school-age children, in Africa measles strikes infants and very young children, in part because they are carried to crowded markets on their mother's backs. His models suggested two approaches that might interrupt measles transmission in West Africa: annual mass vaccination reaching at least 90% of susceptible children or an ongoing program in which 75% of all susceptible children were vaccinated as they became susceptible at approximately six months of age. Absence of the necessary infrastructure necessitated the choice of mass campaigns. Using this approach, measles transmission was interrupted for varying periods. In the Gambia, a zero incidence of indigenous measles was achieved and maintained for more than two years.

The recent success of poliomyelitis eradication programs using mass vaccination aproaches and favorable results from pilot mass measles vaccination programs has reawakened interest in the simultaneous mass vaccination approach to interrupt transmission. In the English-speaking Caribbean countries a goal has been set to eliminate indigenous transmission of measles by 1995 using a two-pronged approach: achieving and maintaining coverage of at least 95% among children 12–15 months of age along with at least one mass vaccination campaign to reach older susceptibles [18].

Whatever approach is used, it is clear that a high level of population immunity must be achieved if measles is to be eradicated. The high communicability of measles suggests that termination of transmission using a single dose strategy may be difficult. Table 1 depicts immunity levels achieved with a single dose of vaccine given varying levels of efficacy and coverage. 95% coverage with a 95% effective vaccine would give immunity levels of 90.2%, leaving 9.8% of the population susceptible. Levels of 96% immunity would only be achieved with a 98% effective vaccine given to 98% of the population. Even with 100% coverage, susceptibility levels could be as high as 10% if the vaccine was only 90% effective.

Table 1. Immunity level after one dose under differing states of efficacy and coverage

Coverage	Efficacy		
	90%	95%	98%
	% Immune		
90%	81.0	85.5	88.2
95%	85.5	90.2	93.1
98%	88.2	93.1	96.0

Table 2. Immunity level after 2 doses under differing states of efficacy and coverage

Coverage	Efficacy		
	90% % Immune	95%	98%
90%	89.1	89.8	90.0
95%	94.0	94.8	95.0
98%	97.0	97.8	98.0
100%	99.0	99.8	100.0

To achieve higher immunity levels, a two dose strategy has been recommended. It assumes that primary vaccine failure is essentially a random event and that someone who did not respond to a first dose would be as likely as anyone else to respond to a second dose. Table 2 shows the immunity levels that would be achieved with two doses of vaccine if the efficacy is the same for both doses. The coverage figures represent the coverage in the general population for the first dose and complete vaccination of first dose recipients with the second dose. For example, the entry in the top left corner shows that 89.1% immunity would be achieved if 90% of the population receives one dose of a 90% effective vaccine and all of those recipients receive a second dose. If 100% of the population receives both doses, even with a vaccine of only 90% efficacy, 99% immunity levels are achieved. Thus, a two dose schedule is attractive, particularly if the coverage achieved can be high. At any coverage, virtually all of the vaccine failures are eliminated and the remaining susceptibles are essentially all unvaccinated.

Two dose schedules have been adopted in many countries which have established elimination targets, including the former Czechoslovakia, the Scandinavian countries, and the United States.

Practical experience with measles eradication

Several countries, including Canada, Cuba, Finland, the former German Democratic Republic and Czechoslovakia, Norway, Sweden, and the United States have individually established targets for elimination of measles and the countries of the European Region of the World Health Organization have adopted a Regional target of elimination of indigenous measles by the year 2000 [19]. Some of us have called for global eradication of measles [10], although others, such as D. A. Henderson, have felt this was unrealistic and potentially dangerous to the overall public health effort [21]. The World Health Assembly and the World Summit for Children set a somewhat more conservative goal: "Reduction by 95% in measles deaths and reduction by 90% of measles cases compared to preimmunization levels by 1995 as a major step to the global eradication of measles in the longer run [22]".

Another paper in this volume has summarized the United States experience with measles elimination [23]. Among the industrialized countries, considerable success has been seen in the former German Democratic Republic and Czechoslovakia and the Scandinavian countries, all of which (except GDR) use a two dose strategy. Among the developing countries, the Gambia had success with a cycled mass vaccination approach in the 1960's. Cuba and the English-speaking Caribbean countries are close to elimination using a modified single-dose approach with supplemental doses provided in mass campaigns reaching a very high proportion of children.

It is clear that with current vaccines and current strategies, reductions in the toll caused by measles can be brought about worldwide and control of measles can be achieved in most countries. In many countries, elimination can be achieved although this may require two-dose strategies. In some areas regional eradication may be achieved. Prospects for global eradication are less clear. No major geographical area has yet successfully eliminated measles and maintained that elimination. There is not yet consensus on whether eradication can be achieved, and if so, what the appropriate strategies are to achieve eradication. Mass campaign approaches, particularly during periods of natural low transmission, may offer some hope of rapidly interrupting transmission. In this regard, it will be very interesting to see the results of efforts to eliminate measles in a number of countries in the Americas in the next few years [17].

In our estimation, there are two major technical impediments to eradication:

1) the need to achieve very high levels of immunity before children are exposed to measles. With current vaccine efficacy in the field, this may not even be possible with 100% coverage with a single dose or 90% coverage with two doses. Use of a mass vaccination approach with somewhat lower coverage might temporarily interrupt transmission but whether twice-yearly mass campaigns could successfully maintain a sizeable area measles-free is unknown.

   There are two components to achieving the necessary levels of immunity – the delivery system and the vaccine itself. The delivery system must be capable of getting children vaccinated at the earliest possible age. In the United States we have been very successful in ensuring that children are immunized before they first enter school – immunization levels among school enterers are 98% nationwide. In much of the country this has successfully interrupted transmission. However, in some areas, notably inner cities, conditions of transmission may be more like those in the developing world, with the median age of infection between one and two. In many such cities the delivery system has not been capable of reaching more than 50–60% of children before their second birthday with the result that transmission continues [24]. We are working to improve the delivery system in these areas. Obviously, if the vaccine was 100% effective it would be more feasible to reach the needed immunity levels with a single dose strategy. Although there is some evidence that a few individuals may lose vaccine-induced immunity over time, it does not seem likely this will pose a major problem.

2) the inability of current vaccines to effectively overcome maternally-derived passive immunity in infants. Since this immunity persists for variable periods, we must delay vaccination until a majority of infants can respond (approximately 9 months in most of the world). This has two major implications. The first is that many children will be susceptible to measles for varying periods before vaccine is given at an age when they are most vulnerable to complications and death. The second is that, since coverage levels generally decline with increasing age at administration of vaccine, it is harder to reach the high proportion of children needed than would be the case if vaccine could be administered in the newborn period.

Development of a 100% effective vaccine that could be given at birth would go a long way to resolving these two impediments but better vaccine alone will not accomplish eradication. Improvements in the delivery systems are also needed. Although measles is theoretically eradicable, realization of the goal may require improvements in vaccines as well as improvements in vaccine delivery systems. Perhaps most importantly, the professional and political will must be developed if eradication is to be achieved.

References

1. World Health Organization Expanded Programme on Immunization Information System. WHO/EPI/CEIS/92.1, Geneva, 1992; 1–163
2. Hinman AR (1984) Prospects for disease eradication or elimination. NY St J Med 84: 501–506
3. Oxford English Dictionary (1971) New York, Oxford University Press
4. Cockburn TA (1961) Eradication of infectious diseases. Science 133: 1050–1058
5. Soper EL (1962) Problems to be solved if the eradication of tuberculosis is to be realized. Am J Public Health 52: 734–745
6. Andrews JM, Langmuir AD (1963) The philosophy of disease eradication. Am J Public Health 53: 1–6
7. Hinman AR, Nieberg PI, Brandling-Bennett AD (1979) The opportunity and obligation to eliminate measles from the United States. JAMA 242: 1157–1162.
8. Orenstein WA, Bart KJ, Hinman AR, Preblud SR, Greaves WL, Doster SW, Stetler HC, Sirotkin B (1984) The opportunity and obligation to eliminate rubella from the United States. JAMA 251: 1988–1994
9. Hedrich AW (1933) Monthly estimates of the child population "susceptible" to measles. 1900–1931, Baltimore, MD. Am J Hyg 17: 613–636
10. Sencer DJ, Dull HB, Langmuir AD (1967) Epidemiologic basis for eradication of measles in 1976. Public Health Rep 82: 253–256
11. Yorke JA, Nathanson W, Pianigiani G, Martin J (1979) Seasonality and the requirement for perpetuation and eradication of viruses in populations. Am J Epidemiol 109: 103–123
12. Anderson RM, May RM (1982) Directly transmitted infectious diseases: control by vaccination. Science 215: 1053–1060
13. Hethcote HW (1983) Measles and rubella in the United States. Am J Epidemiol 117: 2–13

14. Choi K, Millar JD, Young C (1983) Measles outbreak in a community of 800: computer simulation study. Presented at the Annual Meeting of the American Epidemiological Society. Tucson AZ, March 24–25, 1983
15. Fox JP, Elveback L, Scott W, Gatewood L, Ackerman E (1971) Herd immunity: Basic concept and relevance to public health immunization practices. Am J Epid 94: 179–189
16. Miller JD (1940) Theoretical and practical problems in measles control. Proceedings: Seminar on Smallpox Eradication and Measles Control in West and Central Africa. Proceedings of a meeting in Lagos, Nigeria, May 13–20, 1969. P.H.S.-S.E.P. Rep. IV.2, January 30, 1970
17. Thacker SB, Millar JD (1991) Mathematical modeling and attempts to eliminate measles: A tribute to the late Professor George MacDonald. Am J Epidem 133: 517–525.
18. —— (1990) Plan to eliminate indigenous transmission of measles in the English-speaking Caribbean countries. Bulletin PAHO 24: 240–246
19. World Health Organization Regional Office for Europe (1984) Summary report of Second Conference on Immunization Policies in Europe, Karlovy Vary, December 10–12, 1984. 1985; ICP/EPI 001/m01(S)
20. Hopkins DR, Hinman AR, Koplan JP, Lane JM (1982) The case for global measles eradication. Lancet i: 1396–1398
21. Henderson DA (1982) Global measles eradication (letter). Lancet ii: 208
22. World Declaration on the Survival, Protection and Development of Children and Plan of Action for Implementing the World Declaration on the Survival, Protection and Development of Children in the 1990s, World Summit for Children, United Nations, September 30, 1990
23. Orenstein WA, Markowitz LE, Atkinson WL, Hinman AR (1993) The experience with measles in the United States. In: Kurstak E (ed.) Measles and poliomyelitis: Vaccines, Immunization, and Control, Springer Verlag Wien New York, pp. 25–35.
24. Centers for Disease Control (1992) Retrospective assessment of vaccination coverage among school-aged children – selected U.S. cities. Morbid Mortal Weekly Rep 41: 103–107.

6

Safety of measles vaccines

P. E. M. Fine

Communicable Disease Epidemiology Unit, Department of Epidemiology and Population Sciences, London School of Hygiene and Tropical Medicine, London, United Kingdom

Summary

Standard dose further attenuated measles vaccines, widely used throughout the world for the past two decades, are among the safest vaccines in use today. Local reactions at the site of injection are minimal. The dominant reactions are associated with a mild, measles-like syndrome which occurs in from 2 to 30% of recipients approximately a week after vaccination. Febrile convulsions may occur in approximately one per 1000 vaccinees, but generally resolve without sequelae. There is some evidence that measles vaccines may be associated with encephalitis or encephalopathy, but causality is still unconfirmed. The attributable risks may be on the order of one per 100 000 vaccinations. If SSPE is associated at all with measles vaccines its risk is ten-fold less frequent than after wild measles infection, thus no higher than 1 per million vaccinees. All such risks pale in comparison to those associated with natural measles, which still remains endemic throughout the world.

Evidence that high titre measles vaccines may be associated with increased long term mortality has led to the withdrawal of their use.

Introduction

It is a truism that, although modern vaccines provide some of our most powerful and cost-beneficial tools in public health, no vaccine is perfectly effective or perfectly safe. That said, the vaccines included in current EPI programmes have all passed stringent scrutiny, in particular with regard to their safety. Given that no vaccine is totally without risk, the decision to use it reflects a cost – benefit calculation balancing costs and risks associated with vaccination against costs and risks associated with the natural disease – or with the unvaccinated state. Inevitably the latter is a function of the severity and the frequency of the target disease in the population. This audience needs no remainder of the importance of measles as a cause of morbidity and mortality.

The subject of measles vaccine safety is complicated by the variety of vaccines and vaccination policies in use. Intensification of measles control efforts in recent years has led to major innovations in the use of different vaccine virus strains, in recommended vaccine potency, and in the recommended ages and number of doses for measles vaccination [21,32]. Though these changes have been driven mainly by the desire to maximize efficacy, or numbers of cases prevented, they may have implications for the risk of untoward events associated with vaccination. In addition, the introduction of combined MMR vaccines in developed countries complicates the attribution of risks to the measles component alone. Given this changing situation, it is necessary continually to reappraise the balance between benefits and risks.

The need to appraise risks is a broad obligation for vaccination programmes. Vaccines are unusual among medical interventions in that they are generally given to healthy, well individuals – to whom we have an obligation to at least do no harm. Many countries are facing increasingly the issue of compensation for adverse outcomes attributable – or claimed to be attributable – to vaccines. It thus behooves those responsible for vaccination programmes to examine closely the safety of vaccine preparations and procedures in order to take appropriate measures to minimize any associated risks. There is also the pragmatic concern that real or perceived vaccine – associated adverse events reflect ill upon a programme, and may led to undermining of the public support upon which it relies. The spectre of events associated with pertussis vaccines in recent decades, in many countries, shades – or illumines – considerations of the safety of all vaccines [10,26].

We must also appreciate the difficulty of evaluating risks of adverse reactions to vaccines. In general we are dealing with rare events, of a sort which may just as well occur in the absence of vaccination. Much of the literature on the safety of vaccines relates to case reports of one or another adverse event having occurred in temporal association with – and hence possibly attributable to – a vaccination. Attribution of cause in such circumstances is a difficult business. Individuals who receive vaccines are in general different from those who do not, in a variety of ways which may in turn reflect their risk of adverse outcomes or of such outcomes coming to the attention of the health services. Such confounding factors introduce major difficulties to attempts to evaluate the safety of vaccines within routine programmes [9]. Ideally one needs properly controlled trials in which vaccinated and unvaccinated individuals can be compared directly. In this context, the situation with measles vaccines has some advantages, insofar as the recent developments of new vaccines and procedures have encouraged several excellent controlled trials which do allow evaluation of at least the high frequency risks which may be associated with such vaccines.

Most of the experience accumulated over three decades of measles vaccination programmes relates to further attenuated vaccine strains (in particular Moraten and Schwarz) given between 9 and 24 months of age. This experience has given measles vaccines a reputation for a high degree of safety. But the introduction of higher titre vaccines, and recommendations for their use below nine months of age, raised potentially new situations, and deserves particular attention.

Review of the evidence

Hazards associated with measles vaccines may be divided into those which are attributable to errors on the part of the programme or vaccinators – for example contaminated vaccines or injection equipment, or improper injection technique – and those attributable to responses to the measles vaccine itself. The first category need not concern us here, though it should not be forgotten. Such problems arise with all vaccines, but should be rare with a well trained staff and an efficient programme. It is the latter category which concerns us – untoward or adverse outcomes attributable to biological responses to measles vaccines properly constituted and administered according to a programme's recommendations.

These may be categorized in a variety of ways, for example: local problems at the injection site; systemic reactions occurring soon after vaccination; and delayed or long-term effects.

Local injection-site reactions

Several investigators have examined the injection sites and reported reactions in terms of local redness, swelling or tenderness. These are in general minimal, and not a cause for concern. Experience with new high-titre preparations and with vaccination of children in early infancy – before 9 months of age – shows no evidence that the new regimens raise particular problems in this regard. Sore arms are not a significant problem with measles vaccines.

Systemic reactions occurring soon after vaccination

Measles-illness-like reactions

Measles vaccines are attenuated live virus preparations which act by inducing a mild "case" of measles in recipients. Inevitably there is a range of clinical responses to these induced infections, and a proportion of recipients manifest symptoms reminiscent of classical measles: fever, rash, cough, conjunctivitis and diarrhoea. These symptoms generally have onset between 5 and 10 days after injection, and are milder and of shorter duration than those associated with infection with wild virus.

Given that such symptoms are not uncommon among children of measles vaccination age, being associated with a wide variety of infections, the measurement of risks attributable to measles vaccination requires comparisons with an appropriately matched control group. Several such controlled trial comparisons have been carried out. The studies have revealed a considerable range of risks associated with measles vaccination, a range which is attributable to differences in definitions and methods of ascertainment of the responses in the various investigations, as well as to the vaccines used [6,30]. Observed attributable risks of fever and rash range from nil to 30%. Of particular contemporary interest are several studies comparing risks of such

Table 1. Systemic adverse reactions observed in randomized controlled trials of high titre Edmonston-Zagreb (EZ) measles vaccines compared with conventional Schwarz (SW) vaccines

Whittle et al. (Gambia) $10^{4\,6}$ EZ vaccine at 4 months [37]

	Number	> 38°	Rash	Cough	Diarrhoea
Control	63	8%	0%	75%	35%
Hi EZ @ 4 mo	62	11%	0%	53%	29%

Markowitz et al. (Mexico)   Lo ($10^{3\,6}$) vs Hi ($10^{5\,6}$) EZ vs SW at 6 or 9 months [22]

	Number	≥ 38°	Rash	Conjunct.	Diarrhoea
Control	63	11%	2%	10%	10%
Lo EZ @ 6 mo	62	5%	8%	18%	42%
Hi EZ @ 6 mo	62%	2%	18%	15%	39%
Lo SW @ 9 mo	109	8%	15%	18%	34%

Titres expressed as Plaque Forming Unit counts. Percents reflect the proportions of recipients manifesting the specified symptom

events between children receiving high titre vaccines before 9 months of age and unvaccinated controls or children receiving conventional Schwarz-type vaccines [22,37]. Observed differences in reaction rates have been greater between studies than between groups within studies; but the results in general suggest that the new high titre vaccines and schedules do not pose significant acute hazards to recipients (Table 1).

Uncommon/rare systemic adverse events

A variety of more severe "adverse events" have been reported as possibly associated with measles vaccination, in particular convulsions, encephalitis and encephalopathy, and conditions such as subacute sclerosing panencephalitis (SSPE), Guillain Barre syndrome or Reye's syndrome [32]. Because of their rarity, the study of such conditions, and of their possible association with measles vaccination, has been extremely difficult. Rough estimates of the risks of such events associated with measles vaccines are summarized in Table 2. We comment briefly upon the evidence behind such estimates.

Only convulsions occur with sufficient frequency – approximately 1 per 1000 vaccinations – to have been measurable directly in even the largest vaccine trials, such as British Medical Research Council trial in the 1960s. In that trial, 18 physician-confirmed convulsions occurred within three weeks of vaccination in 9577 recipients of Schwarz-type vaccine, compared with 5 convulsions in 16 237 unvaccinated controls, providing an attributable risk of 0.0016 [24]. All but one of these convulsions was associated with fever. This is the only controlled trial estimate we have, and is consistent with follow-up studies of children vaccinated in routine programmes [25].

Risk estimates for rarer adverse events potentially associated with vaccines have had to be derived by less direct means. Some evidence comes from routine surveillance. For example CDC/Atlanta received reports of 59 and 5 measles vaccine-associated encephalopathies during the years 1963–71 and 1979–84, respectively, when some 50.9 and 18.1 million doses of measles vaccines were administered, giving much quoted estimates of 0.3 to 1 per million doses [4,5,19]. Given that ascertainment by routine surveillance is likely to be incomplete, and that such systems have no means to distinguish attributable from coincidental events, such estimates are difficult to interpret. The frequency estimated by this method is similar to the expected background incidence in the population, and thus one is left with just a vague reassurance that whatever the risk of measles vaccine-attributable encephalopathy or encephalitis may be, if it exists at all, it is very small. Evidence that the risk is not zero is provided by a small number of case reports of measles viruses (probably vaccine type) having been isolated from individuals with postvaccination encephalitis [11] or encephalopathy [35]; but such reports are very few, considering the hundreds of millions of doses of measles vaccines which have been administered.

The largest and best known controlled observational study of vaccines as risk factors for encephalopathies has been the National Childhood Encephalopathy Study (NCES), a large case control investigation carried out in the United Kingdom during the late 1970's [2]. Though this study is best known for its findings related to pertussis vaccines, it also revealed a significant association (relative risk 2.3; $p < 0.05$) between convulsions or encephalopathy (9 cases with simple or febrile convulsions [one with transient hemiplegia] and 5 with encephalitis or encephalopathy) and the receipt of measles vaccine between 7 and 14 days prior to onset. The authors derived an estimate of attributable risk – "of the onset of a serious neurological disorder within *fourteen* days of an immunization with measles vaccine ... for previously normal children and irrespective of eventual clinical outcome" – of 1 in 87 000 vaccinations (95% confidence limit 1/830 000 to 1/25 000 vaccinations). This is the only available controlled estimate for measles-vaccine-attributable risk of "serious neurological disorder"; but it should be appreciated that the study faced formidable methodological difficulties, and so even this estimate must be interpreted with extreme caution. (Of the 9 vaccine-associated cases in this study who were classed as "simple/febrile convulsions", all were considered normal when examined 12 months later. Of the 5 originally diagnosed as encephalitis/encephalopathy, 3 recovered and 2 had mild persisting defects one year later. The risk cited in Table 2 is thus derived as $(5/14)/87\,000$).

Unlike convulsions, encephalitis and encephalopathies, SSPE is unique in that the vast majority of cases are apparently attributable to latent or persistent measles virus infection. When live virus measles vaccines were introduced, there thus was a concern that risk of SSPE associated with measles vaccine could be equal or even greater than that associated with wild virus infection. However, a dramatic decline in SSPE observed several years after the decline in measles showed this not to be true, and indicates that

Table 2. Rare neurological adverse events associated with conventional measles vaccines

Adverse event	Risk associated with	
	Measles disease	Measles vaccine
Convulsion	5–10 per 10^3	1.6 per 10^3[24,25]
Encephalitis/encephalopathy	5–40 per 10^4	0.003–0.01 per 10^4[4,5,19] 0.04 per 10^4[2]
SSPE	5–20 per 10^6	< 1 per 10^6[12,28]

the risk of SSPE associated with conventional further attenuated measles vaccines is at most a tenth that associated with wild virus infection, thus on the order of 0.7 per million doses [28,32]. Whether there is in fact any risk at all has not been confirmed. A few cases of SSPE have occurred in individuals with a history of vaccination but not of measles diasease – but it is possible that they had experienced an unrecognized infection with the wild measles virus at some time. Though one virus strain isolated from an SSPE case has characteristics of a vaccine-type virus, there is some concern that this isolate may reflect a laboratory contaminant.

There has not yet been sufficient experience with high titre vaccines or with vaccines given before nine months to evaluate their implications for such rare potential adverse events.

Adverse events in high risk groups – the problem of contraindications

Concern over potential risks associated with measles vaccines has led to the consideration of certain groups which might be particularly prone to such adverse responses. Chief among these are children who are already ill, in particular with febrile disease or tuberculosis, malnourished children, pregnant women, those with allergies to neomycin or egg protein, immunosuppressed individuals, and those infected with HIV. Each of these conditions has been considered a contraindication for measles vaccination in some contexts.

Intercurrent infections

The inclusion of febrile illness as a contraindication was in part because of concern that intercurrent infection might impair efficacy of the vaccination procedure, and in part over a concern that vaccine reactions might worsen the clinical condition of the child. Several recent studies have now shown that children with a variety of illnesses respond well to standard further attenuated vaccines [14,18,27] and that the vaccines are not harmful to them [15,27]. Concerns over tuberculosis in particular have now been discarded by most programmes – though measles vaccines may depress tuberculin responsiveness, there is no evidence that such vaccination increases the risk of clinical tuberculosis [32,39]. On this basis the EPI has recommended that febrile illness should not be considered a contraindication unless a child is so ill as to require hospitalization – in which case the decision on measles vaccination is deferred to the hospital authority (who may well recommend vaccination in order to protect the child from nosocomial transmission). There is little documented experience with high titre vaccines in clinically ill infants.

Malnutrition

Arguments to withhold measles vaccines from malnourished children stemmed from fears over inadequate immune responsiveness, and theoretical concerns that the mild measles-type syndrome induced by measles vaccine virus might exacerbate the clinical condition of the child. These concerns have been abandoned, after several studies showing adequate seroconversion rates in such children, and evidence that measles vaccination does not lead to worsening of the clinical condition [14,15,27]. Given the high risk of measles infections in situations where malnutrition occurs, and the serious effects of wild measles in malnourished children, malnutrition is now generally considered more an indication than a contraindication for measles vaccine. Limited data on high titre vaccines and young age groups show no reason to change this recommendation.

Pregnancy

Pregnancy is generally considered a contraindication for live virus vaccines, including measles. This is based on theoretical grounds, however, insofar as measles vaccine virus – unlike rubella and mumps vaccine viruses – has not been shown to cross the placenta [32].

Allergies

Neomycin allergy and severe allergy to egg protein are generally recognized as contraindications to measles vaccines. Though anaphylactic reactions have been reported in association with measles vaccines (e.g. 9 per 170 000 vaccinations monitored by [31]), perhaps attributable to such preconditions, they are rare. Recommendations to skin test egg-protein allergic children prior to vaccination are controversial [16,34]. Ideally, vaccination clinics should be equipped, and staff trained, to handle rare anaphylactic reactions to any vaccine.

Immunosuppression

It is generally accepted that live measles vaccines are not safe in individuals who are immunosuppressed either because of underlying illness (e.g. congenital or acquired immune deficiency, leukaemia, lymphoma or generalized malignancy), medication (e.g. high dose steroids) or therapy (e.g. radiation) [32].

HIV infection

Early concerns that measles vaccines might be ineffective, or cause serious illness, in children with symptomatic or asymptomatic HIV infection have not been substantiated. Careful observation of several hundred HIV positive individuals subsequent to receipt of standard dose further attenuated measles vaccines have revealed no cause for concern [17,23,29,36]. As with malnutrition, current opinion favours that HIV infected children are at particular need of measles vaccination, as wild measles virus infections can be particularly severe in such individuals. Experience of high titre vaccines in HIV infected infants less than 9 months of age is still limited, but has revealed no particular risks [7,20].

Delayed effects of measles vaccination

Three different situations have arisen suggesting that certain measles vaccination policies could have unexpected, delayed, inopportune effects upon the immune system. Two of these situations have now been more or less resolved, but they serve to remind us of the potential complexity of such interventions, and of the need for careful vigilance in the full evaluation of vaccination programmes.

Killed vaccines and atypical measles

Among the first measles vaccines licensed in the 1960's were formalin-inactivated virus preparations. Follow-up of recipients of these vaccines revealed that some developed an aberrant and sometimes severe clinical "atypical" disease when exposed to wild measles virus. Investigations revealed a mechanism – the inactivation process destroyed immunogenicity of the F (fusion factor = hemolysin) antigen, one of the 6 major structural proteins of the measles virus. Recipients of such vaccines thus developed an immune response emphasizing antibodies to the heamagglutinin but not the fusion protein, a profile which predisposed to the occurrence of Arthus-type delayed hypersensitivity reactions on exposure to wild measles virus [3,32]. This situation is now mainly of historical interest, as production of such inactivated vaccines was discontinued more than two decades ago. Its current relevance is that individuals who did receive such inactivated measles vaccines should have or are recommended to receive live vaccine..., and as a reminder of the potential for long term inopportune effects of some vaccines.

Low HI titres and two-dose regimens

The second instance arose in early studies of two-dose measles regimens. [38] noted that individuals who were vaccinated first at 6–10 months and then revaccinated at over 12 months had lower HI antibody titres than individuals who were vaccinated only once at over 12 months. Subsequent studies have confirmed this effect, though its mechanism remains unclear [8,33]. On the other hand, and importantly, there is no evidence that these lower HI titres in revaccinated individuals are reflected in lower protection. The only reason for mentioning them in this context is as a reminder of inadequately understood immunological implications of certain vaccination schedules, in particular those involving vaccination in the presence of maternal antibody.

Long-term effects on mortality

More recently, evidence has arisen that high titre measles vaccines may be associated with increased mortality risks lasting for up to three years after vaccination. This effect was noted first by Aaby in data from rendomized trials of high titre vaccines in Guinea Bissau (*written communication, Aaby to EPI, 1990*), and later observed in data from Senegal [1,13]. Detailed analyses have suggested that the effect may be restricted largely if not entirely to females. Absence of consistent evidence for increased mortality associated with high titre vaccines in populations with low background mortality suggests that the effect may be mediated by increasing slightly the susceptibility to prevailing causes of death in the community, rather than by directly inducing death by any specific cause. Though the evidence for increased mortality observed to date has not been conclusive, it has been considered sufficiently strong for EPI to withdraw its 1989 recommendation that high titre vaccines be given to young infants in areas where measles is a significant cause of death before nine months of age (*to be published in Weekly Epidemiological Record*).

Conclusions

Standard dose further attenuated measles vaccines, widely used throughout the world for the past two decades, are among the safest vaccines in use today. Local reactions at the site of injection are minimal. The dominant reactions are associated with a mild, measles-like syndrome which occurs in from 2 to 30% of recipients approximately a week after vaccination. Febrile convulsions may occur in approximately one per 1000 vaccinees, but generally resolve without sequelae. There is some evidence that measles vaccines may be associated with encephalitis or encephalopathy, but causality is still unconfirmed. The attributable risks may be on the order of one per 100 000 vaccinations. If SSPE is associated at all with measles vaccines its risk is ten-fold less frequent than after wild measles infection, thus no higher than 1 per million vaccinees. All such risks pale in comparison to those associated with natural measles, which still remains endemic throughout the world.

Evidence that high titre measles vaccines may be associated with increased long term mortality has led to the withdrawal of the recommendation for their use in infants below nine months of age.

References

1. Aaby P, Samb B, Simondon F, Whittle H, Seck AMC, Knudsen K, Bennett J, Markowitz L, Rhodes P (1991) Child mortality after high titre measles vaccines in Senegal: the complete data set. Lancet 338:1518
2. Alderslade R, Bellman MH, Rawson NSB, Ross EM, Miller DL (1981) The National Childhood Encephalopathy Study. In: Whooping Cough: Reports from the Committee on Safety of Medicines and the Joint Committee on Vaccination and Immunisation. Her Majesty's Stationery Office, London.
3. Annunziato D, Kaplan MH, Hall WW, Ichinose H, Lin JH, Balsam D, Paladino VS (1982) Atypical measles syndrome: Pathologic and serologic findings. Pediatrics 70: 203–209
4. Centers for Disease Control (1984) Adverse events following immunization: Surveillance report number 1, 1979–1982. Issued August 1984
5. Centers for Disease Control (1986) Adverse events following immunization: Surveillance report number 2, 1982–1984. Issued December 1986
6. Cockburn WC, Pecenka J, Sundaresan T (1966) WHO – supported comparative studies of attenuated live measles virus vaccines. Bull WHO 34: 223–231
7. Lepage P, Dabis F, Msellati P, Hitimana D-G, Stevens AM, Mukamabano B, Van Goethem C, Van de Perre P (1992) Safety and immunogenicity of high-dose Edmonston-Zagreb measles vaccine in children with HIV-1 infection. A cohort study in Kigale, Rwanda, Am J Dis Child 146:550–555.
8. Desada-Tous J, Cherry JD, Spencer MJ, Welliver RC, Boyer KM, Dudley JP, Zahradnik JM, Krause PJ, Walbergh EW (1991) Measles revaccination. Persistence and degree of antibody titre by type of immune response. Am J Dis Child 132: 287–290
9. Fine PEM, Chen RT (1991) Confounding in studies of adverse reactions to vaccines. Am J Epidem 136: 121–135.
10. Fine PEM, Clarkson JAC (1986) Individual versus public priotities in the determination of optimal vaccine policies. Am J Epidem 124: 1012–1020
11. Forman ML, Cherry JD (1967) Isolation of measles virus from the cerebrospinal fluid of a child with encephalitis following measles vaccination. Presented at the 77th Annual Meeting of the American Pediatric Society, Inc. April 26–29 1967 (Abstract)
12. Galazka AM, Lauer BA, Henderson RH, Keja J (1984) Indications and contraindications for vaccines used in the Expanded Programme on Immunization. Bull WHO 62: 357–366
13. Garenne M, Leroy O, Beau JP, Sene I (1991) Child mortality after high titre measles vaccines: Prospective study in Senegal. Lancet 338: 903–907
14. Halsey NA, Boulos R, Mode F, Andre J, Bowman L, Yaeger RG, Toureau S, Rohde J, Boulos C (1985) Response to measles vaccine in Haitian infants 6 to 12 months old: Influence of maternal antibodies, malnutrition and concurrent illnesses. N Eng J Med 313: 544–549
15. Harris MF (1979) The safety of measles vaccine in severe illness. South African Med J 55: 38
16. Kemp A, Van Asperen P, Mukhi A (1990) Measles immunization in children with clinical reactions to egg protein. Am J Dis Child 144: 33–35
17. Krasinski K, Borkowsky W (1989) Measles and measles immunity in children infected with human immunodeficiency virus. J Am Med Ass 261: 2512–2516
18. Krober MS, Stracener CE, Bass JW (1991) Decreased measles antibody response after measles-mumps-rubella vaccine in infants with colds. J Am Med Ass 265: 2095–2096
19. Landrigan PJ, Witte JJ (1973) Neurologic disorders following live measles virus vaccination. J Am Med Ass 223: 1459–1462
20. Mandala K, Mayala B, Cutts F, Brown C, Davachi F, Deforest A, Behets F, Kamenga M, Quinne TC, Markowitz L, Oxtoby M, St Louis ME (1991) Edmonston-Zagreb measles vaccination in HIV infected children. Abstract WB 2040, 7th International Conference on AIDS. Florence, Italy June 1991

21. Markowitz L (1990) Measles control in the 1990's: Immunization before 9 months of age. WHO/ EPI/GEN/90.3

22. Markowitz L, Sepulveda J, Diaz-Ortega JL, Valdespino JL, Albrecht P, Zell ER, Stewart J, Zarate ML, Bernier RH (1990) Immunization of six month old infants with different doses of Edmonston-Zagreb and Schwarz measles vaccines. N Eng J Med 332: 580–587

23. McLaughlin M, Thomas P, Onerato I, Rubenstein A, Oleske J, Nicholas S, Krasinski K, Guigli P, Orenstein W (1988) Live virus vaccines in human immunodeficiency virus infected children: A retrospective survey. Pediatrics 82: 229–233

24. Measles Vaccine Committee of the Medical Research Council (1966) Vaccination against measles: A clinical trial of live measles vaccine given alone and live vaccine preceded by killed vaccine. Br Med J 1: 441–446

25. Miller CL (1982) Surveillance after measles vaccine in children. Practitioner 226: 535

26. Miller DL, Alderslade R, Rose EM (1982) Whooping Cough and Whooping Cough vaccine: the risks and benefits debate. Epidemiol Rev 4: 1–24

27. Ndikuyeze A, Munoz A, Stewart J (1988) Immunogenicity and safety of measles vaccine in ill African children. Int J Epidemol 17: 448–455

28. Okuno Y, Nakao T, Ishida N, Konnot, Mizutani H, Fukuyama Y, Sato T, Isomura S, Ueda S, Kitamura I, Kaji M (1989) Incidence of subacute sclerosing panencephalitis following measles and measles vaccination in Japan. Int J Epidemiol 18: 684–689

29. Onorato IM, Markowitz LE, Oxtoby MJ (1988) Childhood immunization, vaccine-preventable diseases and infection with human immunodeficiency virus. Pediatr Infec Dis J 7: 588–595

30. Peltola H, Heinonen OP (1986) Frequency of true adverse reactions to measles, mumps, rubella vaccine. A double-blind placebo-controlled trial in twins. Lancet i: 939–942

31. Pollock TM, Morris J (1983) A 7-year survey of disorders attributed to vaccination in North West Thames Region. Lancet i: 753–757

32. Preblud SR, Katz SL (1988) Measles vaccine, In Vaccines edited by SA Plotkin and EA Mortimer. WB Saunders, Philadelphia

33. Stetler HC, Orenstein WA, Bernier RH, Herrmann KL, Sirotkin B, Hopfensperger D, Schuh R, Albrecht MD, Lievens AW, Brunell PA (1986) Impact of revaccinating children who initially received measles vaccine before 10 months of age. Pediatrics 77: 471–476

34. Stiehm ER (1990) Skin testing prior to measles vaccination for egg-sensitive patients. Am J Dis Child 144: 32

35. Valmari P, Lanning M, Tuokko H, Kouvalainen K (1987) Measles virus in the CSF in post vaccination immunosuppressive measles encephalopathy. Pediat Infect Dis J 6: 59–63

36. Von Reyn CF, Clements CJ, Mann JM (1987) Human immunodeficiency virus infection and routine childhood immunization. Lancet ii: 669–672

37. Whittle HC, Hanlon P, Hanlon L, O'Neill K, Marsh V, Jupp E, Aaby P (1988) Trial of high dose Edmonston-Zagreb measles vaccine in the Gambia: antibody response and side effects. Lancet ii: 811–814

38. Wilkins J, Wehrle PF (1979) Additional evidence against measles vaccine administration to infants less than 12 months of age: altered immune response following active/passive immunization. J Pediatr 94: 865–869

39. Zweiman B, Pappagianis D, Maibach H, Hildreth EA (1971) Effect of measles immunization on tuberculin hypersensitivity and in vitro lymphocyte reactivity. Int Arch Allergy 40: 834–841

7

Measles vaccine – one versus two doses: why and when

J. Thipphawong, R. Wittes, and **O. Van-Ham**

Connaught Laboratories Limited, Willowdale, Ontario, Canada

Summary

The decision to choose a two-dose measles vaccination schedule depends on whether elimination rather then control is the goal. Both mathematical models and epidemiologic observations suggest that elimination is not possible with only one dose of current measles vaccines. A two-dose schedule is clearly a minimum requirement for elimination. The decision to implement a second dose requires social and governmental acceptance of the need for elimination. The country must have both the desire to commit economic resources and the health care infrastructure to deliver the second dose. Decision makers must realize that it may require ten or more years to achieve elimination. A two-dose schedule will not prevent outbreaks, unless programs are in place to ensure high coverage with both doses. Poorly implemented programs will result in continued outbreaks and loss of public confidence in the strategy. It may be better to identify and vaccinate high-risk groups and thus ensure high first-dose coverage at an appropriate age, rather than to give 90% of the population two doses and 10% of the population no doses.

Introduction

The optimal measles vaccination schedule is dependent on a number of scientific factors, including the total epidemiology of measles, the vaccine that is used, and such host factors as age, concurrent illness, and malnutrition. In addition, regional and national schedules must take into account such operational factors as the cost and availability of vaccines, the type of health care delivery system, and provider and societal acceptance of the benefits and risks of vaccination. For more than a decade, a number of developed countries have targeted measles for elimination. (Elimination refers to the complete interruption of transmission in a defined geographic area). The initial success of programs aimed at controlling measles suggested that this was a feasible objective. With the resurgence of measles in the latter half of the past decade, it has become clear that elimination is a much more elusive target. Due to persistent outbreaks, the United States implemented a two-dose vaccination schedule in 1989, following the example of a number of European countries. Controversy exists over when and where it is preferable to institute a two-dose measles vaccination schedule over a one-dose schedule. We outline some of the reasons for the two-dose schedule, and discuss factors that must be considered before implementing a two-dose schedule.

Measles disease and epidemiology

Measles is an acute, infectious disease characterized by fever and an erythematous maculopapular rash. The rash is preceded by a prodromal period with cough, coryza and conjunctivitis. The virus is transmitted in droplets or by direct contact of the susceptible with nasal or throat secretions of infected individuals, and is transmissible prior to the onset of the typical rash (and thus prior to the ability to diagnose measles).

Measles is one of the most contagious diseases known. In outbreaks, nearly 100% of non-immune subjects are rapidly infected. This has been illustrated by outbreaks in islands: all but 5 individuals in a population of 4000 were infected within 6 weeks in a 1961 Greenland epidemic [30]. Due to its high infectivity, measles transmission cannot be sustained in small isolated populations. A minimum population of 200 000–500 000 [30,61] or an annual birth cohort of 5000–10 000 is required for continuous recirculation of the disease [30].

Prior to the introduction of measles vaccines, nearly the entire birth cohort world-wide was infected with measles. A one-dose vaccine strategy reduced the incidence of measles by 99% in the United States [4]. In countries with effective vaccination programs, measles becomes a sporadic disease, with most cases occurring in scattered outbreaks [59]. The pre-vaccination era epidemic peaks that occurred every 2–5 years have become progressively smaller and less frequent [30].

The average age of infection varies. In developed countires, exposure is generally delayed until school entry [30]. Prior to immunization programs, the average age of infection in North America was 5–9 years of age [45]. In developing countries, it may be as low as 1 to 3 years [89]. In parts of Africa, nearly 100% of children are infected with measles by 4 years of age [102]. Vaccination causes a gradual upwards shift of the average age of infection. In the United Kingdom, the peak incidence shifted during the 1970's from the 5–9 year age group (mainly 5 and 6 year olds) to the 10–14 year age group [51]. The remaining susceptibles (vaccine failures and unvaccinated) are less likely to encounter measles until an older age due to the reduction in incidence.

Complications of measles include otitis media, pneumonia, encephalitis, sub-acute sclerosing pan-encephalitis, and death [102]. Mortality in developed countries has been quoted to range from 1 per 3000 to 1 per

10 000 cases [85,102]. Measles-associated mortality is usually higher among the very young and the very old [30,102]. Mortality in developing countries may be as high as 10% due to one or several factors, including the early age of infection, malnutrition, diarrhoea, concomitant or secondary bacterial infections, and lack of access to good medical care [120]. In populations with no previous exposure to measles, case-fatality rates may exceed 20% [30]. Recent outbreaks in the United States have demonstrated unusually high case-fatality rates. In 1990, there were 89 deaths reported among 27 672 cases, for a case-fatality rate of 3.2 per thousand cases [71]. This high rate may partly be explained by the high number of infant cases. Since the first introduction of measles vaccine in 1960, we are just now witnessing the experience of cohorts of infants born to vaccinated (as opposed to naturally-infected) mothers, who have also had limited exposure to wild measles. The lower levels of maternal antibodies in these infants [93] may result in susceptibility to infection at a younger age and possibly in more severe disease once infected. Additional contributing factors to the higher case-fatality rate include underlying diseases such as HIV infection, factors related to poverty such as poor nutrition, and a higher number of cases in unvaccinated individuals [27]. Less severe cases may also have been under-reported [27,71].

Measles immunity

Immunologic response to measles infection and vaccines

Natural measles infection appears to induce lifelong immunity. This was illustrated by Panum's observation, in an 1846 Faroe Islands measles outbreak, that individuals infected during a previous outbreak 60 years prior were still immune despite high attack rates among those born in the intervening period [79]. Very rare documented cases of clinically-apparent measles reinfection have been reported [102]. Natural measles infection produces higher antibody levels and longer persistence of antibodies than do measles vaccines [68]. Until recently, the immunity derived from measles vaccines was assumed to be lifelong. (Because the oldest cohort to be vaccinated against measles is currently only in its fourth decade of life, it will be many years before an empiric answer is known).

Humoral versus cellular immunity

Both natural measles infection and measles vaccination induce both cellular and humoral immune responses. Children with congenital agammaglobu-linemia appear to be immune to reinfection after primary measles infection; by contrast, individuals with defects of cellular immunity are not protected from infection by immune serum globulin. Immunocompromised individuals, particularly those with defects of the cellular immune system, are at increased risk of giant cell pneumonia, a severe complication of measles. These observations suggest that cellular immune response is crucial for protection against measles [30,90]. However, neonates (who have a very immature cellular immune system) appear to be protected from infection by maternal antibodies that were transmitted *in utero*. Chen et al [38] were able to correlate levels of measles antibodies, as measured by the sensitive plaque neutralization technique, with protection from disease. Some individuals with measurable but low concentrations of antibodies in this study were not protected from disease. It is not clear from these observations whether either arm of the immune system alone, or rather both, are necessary or sufficient for protection against measles infection. Unfortunately, cellular immune studies are difficult to perform and to standardize. Most studies of measles have relied on humoral responses as evidence of infection or immunogenicity of the vaccine.

Measurement of humoral response

Three methods commonly used to measure serologic response to measles vaccination and infection are Hemmaglutination Inhibition (HI), neutralization assays such as Plaque Reduction Neutralization (PRN), and Enzyme Immunoassays (EIA) [33]. HI is the most widely used but the least sensitive. Clinically significant amounts of antibody may be missed by this method; for example, sufficient maternal antibodies may still be present in HI-defined seronegatives to inhibit seroconversion to measles vaccine [6]. Most individuals who seroconvert on HI are considered to be protected from infection. Neutralization assays are the most sensitive, and probably detect the most clinically relevant antibodies, but are difficult to perform. EIAs are relatively sensitive and easy to perform, but may lack specificity in the detection of clinically important antibodies.

Measles infection and vaccination induce IgG, IgM and IgA responses in both serum and nasal secretions. IgA predominates in nasal secretions, whereas first IgM and then later IgG predominate in serum [102]. The timing of specimen collection is important for the detection and determination of absolute antibody levels. Measles IgM, for instance, may be difficult to detect in the first few days of measles infection [73,105]. It is often difficult to compare results from different studies due to the different serologic methods used and the lack of standardization between laboratories in carrying out each of these methods [33].

Measles vaccines

Measles vaccine strains

Measles virus was first isolated in 1954 by Enders and Peebles from a boy named Edmonston [47]. This clinical isolate was the source of most of the measles vaccine used in the world today [65]. These include the

Moraten, Edmonston-Zagreb, and Connaught strains derived from the Edmonston B derivative, and the Schwarz strain derived from the Edmonston A derivative. Another Edmonston derivative of recent interest is the AIK-C strain [75,111]. Most countries in Eastern Europe use derivatives of the Leningrad-16 measles strain [98]. A number of different strains have been developed and are in use in China [44,121]. All measles vaccine strains have been attenuated from the clinical isolates by serial culture in cells such as chick embryo fibroblasts or human diploid cells. The serial passages result in safer viruses that are less likely to cause clinical illness but may also be less immunogenic.

Immunogenicity

The immunogenicity of a measles vaccine is dependent on both the strain and the titre (i.e., the quantity of attenuated virus particles) of the vaccine. Most vaccines typically induce seroconversion rates of 90% to 95% when administered at 12–15 months of age. Seroconversion is much lower with administration under one year of age due to presence of maternal antibodies against measles. (Maternal antibodies against measles are transferred transplacentally to the fetus. These antibodies protect newborns from measles during the first 6 months of life, but also interfere with the response to measles vaccines). Due to the high morbidity and mortality in developing countries from measles in infants less than one year of age, considerable research has attempted to identify an effective vaccine in infants as young as 6 months of age. Attempts to increase immunogenicity in younger age groups have focused on the development of new vaccine strains such as the Edmonston-Zagreb [29,79,116,117], and AIK-C strains [111].

Higher immunogenicity may also be achieved by increasing the titre of the vaccine. Measles vaccines licensed in North America must contain at least log 3.0 TCID [50]. ("Log" = log 10; thus, log $3.0 = 10^{3.0} = 1000$. $TCID_{50}$ and PFU are quantitative assays of the potency of measles vaccines). The World Health Organization (WHO) recommends that measles vaccines used in countries where vaccination is given at 9 months of age contain at least log 4.0 infective particles ($TCID_{50}$ or PFU) [15]. "High-potency" vaccines were initially defined by the WHO as containing greater than log 5.0 [15]; subsequently, this was reduced to log 4.7 [26]. Although the level of serologic immune response may be a predictor of subsequent risk of measles vaccine failure [83], the clinical importance of differences in immunogenicity with different vaccines has yet to be determined. There is some evidence that certain strains may be more efficacious at a younger age [2,18].

Safety

Measles vaccines have always had an excellent safety profile until recent observations of increased mortality among cohorts receiving high-potency vaccines at ages <12 months [8,53]. The increased mortality was only apparent after two or more years of follow-up and was greater among females than in males; its cause is not yet understood. As a result, vaccines with titres greater than log 4.7 infective particles are no longer recommended by the WHO [26]. It is still not clear whether the increased mortality associated with high-potency vaccines would be found with highly immunogenic strains when administered at a standard titre.

Primary versus secondary vaccine failure

Primary vaccine failure is the failure to serologically respond to the measles vaccine. In contrast, secondary vaccine failure (or waning immunity) may be defined as the loss of vaccine-induced protection in an individual who has previously seroconverted to vaccination [58]. The most common factor contributing to primary vaccine failure is vaccination at too young an age, when sufficient maternal antibodies still exist to interfere with response to vaccination. Additional factors causing primary vaccine failure are related to vaccine potency; these include vaccine production problems, failure of the cold chain, and inappropriate administration techniques. Individuals with immunodeficiency (biologic or secondary to immunosuppressive therapy), or with poor health due to concurrent illness or malnutrition, may also have sub-optimal response to vaccination. A recent paper suggests that a mild respiratory illness at the time of vaccination may also reduce the response to measles vaccine [67]. Some individuals who fail to seroconvert have no identifiable risk factors.

The existence of secondary vaccine failure has been well documented in several outbreaks [46,83,104]. At the current time, however, the evidence does not suggest that secondary vaccine failure is a major factor in the recent resurgence of measles in North America. The causes of secondary vaccine failure are not well understood, although it is known that individuals who received either inactivated measles vaccine or else immune serum globulin concurrently with measles vaccine may lose their immunity after several years [102]. The lower incidence of measles in developed countries results in less exposure to measles and consequently less natural boosting of individuals' antibody levels. This may contribute to secondary vaccine failure.

Duration of immunity

As previously mentioned, measles vaccines appear to induce antibody levels lower than those observed following natural infection. Vaccine-induced antibodies also appear to decline at a faster rate than after natural infection [68]. Additional information may be found in a recent review by Markowitz et al [80]. Information is also needed to compare duration of antibodies between different measles vaccine strains [44,80,84,111,118].

Response to a second dose of vaccine

A number of studies suggest that vaccination prior to twelve months of age not only results in lower overall seroconversion rates, but also affects the response to subsequent revaccination [74,102,108]. This may be related to the presence of maternal antibodies that alter the infant's immunologic response to vaccination. If so, then lower infant maternal antibody levels resulting from vaccinated (versus naturally-infected) mothers and the lack of boosting from exposure to measles may alter this risk, allowing for vaccination at a younger age.

Adults who are seronegative some time after vaccination (whether due to primary or secondary vaccine failure) appear to respond to revaccination [32,81]. It appears that individuals who failed to respond to the first shot (the primary vaccine failures) will respond later with a primary immune response. Some initial responders who have low antibody levels on long term follow-up, respond to revaccination with a secondary immune response. Many of these individuals tend to rapidly return to their pre-existing levels of antibody [44,81]. Revaccination will therefore, successfully immunize nearly all adult primary vaccine failures, but may have less benefit for those who have low antibody levels after vaccination. Since immunity also seems to depend on cellular immune response, many of these individuals may still be protected from infection. There is little data on the required time interval between vaccinations to obtain optimal results [44]. Additional research is needed on this subject to guide policy.

Co-administration with other vaccines

Measles vaccine may be administered alone or in combination with mumps and rubella vaccine as MMR. The combination of measles vaccines with the other two antigens does not significantly affect its immunogenicity or reactogenicity when compared to the separate administration of antigens [102]. Although monovalent measles vaccine is cheaper than MMR, the cost of vaccine is not as important as the program cost of delivering the vaccine. The incremental benefits of reduction in mumps and rubella incidence and reduction in delivery costs from MMR far outweigh the increased cost of vaccine in developed countries [28]. The benefits of a second dose of measles vaccine would also be incrementally improved by the reduction in congenital rubella syndrome and mumps-associated complications if given as MMR. MMR may also be administered at the same time as other nonliving vaccines such as DTP and *Haemophilus influenza* type b conjugate vaccines [4,95].

The goal of measles elimination

Control of a disease may be defined as the reduction in transmission to the point where disease incidence no longer represents a major public health problem. Elimination is the complete cessation of transmission in a defined region, while eradication is a complete worldwide cessation of transmission. Yekutiel has reviewed preconditions required for the elimination of a disease [122]. The epidemiologic characteristics of measles suggest that its elimination may be a desirable and achievable goal. These characteristics include:

1. Without immunization, measles is associated with high morbidity and mortality. Even with primary one-dose immunization, considerable ongoing efforts are required to control the disease. Measles is, therefore, an important enough public health problem to warrant consideration for elimination.
2. Man is the only known reservoir for the disease.
3. There is no human carrier state.
4. Except for rare exceptions, infection or vaccination (be it with one, two or periodic doses) confers lifelong immunity.
5. There are no animal or insect vectors.
6. The disease is relatively easily recognized.
7. A suitable intervention – i.e., vaccination – exists. Outbreaks throughout the world have consistently shown that a single dose of vaccine protects 90–95% of appropriately vaccinated individuals.
8. Although minor antigenic variations in circulating measles virus may exist, no evidence to date suggests that current measles vaccines will not protect individuals throughout the world.

In spite of the characteristics listed above, closer examination reveals that significant barriers to elimination or eradication exist [48]. Most importantly, these include the extreme infectivity of measles and its infectivity during the prodromal period (before diagnosis is possible), making outbreak control more difficult. The significance of secondary vaccine failure for elimination has not yet been determined [83]. However, the principle barrier to elimination has been the failure to implement existing control strategies [4].

Theoretical considerations for elimination

Mathematical models can help predict the proportion of the population that must be vaccinated in order to prevent transmission of disease. All mathematical models must incorporate parameters derived from empirical observations. They are therefore limited by the quality of data and the assumptions used to create the models.

Central to the understanding of what proportion of the population must be immune in order to interrupt disease transmission is R_0, the basic reproductive rate. R_0 is defined as the average number of secondary cases produced by one primary case in a wholly susceptible population, in the absence of any medical or public health intervention [8]. R_0 is dependent in part on population density. The higher the population density, the higher the value of R_0. Measles, a highly infectious disease, has been calculated to have an R_0 of between 10–20 [8]. To achieve elimination, the proportion of the population immune to measles must clearly be much higher than was the case with smallpox, the one infectious disease successfully eradicated, which has an R_0 of 2–4.

Measles outbreaks may occur despite measles vaccine coverage approaching 100% [38,87]. The spread of measles depends not on the proportion with immunity, but on the total number of susceptibles and their proximity to each other. In special settings such as schools or colleges or in crowded urban settings, susceptibles are in close proximity to each other. Choi demonstrated in a mathematical simulation that a prevalence of 4% susceptibility in a population would allow continued propagation of measles [110]. In fact, transmission of measles in a school has been demonstrated despite serologic evidence of 96% measles immunity [56].

Mathematical models show that measles immunity of 93–96% in the population at large is required to halt transmission [60,110]. These numbers should be regarded as a minimum. The models assume that susceptibles are homogeneously distributed through the population and are thus less likely to come into contact with a case. As demonstrated in school outbreaks, even in the presence of high vaccine coverage and hence population immunity, transmission may rapidly occur if these susceptibles cluster.

Although a one-dose vaccination policy will reduce the incidence of measles as compared to the absence of measles vaccination, it may not reduce the total number of susceptibles in the population [8]. Before mass vaccination, nearly 100% of adults were infected as children and were thus permanently protected. Depending on the age of vaccination, one dose of measles vaccine will allow 5–10% of individuals to remain susceptible. Immediately after the implementation of a mass measles vaccination program, the total number of susceptibles will decline as children are vaccinated before infection. The cohort of naturally infected adults will, however, eventually be replaced by the cohort of vaccinated adults with a lower population immunity level. This will result in a gradual rise in the number of susceptibles, potentially to levels observed prior to introduction of vaccination. Thus, the current one-dose policy may not reduce the total number of population susceptibles in the long run, but merely shift them to an older age [72].

Even with a single-dose coverage of 95%, a vaccine failure rate of less then 5% will allow the gradual accumulation of sufficient susceptibles to sustain periodic epidemics [72,110]. If elimination is the goal of measles vaccination programs, then a two-dose measles vaccination schedule will be required. Even a two-dose measles vaccine schedule will require high vaccination coverage of both doses to prevent the accumulation of susceptibles and therefore to prevent outbreaks (see Table 1).

Due to the highly infectious nature of measles, elimination is difficult if not impossible to maintain without control of the disease in adjoining regions. High coverage rates will need to be maintained for years. A cohort of adolescents that is completely vaccinated with two doses would not be realized for 10–14 years if the second dose is given at school entry [76]. Options

Table 1. Immunity levels achieved with different vaccine overage and efficacy of one and two doses of measles vaccine*[†]

Coverage	Efficacy					
	90%		95%		98%	
	one	two	one	two	one	two
			% immune			
90	81	89	86	89	88	90
95	85	94	90	95	93	95
98	88	97	93	98	96	98

Source: Markowitz LE, Orenstein WA (1990) Measles Vaccines. Pediatric Clinics of N.
    America 37(3): 603–625
*It is assumed that all first-dose recipients receive a second dose, that the second dose is
    limited to first-dose recipients, and that the efficacy of the second dose is equivalent to that
    of the first dose
[†]Copied with permission of the authors

to accelerate this process would include mass revaccination of selected populations.

Elimination in developed countries

The success of vaccination programs in reducing measles incidence by over 90% in Europe and North America prompted some experts to call for the elimination of measles [52,62,85,115]. In October, 1978, the United States set a goal for the elimination of measles by October, 1982 [52]. In September, 1984, the WHO European Regional Committee established the following two elimination targets:

1. Elimination of measles by 1990 in countries with already effective programs.
2. Elimination of measles by 1995 in the remaining countries of the European region.

In December, 1984, the Second Conference on Immunization Policy, held in Karlovy Vary, Czechoslovakia, identified strategies to achieve these objectives. These included targets for immunization coverage, disease surveillance, and outbreak control [9]. At the current time, it appears that these targets have not been met. Failure to change national vaccination policies or to achieve vaccination coverage targets have been the main problems cited for this failure.

Former Czechoslovakia claimed to have eliminated measles for at least 8 months [107]. Albania has also claimed to have interrupted transmission [103]. To date, except for limited time periods in regions with high vaccine coverage and either geographic isolation or a two-dose schedule, no developed country has achieved and maintained the documented elimination of measles.

Revised targets for measles elimination are soon to be issued by the European Regional offices of the WHO and the EPI based on meetings held in February, 1992 in Langen, Germany and in June, 1992 in Milan, Italy.

Control in developed countries

Measles control programs in developed countries are based on the triad of delivering one dose of measles vaccine (usually as MMR) to as many children as possible, surveillance of disease, and vigorous outbreak control [3,4]. Outbreak control is a time-consuming and labour intensive process, consisting of the identification and exclusion of and/or revaccination of susceptibles. The definition of "susceptible" varies, from anyone who received no vaccine or inappropriately-administered vaccine, to anyone who did not receive the second dose of vaccine at an appropirate age. Canada has included in its definition of "susceptible" anyone born after 1957 who received vaccine prior to a change in vaccine formulation in 1980 [3]. This aggressive outbreak control strategy has been found to be very disruptive in school settings. It is also unclear whether this strategy is effective in preventing the spread of measles [92]. By the time the outbreak has been identified, many of the susceptibles have already been infected. These efforts tax already limited public health resources and divert time and effort away from other programs. Despite vaccine coverage exceeding 98% in the 1970's in Czechoslovakia [107] and in the late 1980's in Hungary [14], both countries continued to have outbreaks. Failure to contain outbreaks despite aggressive control programs resulted in the decision to implement two-dose strategies.

Developed countries with two-dose schedules

A number of European countries have implemented two-dose schedules since the early 1980's (see Table 2) [34]. These include the former Czechoslovakia and German Democratic Republic, Sweden, Finland, Norway, the Netherlands, and Bulgaria. It is worthwhile to review some of their experience.

Czechoslovakia introduced mass revaccination of children on school entry in 1975 because of a poorly-immunogenic vaccine in use prior to 1974. Despite vaccine coverage of 98% [106], measles outbreaks continued to occur in the late 1970's. By 1982, it was decided to introduce a routine second dose of vaccination six to ten months after the first dose [107]. It appears that this program was successful in interrupting measles transmission for at least eight months [107]. However, measles rates increased again in 1984 [12].

Sweden implemented a two-dose strategy in 1982 [31,40]. The net effect has been a substantial reduction in the rate of measles from 76 per 100 000 individuals in 1982 to 1.2 per 100 000 in 1988 [41]. Finland has had a national two-dose schedule since 1982 [97]. Coverage was initially low (only 80.5%), but an intense education programme raised the coverage for both doses to over 96% by 1986 [96]. There has also been a marked decline in the incidence of measles. Norway has also had good success with implementation of a two-dose schedule since 1983.

Table 2. Countries with two-dose measles schedules

Country[a]	Age of first dose	Age of second dose[b]
Czechoslovakia (107)	12 months	6–10 months after first (1982)
Bulgaria (79)	12 months	4 years
France (100)	12–15 months	5–6 years
Germany (49)	15 months	6 years (1991)
Finland (79)	18 months	6 years (1982)
Hungary (14)	14 months	6–7 years (1989)
Israel (54)	15 months	6–7 years (1990)
Poland (21)	13–15 months	7–8 years (1991)
The Netherlands (113)	14 months	9 years (1983)
New Zealand (24)	12–15 months	11–12 years (1991)
Denmark (35)	15 months	12 years
Sweden (79)	18 months	12 years (1982)
Norway (79)	18 months	13 years (1983)
United States (3,5)	15 months	4–6 years/11–12 years (1989)[c]

[a] Number in parenthesis refers to source reference in text
[b] Year that the second dose was implemented, where available
[c] 4–6 years – Recommendation of U.S. Immunization Practices Advisory Committee (ACIP); 11–12 years – Recommendation of American Academy of Pediatrics

In 1990–91, Bulgaria, a country which had implemented a two-dose schedule since at least 1985 [34], reported a major measles outbreak with over 2000 cases and two deaths [26]. The outbreak started in an unvaccinated ethnic community. Over 70% of the cases were unvaccinated or inadequately-vaccinated, including many infants. Additional investigations are necessary to rule out other potential causes, including problems with the vaccine.

The Netherlands implemented a two-dose schedule in 1983 [113]. In 1987–8, a measles epidemic occurred mainly in religious objectors to vaccination.

Other European countries have recently implemented two-dose schedules, including Hungary [14], Poland [21], Denmark [35], Germany [49], and Malta [21]. Israel [12] and New Zealand [24] have also recommended two-dose schedules.

The decline of measles in countries with a two-dose schedule may not be solely attributable to the change in strategies, since coverage levels with at least one dose increased at the same time.

The outbreaks that occurred in countries with two-dose schedules should not be considered as failures of this strategy. It may take ten or more years before the full impact of a two-dose schedule is realized. However, they do illustrate that measles outbreaks can occur, despite high vaccine coverage and a two-dose schedule if pockets of inadequately immunized individuals exist. A two-dose schedule cannot replace or compensate for weaknesses in measles control programs that fail to deal with target groups who refuse vaccination.

Some countries have not implemented two-dose schedules [24,69,85], choosing to optimize the one-dose schedule at the current time (see Table 3).

Table 3. Countries with one-dose measles schedules

Country*	Age of administration
Australia (24)	12 months
Canada (85)	12 months
Hong Kong (69)	12 months
Singapore (25)	12 months
Austria (11)	14 months
Belgium (19)	15 months
Italy (101)	15–18 months
Switzerland (99)	15–24 months
U.K. (70)	second year

*Number in parenthesis refers to source reference in text

Elimination and control in developing countries

The objective of vaccination in most developing countries is the reduction of measles-associated morbidity and mortality [120]. Measles incidence and mortality is typically very high and affects very young children in developing countries [30,109]. The goal therefore is to vaccinate children at the earliest possible age to prevent infection. The EPI/WHO recommends measles vaccination at 9 months of age. In countries where significant measles incidence and mortality occurs under 9 months, vaccination may be given as early as 6 months of age, with a subsequent second dose [26].

The optimal time to administer vaccine is when the greatest proportion of infants have lost their maternal antibodies (making them susceptible to infection) [88], but before being infected [7,8]. In countries with high disease incidence, a large number of infants are infected very soon after losing the protection from maternal antibodies. The proportion that seroconvert following vaccination at this early age may not be sufficient to block transmission. This differs from developed countries where the average age of infection is older; it is possible to vaccinate at an older age without significant risk of disease among infants, allowing for a higher seroconversion rate. For developing countries, two possible solutions exist. One is to find a vaccine that induces seroconversion in the presence of maternal antibodies, and the other is to institute a two-dose vaccine policy. In a two-dose policy, the first dose is given at 6 months to prevent disease in those children who are already susceptible, thus reducing transmission. The second dose is given at 9 months to achieve sufficiently high seroconversion to block transmission. Although scientifically desirable, the practical constraints of delivering two doses of potent vaccine in developing countries are considerable.

Nokes et al [89] used a mathematical model to demonstrate that an isolated policy of lowering the age of vaccination with current vaccines from 9 to 6 months of age in countries with high rates of disease between 6 and 9 months of age, without a subsequent second dose, would actually increase the total number of measles cases. This model predicted that vaccination at 9 months with selective vaccination at 6 months as well would be beneficial.

Gambia claimed to have interrupted measles transmission in 1963 and 1970. This was achieved by a mass mobile vaccination campaign of all infants six months to 4 years of age from 1967 to 1972. Gambia is a small country with many isolated villages. The epidemiologic conditions may have been conducive to short-term elimination. This success was not maintained when the program was terminated due to a lack of vaccine, the inability to identify and vaccinate new susceptibles, and the lack of skills and resources to maintain the mobile vaccination program [110,120].

Some Caribbean islands also claim to have eliminated measles [94]; however, because of their geographic isolation, elimination of measles on islands cannot be extrapolated to other countries.

Developing countries in the Americas have successfully implemented mass vaccination days to control and eventually eliminate polio. The Pan American Health Organization also plans to use this approach for measles [94]. The campaigns consist of mass vaccination of all children from the age of 1–14 years followed by intensive surveillance, outbreak control, and routine universal vaccination.

Delivery issues

Problems associated with the delivery of vaccines

Because of the extreme infectiousness of measles, even a 100% efficacious vaccine would not eliminate measles unless coverage were over 95%. Even if a two-dose schedule were implemented, if the same people who failed to receive the first dose also failed to receive the second dose, then an overall coverage of 95% would still fail to eliminate measles! Therefore any consideration of one versus two doses of measles vaccine must include, in addition to scientific considerations, a discussion of the factors that affect coverage.

Although elimination has not yet been realized, many excellent measles control programs exist in European countries, such as in the Scandinavian nations, the Netherlands, and Czechoslovakia. Progress has been steady in the United Kingdom; in 1991, measles vaccine coverage reached 91%, having been as low as 53% during the previous decade [70]. Despite the stated objective of measles elimination by the European region of the WHO, many other large European countries have failed to achieve high coverage. It is estimated that the coverage in the former Federal Republic of Germany (FRG) was only 64–68% in 1984, although this may have increased in recent years [49]. Coverage in the FRG has been stated to be as low as 50% in 1987 for children under 2 years of age [119]. Coverage for MMR in France is estimated to be 59% [82]. Many developed countries do not even have national surveillance systems and are unable to provide national and regional data on disease incidence, let alone assess regional vaccine coverage. Without this information, it is difficult if not impossible to identify program needs or to evaluate existing control programs.

Other problems exist in countries where national objectives and control programs exist. Analysis of the resurgence of measles in the United States has identified many barriers to delivering measles vaccine [44,64,86]. These may be divided into failures of the health care system, barriers to accessing existing health care services, and lack of public and provider knowledge. The socially disavantaged are particularly at risk of missing vaccination. Failures of the health care system include the lack of adequate health care insurance, lack of health care facilities and providers, and economic hardship, including the inability to purchase vaccines. Difficulties in accessing existing facilities may be physical (lack of transportation, inconvenient clinic hours), bureaucratic (ineligibility for assistance programs), or social (language or fear of public/ governmental agencies). Parents often forget or do not understand the need for vaccination. Physicians and other health care providers are poorly-educated about vaccination. They often delay vaccination inappropriately because of mild ailments, or miss opportunities to vaccinate when patients visit for other reasons.

Other barriers may be more difficult to overcome. In the 1987–8 measles epidemic in the Netherlands, two thirds of cases occurred in religious objectors who had refused vaccination [113]. Certain groups such as the homeless or illegal immigrants may be particularly difficult to vaccinate. These problems may become more accute in view of the major social changes occuring in Eastern Europe.

The main reason why existing programs fail to control measles is their failure to deliver a single dose of measles vaccine to a high proportion of infants at the appropriate age.

Considering the difficulties involved in delivering one dose of vaccine, opponents to two-dose measles vaccine schedules argue that the efforts required to deliver the second dose may divert resources away from achieving high coverage with the first dose. What requirements are necessary for an effective two-dose strategy?

National co-ordination

Minimum requirements to achieve high coverage should include a nationally organized control program. Measles does not recognize national or regional borders, so continuous elimination in one area is difficult to maintain without control in adjacent areas. Optimization of measles control therefore, requires an integrated national program. This should include national goals and objectives and program standards to help focus resources and provide measures for evaluation. National programs should be flexible and innovative enough to respond to local needs. Regional control programs with national coordination may be more appropriate in some countries.

Political and public commitment

In order for decision makers to acknowledge and commit resources, they must appreciate the need for measles control. Public health programs have been under pressure due to diminishing government resources and increasing pressure to rationalize expenses in health care. There is no doubt that existing one-dose programs are cost-effective [28]. New extremely effective vaccines, such as *Haemophilus influenzae* type b conjugate vaccines, have been successfully introduced and implemented. Universal vaccination with Hepatitis B vaccine had also been recommended. Implementation of a second dose of measles vaccine often must compete with these programs as well as with the acute care sector for health care dollars. Other diseases such as AIDS command higher priority in the public eye. The final decision requires a political and social commitment to eliminate measles. Decision makers and the public must be persuaded that the control and ultimately the elimination of measles is socially desirable. Only then will the additional organizational and capital costs of implementing a second dose be successfully applied.

Surveillance and evaluation

National programs should develop the means to collect data such as disease trends and vaccination coverage at the regional level to evaluate vaccination programs, identify impediments to achieving disease control, and identify areas requiring additional resources.

Timely assessment of coverage levels is necessary to evaluate the effectiveness of vaccine delivery systems and identify areas where additional resources must be committed to achieve effective coverage levels. High national coverage figures may conceal local pockets of unvaccinated individuals. These pockets may serve as foci for the initiation of outbreaks when measles is introduced from outside sources. Coverage figures must be provided locally in order to customize local programs. Coverage should be assessed at appropriate ages; e.g. 2-year-old measles vaccine coverage. Evaluation at convenient times (e.g., school entry) may be deceptive; vaccine coverage at school entry is over 90% in most areas of the United States, but area surveys revealed local coverage of 54–71% of 2-year olds in some areas [43].

Specific strategies to increase coverage

Different methods have been used in various countries to increase coverage with varying success [43,70,86]. Some of these include legislation, economic incentives, education and mass media, computerization of record-keeping and notification, alternative delivery system, change in the vaccination schedule, and elimination of various barriers to vaccination.

1. Mandatory vaccination has been used effectively in many countries. Legislation is only effective if a commitment is made to enforce it. It is also useful in enforcing outbreak control measures. Some jurisdictions may perceive this strategy as socially unacceptable. High vaccination rates of over 95% have been achieved in many countries without legislated requirements, such as in Finland [96]. Other countries with different delivery systems and more heterogenous populations may not be able to achieve

this level without legislation. If coverage rates of nearly 100% are necessary, then legislation may need to be considered to reach the remaining hold outs. The acceptability of legislation may be increased if formal procedures are in place to allow conscientious and religious objectors to refuse vaccination.

2. Economic incentives include payments to parents in France (with unproven effectiveness) and structuring of physician reimbursement based on vaccine coverage goals in the United Kingdom (effictive so far).

3. Most parents will have their child vaccinated if they are reminded. A simple reminder to the health care provider or directly to the parent often suffices.

4. There must be a method to identify the birth cohort and to keep accurate vaccination records. This process can be greatly expedited by automation. Automated identification of eligible infants at birth and recording of vaccination status may be coupled to reminders to parents and providers. Automation projects in the United Kingdom [55], the Netherlands [113], and Finland [96] have demonstrated the potential utility of this tool, especially if linked to other educational, social, and health care services.

5. Mass media have been used in Finland [96]. Further research is needed to validate this approach, due to its expense and the potential for adverse publicity [70]; e.g., DTP-associated deaths reduced vaccine acceptance in Sweden, Japan, and the United Kindom.

6. Perhaps the most important method of increasing compliance is to remove economic barriers. Governments should provide free vaccine and the means to receive them. An excellent delivery system exists in Finland, where vaccination is integrated into a comprehensive child health care program.

7. Special programs are required to reach special groups. In American inner cities, vaccination of infants visiting emergency rooms or special community programs are examples of innovative outreach programs.

8. Compliance may be affected by changing the age of vaccination. Parents are more likely to bring their infants for vaccination at a young age. The United Kingdom is experimenting with an accelerated schedule to increase compliance. If vaccination coincides with a regular periodic health check and eliminates an additional visit, compliance may be increased.

Minimum program requirements

It is incumbent on public health officials to persuade policy makers of the need for measles control. Measles control and elimination can only happen if governments and the public believe that this is a necessary objective. Remarkable progress is possible if governments make firm commitments to measles control. China increased measles vaccination coverage among infants 12 months of age from 77% to over 97% in less then a decade [22]. This was accompanied by a 95% reduction in measles incidence. This was attributed to a national commitment to vaccination programs.

Minimum program requirements should include:

1. A national control program with regional (state/provincial) coordinators.
2. Surveillance of disease incidence and vaccine coverage at the local level.
3. Effective outbreak control.
4. Continuous evaluation of existing control methods.
5. Free vaccine and administration of such.

If these minimum requirements cannot be met, then implementing a second dose of vaccine may not be advisable.

Choice of optimal strategy

Cost-benefit analyses of strategies

A one-dose vaccine policy is clearly cost-beneficial [28,115]. Outbreak control is expensive and often difficult to implement. The decision to pay for implementing a second dose of measles vaccine is based on the assumption that this strategy will eventually eliminate measles and reduce costs incurred by outbreak control. However, in the interim, program delivery costs, surveillance, and limited containment efforts will need to continue, since it is unlikely that the risk of importation from developing countries will be eliminated in the foreseeable future.

A cost-benefit analysis of implementing a two-dose schedule in Israel showed that the economic benefits of reduced disease incidence outweighed the costs of a second dose [54,112]. This analysis also demonstrated that mass vaccination of children older then the recommended revaccination age of 6 years of age was not cost-beneficial. This analysis is not, however, applicable to most developed countries, because the authors assumed that there were no program costs of delivering the second dose other than the cost of vaccine itself. In addition, Israel has a high incidence of measles and a coverage of only 90%. Therefore, the incremental benefit of a second dose of vaccine would be less in low-incidence countries. Additional cost-benefit analyses are required. Cost-utility analyses would be useful in comparing two-dose measles programs to other health care programs.

Age at first dose

The age of first vaccination varies from 12 months in Canada and Czecho-slovakia to 18 months in Norway and Sweden [79].

If a one-dose schedule is chosen, then the age of vaccination has implications for outbreaks and risk of disease to the individual. If, for example, seroconversion at 12 months is 90% and seroconversion at 15 months is 95%, then choosing a 12-month vaccination schedule implies that the number of population susceptibles will accumulate at twice the rate as compared to a 15-month vaccination schedule. The critical mass to sustain large outbreaks will be reached earlier and more frequently. On the other hand, delaying vaccination to 15 months exposes infants to disease for an additional 3 months. With an increasing number of children born to vaccinated mothers, the window of susceptibility also increases as these infants become susceptible to infection at an earlier age. The American Advisory Committee on Immunization Practices is now discussing the need to lower vaccination to 12 months of age. Population serosurveys of measles antibody levels would aid in making this decision.

The choice of the age of first vaccination may also be altered if a two-dose schedule is chosen. A lower age of first vaccination, inducing a lower rate of seroconversion, may be acceptable in the context of a two-dose schedule.

Age at second dose

Different countries have chosen different ages for revaccination. These range from Czechoslovakia, where the second dose is delivered 6–10 months after the first dose, to Norway, where the second dose is given at 13 years of age [79]. Recommendations may vary even within the same country. In the United States, the Advisory Committee on Immunization Practices (ACIP) has recommended revaccination on school entry [4], in contrast to the American Academy of Pediatric's (AAP) recommendation of 11 to 12 years [3]. There are inadequate data on the effect of time between doses on seroconversion to guide policy. Delivery issues must guide this decision until additional epidemiologic data are collected.

Many countries choose the age for measles revaccination to coincide with the onset of increased disease incidence in older age groups. Balanced against the epidemiologic criteria are the logistic constraints of delivering and monitoring the second dose of vaccine. It is more cost-effective if the delivery of the second dose coincides with another preexisting health care visit. Additional public health resources are required to monitor coverage of the second dose and increase compliance where indicated.

This is the reason for the difference in AAP and ACIP recommendations. The ACIP school entry recommendation coincides with the preschool DTP booster. All fifty American states have legislation for mandatory vaccination on school entry. Vaccination at this age increases compliance by potentially reducing the number of health care visits because of coadministration of DTP and MMR at the same visit. It allows the public health system to monitor coverage at school entry and to increase compliance through the use of existing legislation. Balanced against this, is the fact that the impact of this strategy will not be realized until the revaccinated cohort reaches the time of peak disease incidence. The AAP choose their recommendation based on the fact that the outbreaks and disease incidence peaked in the 10–14-year age group in the United States. Revaccination at 11 to 12 years would have an immediate impact and boost immunity just at the time of peak risk.

If a two-dose schedule is chosen, then the second dose would reduce the risk of outbreaks that would otherwise result from a more rapid creation of a large pool of susceptibles because of the younger age at first vaccination. This decision must be considered against the risk of disease in the age group between the two doses. Czechoslovakia has minimized this period of susceptibility by vaccinating at the earlier age of 12 months and revaccinating only 6 to 10 months later. Does the protection from such a schedule last to the age when peak disease incidence occurs? Additional data on the duration of antibodies after early revaccination, for example by population serosurveys, might be useful in evaluating this option. In North America, such a program might not have an impact for at least ten years. A program of this nature might; therefore, need to be coupled with a catch-up revaccination program in older groups.

Until additional epidemiologic evidence is obtained from the experience of ongoing programs, each country must choose the optimal age for the second dose based on local disease epidemiology and health care delivery constraints.

One versus two doses

The decision to choose a two-dose schedule depends on whether elimination rather than control is the goal. Mathematical models and epidemiologic observations suggest that elimination is not possible with only one dose of current measles vaccines. A two-dose schedule is clearly a minimum requirement for elimination.

The decision to implement a second dose requires social and governmental acceptance of the need for elimination. The country must have the desire to commit economic resources and the health care infrastructure to deliver the second dose. Decision makers must realize that this is a long-term investment. It may require ten or more years to achieve elimination. A two-dose schedule will not prevent outbreaks unless programs are in place to ensure high coverage with both doses. Poorly-implemented programs will result in continued outbreaks and loss of public confidence in the strategy. Measles elimination programs must also be complemented with methods of assessing disease incidence and vaccine coverage. Methods of responding to and controlling outbreaks must also be in place. It may be better to identify and vaccinate high-risk groups and thus ensure high first-dose coverage at an appropriate age, rather than to give 90% of the population two doses and 10% of the population no doses. Untimely implementation of a two-dose program may divert resources from delivery of the first dose with no major improvements in control.

Unresolved questions

1. What is the optimal age for delivery of both the first and second dose in a two-dose schedule?
2. What effect will the increasing cohort of infants born to vaccinated mothers (with their lower maternal antibody levels) have on disease risk and age of vaccination?

3. Can measles truly be eliminated even with a two-dose schedule? Even if regional elimination is possible, can this be maintained without worldwide eradication?

References

1. Aaby P, Bukh J, Leerhøy J, Lisse IM, Mordhorst CH, Pedersen IR (1986) Vaccinated children get milder measles infection: A community study from Guinea-Bissau. J Infect Dis 154(5): 858–863
2. Aaby P, Jensen TG, Hansen HL, et al (1988) Trial of High-Dose Edmonston-Zangreb Vaccine in Guinea-Bissau: Protective Efficacy. Lancet II: 809–811
3. Advisory Committee on Epidemiology (1991) Guidelines for Measles Control in Canada. Canad Dis Wkly Rep 17(7): 35–40
4. Advisory Committee on Immunisation Practices (1989) Measles Prevention: Recommendations of the Immunization Practices Advisory Committee (ACIP). MMWR 38(S-9): 1–18
5. American Academy of Pediatrics (1989) Measles: Reassessment of the current immunization policy. Pediatrics 84(6): 1110–1113
6. Albrecht P, Ennis FA, Saltzman EJ, Krugman S (1977) Persistence of maternal antibody in infants beyond 12 months: Mechanism of measles vaccine failure. J Pediatr 91(5): 715–718
7. Anderson RM, May RM (1990) Immunisation and herd immunity. Lancet II: 641–645
8. Anderson RM (1992) The concept of herd immunity and the design of community-based immunization programmes. Vaccine 10(13): 928–935
9. Anonymous (1985) Expanded program on immunization: European conference on immunization policies. Wkly Epidemiol Rec 60(22): 165–168
10. Anonymous (1986) Expanded programme on immunization, impact of immunization programme. Wkly Epidemiol Rec 29: 221–222
11. Anonymous (1987) Expanded programme on immunization, immunization coverage survey. Wkly Epidemiol Rec 25: 183–185
12. Anonymous (1988) Expanded programme on immunization, reported trends of EPI disease in Europe. Wkly Epidemiol Rec 12: 81–85
13. Anonymous (1988) Measles, measles elimination by 1990? Wkly Epidemiol Rec 22: 163–164
14. Anonymous (1989) Measles outbreak. (Hungary) Wkly Epidemiol Rec 18: 137–138
15. Anonymous (1990) Expanded programme on immunization, global advisory group, part I. Wkly Epidemiol Rec 2: 5–11
16. Anonymous (1990) Measles surveillance, outbreak of measles in a private international school in french-speaking Switzerland. Wkly Epidemiol Rec 23: 173–175
17. Anonymous (1990) Expanded programme on immunization, measles outbreak. Wkly Epidemiol Rec 49: 379–381
18. Anonymous (1991) Expanded programme on immunization, safety and efficacy of high titre measles vaccine at 6 months of age. Wkly Epidemiol Rec 34: 249–251
19. Anonymous (1991) Measles surveillance 1982–1990. Wkly Epidemiol Rec 43: 314–317
20. Anonymous (1992) Expanded programme on immunization, global advisory group, part II. Wkly Epidemiol Rec 4: 17–19
21. Anonymous (1992) Expanded programme on immunization, measles epidemic, 1989–1990. Wkly Epidemiol Rec 8: 50–54
22. Anonymous (1992) Expanded programme on immunization, programme review. Wkly Epidemiol Rec 15: 109–111
23. Anonymous (1992) Measles outbreak in Bulgaria. Wkly Epidemiol Rec 12: 84–85
24. Anonymous (1992) Measles surveillance 1991-Austria & New Zealand. Wkly Epidemiol Rec 19: 139–142

25. Anonymous (1992) Seroepidemology of measles, mumps and rubella, Singapore. Wkly Epidemiol Rec 31: 231–233
26. Anonymous (1992) Expanded programme on immunization, Safety on high titre measles vaccines. Wkly Epidemiol Rec 48: 357–361
27. Atkinson WL, Orenstein WA (1992) The resurgence of measles in the United States 1989–90. Annual Reviews of Medicine 43: 451–463
28. Bart KJ, Orenstein WA, Hinman AR (1986) The Virtual Elimination of Rubella and Mumps from the United States and the Use of Combined Measles, Mumps and Rubella Vaccine (MMR) to Eliminate Measles. Devel Biol Stand 65: 45–52
29. Berry SB, Hernandez H, Kanoshiro R, et al (1992) Comparison of high titer Edmonston-Zagreb, Biken-CAM and Schwarz measles vaccines in Peruvian infants. Pediatr Infect Dis J 11(10): 822–827
30. Black FL (1989) Measles. In: Evans AS (ed) Viral Infections of Humans: Epidemiology and Control Third Edition. Plenum Medical Book Company, New York and London, pp 451–469
31. Böttiger M, Christenson B, Romanus V, Taranger J, Strandell A (1987) Swedish experience of two dose vaccination programme aiming at eliminating measles, mumps, and rubella. Br Med J 295: 1264–1267
32. Braunstein H, Thomas S, Ito R, et al (1991) Response of seronegative adults to measles immunization. Am J Dis Child 145: 969
33. Brunell PA (1990) Measles control in the 1990s: Measles serology. WHO/EPI/GEN. 1–34
34. Bytchenko BD, Dittmann S (1986) Elimination of diseases from Europe through use of vaccines. Devel Biol Stand 65: 13–21
35. Bytchenko B, Prokhorskas R (1991) Regional immunization programme: Current and perspectives. ICP/EPI 027/20
36. Centers for Disease Control (1983) The feasibility of measles elimination in Europe. MMWR 32: 523–4, 530
37. Centers for Disease Control (1991) Measles – United States 1990. MMWE 40(22): 186–189
38. Chen RT, Goldbaum GM, Wassilak SGF (1989) An Explosive Point Source Measles Outbreak in a Highly Vaccinated Population. Am J Epidemiol 129(1): 173–181
39. Chen RT, Markowitz LE, Albrecht P, et al (1990) Measles antibody: Reevaluation of protective titers. J Infect Dis 162: 1036–1042
40. Christenson B, Böttiger M (1991) Changes of the immunological patterns against measles, mumps and rubella. A vaccination programme studied 3 to 7 years after the introduction of a two-dose schedule. Vaccine 9: 326–329
41. Christenson B, Böttiger M, Heller L (1983) Mass vaccination programme aimed at eradicating measles, mumps, and rubella in Sweden: first experience. Br Med J 287: 389–391
42. Cutts FT (1990) Measles vaccination in early infancy and persistence of antibody (Letter). Lancet I: 913
43. Cutts FT, Zell ER, Mason D et al (1992) Monitoring progress toward U.S. preschool immunization goals. JAMA 267(14): 1952–1955
44. Dai Bin, Chem Zhihui, Wu Ting, et al (1991) Duration of immunity following immunization with live measles vaccine: 15 years of observation in Zhejiang Province, China. Bull WHO 69(4): 415–423
45. Davies JW, Acres SE, Varughese PV (1982) Experience with Measles in Canada and the United States. Canad Med Assoc J 126: 123–126
46. Edmonston MB, Addiss DG, McPherson JT, et al (1990) Mild measles and secondary vaccine failure during a sustained outbreak in a highly vaccinated population. JAMA 263(18): 2467–2471
47. Enders JF, Peebles TC (1954) Propagation in tissue cultures of cytopathic agents from patients with measles. Proc Soc Exp Biol Med 86: 277–286
48. Evans AS (1985) The eradication of communicable diseases: Myth or reality. Am J Epidemiol 122(2): 199–207

49. Fescharek R, Quast U, Maass G, Merkle W, Schwarz S (1990) Measles-mumps vaccination in the FRG: an empirical analysis after 14 years of use. I. Efficacy and analysis of vaccine failures. Vaccine 8: 333–336
50. Fine PEM, Clarkson JA (1982) Measles in England and Wales – I: An analysis of factors underlying seasonal patterns. Int J Epidemiol 11(1): 5–14
51. Fine PEM, Clarkson JA (1982) Measles in England and Wales – II: The impact of the measles vaccination programme on the distribution of immunity in the population. Int J Epidemiol 11(1): 15–25
52. Frank JA, Orenstein WA, Bart KJ, et al (1985) Major impediments to measles elimination. Am J Dis Child 139: 881–888
53. Garenne M, Leroy O, Beau JP, et al (1991) Child mortality after high-titre measles vaccines: prospective study in Senegal. Lancet II: 903–906
54. Ginsberg GM, Tulchinsky TH (1990) Costs and benefits of a second measles inoculation in Isreal, the West Bank, and the Gaza. J Epidemiology Community Health
55. Goodwin S (1990) Preventive car for children: Immunization in England and Wales. Pediatrics 86(6) II: 1056–1060
56. Gustafson TL, Lievens AW, Brunnell PA (1987) Measles outbreak in a fully immunized secondary-school population. N Engl J Med 316: 771–778
57. Halsey NA (1983) The optimal age for administering measles vaccine in developing countries. In: Halsey NA, deQuadros CA (eds) Recent Advances in Immunization: A Bibliographic Review. Scientific Publication No. 451. PAHO: 4–17
58. Hayden GF (1979) Measles Vaccine Failure. Clin Pediator 18(3): 163–167
59. Hersh BS, Markowitz LE, Maes EF (1992) The Geographic Distribution of Measles in the United States, 1980 Through 1989. JAMA 267(14): 1936–1941
60. Hethcote HW (1983) Measles and Rubella in the United States. Am J Epidemiol 117: 2–13
61. Hinman AR, Brandling-Bennett AD, Bernier RH et al (1980) Current features of measles in the United States: Feasibility of measles elimination. Epidemiologic Reviews 2: 153–167
62. Hinman AR, Kirby CD, Eddins DL et al (1983) Elimination of indigenous measles from the United States. Rev Infect Dis 5(3): 538–545
63. Hinman AR, Bart KJ, Hopkins DR (1985) Costs of not eradicating measles. Am J Publ Heal 75(7): 713–4
64. Hinman AR (1990) Immunization in the United States. Pediatrics 86(6) II: 1064–1066
65. Hirayama M (1983) Measles Vaccines Used in Japan. Rev Infect Dis 5(3): 495–503
66. Kaplan LJ, Daum RS, Smaron M, et al (1992) Severe measles in immunocompromised patients. JAMA 267: 1237–1241
67. Krober MS, Stracener CE, Bass JW (1991) Decreased measles antibody response after measles-mumps-rubella vaccine in infants with colds. JAMA 265: 2095–2096
68. Krugman S (1983) Further-attenuated measles vaccine: Characteristics and use. Rev Infect Dis 5(3): 477–481
69. Lau Y, Chow C, Leung T (1992) Changing epidemiology of measles in Hong Kong from 1961 to 1990 – Impact of a measles vaccination program. J Infect Dis 165: 1111–1115
70. Leese B, Bosanquet N (1992) Immunization in the UK: policy review and future economic options. Vaccine 10(8): 491–499
71. Lett S, Scambio E, Spitalny K, et al (1991) Measles vaccination levels among selected groups of preschool-aged children – United States. MMWR 40(2): 36–39
72. Levy D (1984) The Future of Measles in Highly Immunized Populations: A Modelling Approach. Am J Epidemiol 120(1): 39–48
73. Lievens AW, Brunell PA (1986) Specific immunoglobulin M enzyme-linked immunosorbent assay for confirming the diagnosis of measles. J Clin Microbiol 24(3): 391–394
74. Linnemann CC, Dine MS, Rosella GA, et al (1982) Measles immunity after revaccination: Results in children vaccinated before 10 months of age. Pediatrics 69: 332

75. Makino S (1983) Development and Characteristics of Live AIK-C Measles Virus Vaccine: A Brief Report. Rev Infect Dis 5(3):504–505
76. Markowitz LE, Preblud SR, Orenstein WA, et al (1989) Patterns of transmission in measles outbreaks in the United States, 1985–1986. N Engl J Med 320(2):75–81
77. Markowitz LE, Sepulueda J, Diaz-Ortega JL, et al (1990) Immunization of six-month-old infants with different doses of Edmonston-Zagreb and Schwarz measles vaccines. N Engl J Med 322:580–587
78. Markowitz LE (1990) Measles control in the 1990s: immunization before 9 months of age. WHO/EPI/90.3
79. Markowitz LE, Orenstein WA (1990) Measles vaccines. Ped Clinics of N America 37(3):603–625
80. Markowitz LE, Preblud SR, Fine PEM, et al (1990). Duration of live measles vaccine-induced immunity. Pediatr Infect Dis J 9:101–110
81. Markowitz LE, Albrecht P, Orenstein WA, et al (1992) Persistence of Measles Antibody after Revaccination. J Infect Dis 166:205–208
82. Mary M, Garnerin P, Roure C, Villeminot S, Swartz TA, Valleron A (1992) Six years of public health surveillance of measles in France. Int J Epidemiol 21(1):163–168
83. Mathias RG, Meekison WG, Arcand TA, et al (1989) The role of secondary vaccine failures in measles outbreaks. Am J Publ Heal 79(4):475–478
84. Miller C (1987) Live measles vaccine: a 21 year follow up. Br Med J 295:22–24
85. National Advisory Committee on Immunization (1990) Statement on recommended use of measles vaccine in Canada. Canad Dis Wkly Rep 16(2):7–10
86. National Vaccine Advisory Committee (1991) The Measles Epidemic; The Problems, Barriers and Recommedations. JAMA 266(11):1547–1552
87. Nkowane BM, Bart SW, Orenstein WA, et al (1987) Measles outbreak in a vaccinated school population; epidemiology, chains of transmission and the role of vaccine failures. Am J Publ Heal 77(4):434–438
88. Nokes DJ, Anderson RM (1988) Measles, mumps and rubella vaccines: What coverage to block transmission? Lancet II:1374
89. Nokes DJ, McLean AR, Anderson RM, Grabowsky M (1990) Measles immunization strategies for countries with high transmission rates: Interim guidelines predicted using a mathematical model. Int J Epidemiol 19(3):703–710
90. Norrby E, Oxman MN (1990) Measles virus. In: Fields BN (ed) Virology, Raven Press Ltd., New York, 1013–1044
91. Orenstein WA, Markowitz L, Preblud SR (1986) Appropriate age for measles vaccination in the United States. Devel Biol Stand 65:13–21
92. Osterman JW, Melnychuk D (1992) Revaccination of children during school based measles outbreaks: Potential impact of a new policy recommendation. Canad Med Assoc J 146(6):929–936
93. Pabst HF, et al (1992) Reduced measles immunity in infants in a well-vaccinated population. Pediatr Infect Dis J 11(7):525–529
94. Pan American Health organization (1992) Measles elimination in the Americas. Bulletin of PAHO 26(3):271–274
95. Parkman PD, Hopps HE, Albrecht P, et al (1983) Simultaneous Adminstration of Vaccines. Biographic Reviews PAHO. 65–80
96. Paunio M, Virtanen M, Peltola H, Cantell K, Paunio P, Valle M, Karanko V, Heinonen OP (1991) Increase of vaccination coverage by mass media and individual approach: Intensified measles, mumps and rubella prevention program in Finland. Am J Epidemiol 133(11):1152–1160
97. Peltola H, Karanko V, Kurki T, Hukkanen V, Virtanen M, Penttinen K, Nissinen M, Heinonen OP (1986) Rapid effect on endemic measles, mumps, and rubella of nationwide vaccination programme in Finland. Lancet I:137–139
98. Peradze TV, Smorodintsev AA (1983) Epidemiology and specific prophylaxis of measles. Rev Infect Dis 5(3):487–490
99. Personal Communication from Dr. Max Just, Badler Kinderspital, Basel, Switzerland

100. Personal Communication from Odile Leroy, Pasteur Mérieux S.V., Marne La Coquette, France
101. Pontecorvo M (1990) Vaccini Sieri Immunoglobuline: Prontuario per l'uso. VI Edizione. Edizioni Minerva Medica, Torino
102. Prebuld SR, Katz SL (1988) Measles Vaccine. In: Plotkin SA, Mortimer EA (ed) Vaccines. W.B Saunders Company. Philadelphia, pp 182–222
103. Rabo E, Taranger J (1984) Scandinavian model for eliminating measles, mumps, and rubella. Br Med J 289: 1402–1404
104. Reyes MR, De Borrero MF, Roa J, et al (1987) Measles vaccine failure after documented seroconversion. Ped Infect Dis J 6(9): 848–851
105. Rossier E, Miller H, McCulloch B, (1992) Measles in Ontario. PHERO 3(8): 125–128
106. Sejda J (1979) Evaluation of the eight-year period of compulsory measles vaccintion in the Czech Socialist Republic (CSR). J Hyg Epidemiol Microbiol Immunol 23(3): 273–283
107. Sejda J (1983) Control of Measles in Czechoslovakia (CSSR). Rev Infect Dis 5(3): 564–567
108. Stetler HC, Orenstein WA, Bernier RH (1986) Impact of revaccination on children who initially received measles vaccine before 10 months of age. Pediatrics 77: 471
109. Taylor WR, Mambu R, Weinman JM (1988) Measles control efforts in urban Africa complicated by high incidence of measles in the first year of life. Am J Epidemiol 127: 788–794
110. Thacker SB, Millar JD (1991) Mathematical modeling and attempts to eliminate measles: A tribute to the late professor George Macdonald. Am J Epidemiol 133(6): 517–525
111. Tidjani O, Grunitsky B, Guerin N, et al (1989) Serological effects of Edmonston-Zagreb, Schwarz, and AIK-C measles vaccine strains given at ages 4–5 or 8–10 months. Lancet II: 1357–1360
112. Tulchinsky TH, Abed Y, Ginsberg G, Shaheen S, Friedman JB, Schoenbaum ML, Slater PE (1990) Measles in Israel, the West Bank, and Gaza: Continuing incidence and the case for a new eradication strategy. Rev Infect Dis 12(5): 951–958
113. Verbrugge HP (1990) The national immunization program of the Netherlands. Pediatrics 86(6)II: 1060–1063
114. White CC, Koplan JP, Orenstein WA (1985) Benefits, Risks and Costs of Immunization for Measles, Mumps and Rubella. Am J Publ Heal 75(7): 739–744
115. White FM (1983) Policy for measles elimination in Canada and program implications. Rev Infect Dis 5(3): 577–582
116. Whittle HC, Rowland MG, Mann GF, et al (1984) Immunization of 4–6 month old Gambian infants with Edmonston-Zagreb measles vaccine. Lancet II: 834–837
117. Whittle HC, Mann G, Eccles M, et al (1988) Effects of dose and strain of vaccine on success of measles vaccination of infants aged 4–5 months. Lancet I: 963–966
118. Whittle HC, Campbell H, Rahman, S, et al (1990) Antibody persistance in Gambian children after high-dose Edmonston-Zagreb measles vaccine. Lancet II: 1046–1048
119. Williams BC (1990) Immunization coverage among preschool children: The United States and selected European countries. Pediatrics 86(6)II: 1052–1056
120. Williams PJ, Hull HF (1983) Status of measles in the Gambia, 1981. Rev Infect Dis 5(3): 391–394
121. Xiang Jianzhi, Chen Zhihui (1983) Measles Vaccine in the People's Republic of China. Rev Infect Dis 5(3): 506–510
122. Yekutiel P (1981) Lessions from the big eradication campaigns. World Health Forum 2(4): 465–490

8

Cost-benefit analysis of a second dose measles inoculation of children

G. M. Ginsberg[1], and **T. H. Tulchinsky**[2]

[1] Departments of Information, Ministry of Health, Jerusalem, Israel
[2] Personal and Community Preventive Health Services, Ministry of Health, Jerusalem, Israel

Summary

The traditional emphasis on single-dose immunization against measles has failed to meet control or eradication requirements even in the most developed parts of the world. A single dose strategy is limited because of insufficient coverage rates in infancy, schoolage epidemics and primary and secondary vaccination failure. A second measles booster dose in required in order to reduce population susceptibility to sufficiently low levels to allow the goal of measles elimination to be achieved. This paper estimates the costs and benefits of a decision taken in Israel in 1988 to add school age measles immunization to the present 15 month old use of measles vaccine in Israel. The second dose policy of immunizing all Israeli children aged 6 for the years 1988–2008, costs around $1.81 million and has estimated benefits of $14.15 million, yielding a benefit-to-cost ratio of 7.86/1. The vaccination programme was estimated to prevent over a 20 year period, approximately 80 300 simple cases, 9162 hospitalized cases, 21 non-fatal cases of encephalitis, 18 cases of SSPE and to save 85 lives. The break-even point for benefits to the health services alone, occurs if vaccine costs were to rise from $0.42 to $2.59 per dose. Therefore the decision to introduce a second measles dose in Israel seems to be fully justifiable on monetary grounds alone. Extremely high benefit-to-cost ratios for the adoption of a two dose policy, were also estimated for the UK (86.3/1), Spain (76.1/1) and Italy (53.0/1). For the Philippines and Nigeria the benefit to cost ratios were 11.4/1 and 20.6/1 respectively, though in the Philippines the benefit-to-cost ratio to the health services alone was less than unity. Limitations in resources for preventive health care remains a serious problem, even in developed countries. However, our cost-benefit calculations show the two-dose approach to be economically justifiable for many developing and developed countries.

Introduction

Measles and its often debilitating complications remains one of the key childhood infectious diseases causing large numbers of deaths and disabilities with attendant direct and indirect economic costs [1], albeit of differing magnitudes, in both developed and developing countries.

Globally, immunization rates against measles have risen from around 50% in the 1980's [2] to around 80% in children under one years old in 1990 [3]. Current stress in measles elimination strategy is being placed on expanding immunization coverage in susceptible groups and case outbreak control epidemiologic methods [4].

The traditional emphasis on single-dose immunization against measles has failed to meet control or eradication requirements even in the most developed parts of the world. For example, the United Kingdom, which uses one dose of MMR at 12–24 months, has measles immunization coverage rate of 80%, with variation by region from 60–85%, continues to have high rates of measles incidence with as many as 20 deaths per year [5,6].

Current estimates of the coverage required to reach fully protective herd immunity are 94–97% [7–10]. This level of coverage is extremely difficult to reach with a single dose schedule in infancy, even in the most developed countries, but it may be more readily achievable at school age through the adoption of a second compulsory vaccination requirement.

A further rationale for a two-dose strategy is the problem of failure to sero-convert, particularly when the vaccine is administered before the optimal time. Primary vaccination failure occurs at rates of 4–8%, with a further 4–5% of people who were vaccinated against measles later exhibited declining antibody levels (secondary failures) [11–13]. Therefore, even vaccination coverage of higher than 95% of children up to 2 years of age might be insufficient to fully control or eliminate measles [14]. Failure to sero-convert gives rise to the high number of cases occurring among previously vaccinated persons during epidemics as evidenced by in the 1986 epidemic in Israel (with coverage rates of 88%) where approximately 20% of the cases occurred in children with a documented history of past immunization [15] and in a 1988 epidemic in Zimbabwe (with 83% coverage rates), where 59% of the reported cases were previously vaccinated [16].

International agencies recommend early vaccination at 6–9 months because of the high susceptibility of young infants to the disease and its complications. Many children may thus be immunized before maternal antibody levels wears off, causing immunization failure and decreasing herd immunity. Where early immunization is indicated by the epidemiologic picture of high morbidity in infants, the subsequent second dose at a later age would help boost immunity levels and catch those who did not get the primary measles vaccination.

A further argument in favor of a two-dose policy is the phenomena of epidemics in school aged populations with high vaccination coverage. Schools play a role in facilitating the transmission of measles epidemics of infectious diseases as documented in Canada [17–20] and the USA [9,12–13,21–31]. Secondary spread of measles from epidemics which begin in school has been shown to be greater than when the primary case is a child under five years old [14,32–33]. Following these schoolchild epidemics a two-dose policy was recommended in the USA in 1989 [14,32,34–36], which has shifted the focus of future epidemics from previously immunized schoolchildren to infants in crowded urban slum areas [37–39]. On the other hand, the Canadian Advisory Committee on Immunization Practice recommended continuation of the one-dose policy, along with mandatory school age proof of immunization or prior measles, and outbreak control strategies [40–41].

The phenomena of insufficient coverage rates, schoolchild epidemics, primary and secondary vaccination failures, support the call for a policy of a second booster dose [26] in order to reduce susceptibility to sufficiently low levels to allow the goal of measles elimination to be achieved.

Use of two-dose schedules

Of the 11 European countries operating with a single-dose protocol as at 1989, who reported incidence data (GDR, Greece, Iceland, Ireland, Italy, Luxembourg, Poland, Romania, Spain, Turkey, UK), only four experienced a decrease in measles incidence [42] over 50% in the period 1974/77 to 1983/87. In contrast, all of the six countries (Bulgaria, Czechoslovakia, Finland [43], Hungary, Norway and Sweden [43]) who had instituted a two-dose policy prior to 1984, had decreases in incidence over 70% over the period. The greater decline in incidence of the two-dose countries could also be attributed to their higher infancy vaccine coverage rates (92% vs 75% in the one-dose countries). The lower coverage rates of the one-dose countries could be due to organizational inefficiencies or a reflection of the lower priority given to measles, which in turn could explain why they have not yet adopted a two-dose policy. In any case, the one-dose countries low infant vaccine coverage rates, only serve to strengthen the argument for them to adopt a two dose policy. Israel, Denmark and the Netherlands [10], introduced a two-dose policy between 1984 and 1989.

Both developed and especially developing countries suffer from the problem of having limited resources available for their society in general, and for health needs in particular. Despite the aforementioned arguments in favour of a two-dose policy, budgetary constraints can provide a barrier to the implementation of such a program. If a cost benefit analysis can show high benefit to cost ratios for the introduction of a two-dose policy, then these budgetary barriers to implementation can be brought down.

Previous cost-benefit studies of vaccination against measles only related to a first dose situation and found high benefit-to-cost ratios [33,44–48]. Partly as a result of our previous study [49], which prospectively estimated benefit-to-cost ratios to society of the addition of a second measles vaccination, of 4.5/1 in Israel, 5.7/1 in the West Bank and 9.6/1 in Gaza [49–51], a school age vaccination was introduced in addition to the 15 month first dose in these areas.

This paper re-estimates a cost-benefit analysis of the adoption in 1988 of the policy of expanding the existing vaccination programme at 15 months to inoculate schoolchildren with a second dose of measles vaccine in Israel. The incremental costs and benefits take into account reductions in ambulatory care, hospitalization, work absences and mortality from projected reductions in measles incidence, and its complications such as encephalitis (with and without sequelae) and subacute sclerosing panencephalitis (SSPE) cases.

Methods

A vast array of Israeli demographic (age structures and projections, labor force participation, life expectancy), epidemiologic (measles incidence, mortality and transitory probabilities to complications), health service (type and amount of care required for measles cases) and economic (costs of inoculation, costs of caring for measles) data were entered into a computerised spreadsheet model. The major advantage of having the model on a spreadsheet (in comparison to writing the model in program form) is the relative ease of adding in new values as new information became available, changing the model specifications and performing sensitivity analyses.

The basic model reported previously [49], was adjusted to take into account substantial population changes (as a result of the recent waves of immigration from Ethiopia and the former USSR), changes in hospital costs, changes in labor force costs and participation rates, increases in unemployment, a higher provision for vaccine wastage, a decrease in the discount rate from 10% to 7.5%, incidence after seven years falling to only 10% instead of 1% of the baseline incidence, decreases in vaccine costs and an additional cost for nurses time required to carry out the vaccinations in a school setting. In addition, the time horizon used for measuring the costs and benefits was increased from ten to twenty years. The following paragraphs describe the data which was entered into the model.

Cost-benefit model parameters

Childhood immunization in Israel is the responsibility of the Ministry of Health. Measles cases are reported individually based on clinical diagnosis, with or without laboratory confirmation by the attending physician, clinic, hospital or other health care institution. Data are collected in the Epidemiology Department, and the Public Health Divisions. Reporting of infectious diseases, including measles is required by law in Israel.

Immunization against measles of infants at age nine months began in Israel in April, 1967, with live attenuated vaccine. Coverage in 1970 was reported to be 78% of the eligible infant population [8,52]. In 1971, the age of administration was changed to 15 months, and in 1975, the imported vaccine Attenuvax was adopted [52]. Immunization coverage has increased from 82% of children under age two in 1980 to 88% in 1989 [53] and to 95% by the time children enter junior schools [54]. Measles, mumps and rubella (MMR) vaccine replaced the measles vaccine in 1989 in Israel.

Morbidity and mortality

The case fatality ratio of 6/1000 in 1973, rose to a peak of 15.5/1000 in 1983 and declined to 2.5/1000 in 1986/88. Israel experienced an average annual death rate from measles of 0.101/100 000 population during the period 1980–88 [55–56]. The overall epidemiologic picture has been a marked reduction in morbidity, mortality and Subacute Sclerosing Panencephalitis (SSPE) incidence [57] over the years 1968–1988.

The expanded intervention program designed to eliminate the disease in Israel was based on the assumption that adding (to the existing dose given at 15 months) a routine second measles immunization at early school age (children aged 6) would, after seven years reduce measles incidence by to a level of just 10% of the 1983–1988 incidence level. The projected costs and benefits which would follow the adoption of the two-dose policy were estimated by calculating the reduction in incidence expected to occur during the period from 1988 to 2008 inclusive.

Annual incidence rates in Israel declined by 45.5% over the period 1968–1987 from 50.1 to 26.5/100 000. Baseline incidence rates were calculated by dividing the total number of reported cases in period 1983–1988 by the relevant population at risk [53]. This resulted in rates of 21.7/100 000 for Israeli Jews, 26.5/100 000 for Israeli non-Jews, giving an overall rate in Israel of 22.5/100 000. Estimates of notification rates [57] of 29% of cases for Jews and 17% of cases for non-Jews in Israel were used to adjust the reported incidence rates to estimates of the true incidence rates.

These rates were applied to national population estimates to simulate incidence rates of measles for the years 1988–2008 in the absence of any second measles vaccination programme.

Table 1. Costs of caring for measles

	Jews	Non-Jews
Simple case		
prob per case	1	1
−ambulatory costs	$12	$12
−work losses	$151	$62
Total cost	$162	$73
Hospital cases: (100% reported)		
prob per case reported	0.3209	0.74
−ambulatory costs	$16	$16
−hospital costs	$1242	$894
−work losses	$210	$74
Total cost	$1468	$983
Measles encephalitis		
prob per case	0.00021	0.00021
−ambulatory costs	$85	$85
−hospital costs	$4079	$4079
−work losses	$421	$147
Total cost	$4586	$4312
Measles encephalitis + sequelae		
prob per case	0.0000875	0.0000875
−ambulatory costs	$16	$16
−hospital costs	$37 146	$37 146
−work losses	$631	$221
Total cost	$37 793	$37 383
SSPE		
prob per case	0.000232	0.000657
−ambulatory costs	$16	$16
−hospital costs	$11 775	$11 775
−work losses	$457	$160
Total cost	$12 248	$11 951
Mortality − gnp	$142 069	$141 833
Non-SSPE mortality costs		
prob per case	0.00093301	0.01126146
Mortality − gnp	$125 991	$134 189

In such a situation it is estimated that there would be 101 674 cases of measles in the Israeli population (72 851 among Jews) during the period 1988–2008. As a result of giving a second dose to children age 6, assuming a 95% compliancy rate, the number of measles cases would decrease by 78.1%, to 21 378 in Israel (Table 1).

The number of hospitalized cases was calculated by applying the age and ethnic group specific probabilities of a reported case being hospitalized in 1979, being 0.35 for Jews and 0.68 overall for non-Jews [58], to the estimated number of reported cases. If, the two dose option were adopted, then the number of hospitalized cases would drop from 11 616 to 2454 during 1988–2008.

A notified case fatality rate of 0.275% from non-SSPE measles related deaths were calculated by dividing the 38 deaths (10 among Jews and 28 among non-Jews) for the period 1981–1988 by the total number of notified cases during the same time period. The notified case fatality rate of 0.088% for Israeli Jews, was considerably lower than the rate of 1.20% among non-Jews in Israel. This difference in case fatalities might not only arise from differences in home or health service care, but may also be due to differences in reporting rates in the different communities. During the study period, non-SSPE mortality in Israel would fall from 84.9 to 18.4 cases with a two dose option.

Secondary illnesses

Measles is usually characterized by a transient febrile illness; however, cases can be complicated by pneumonia, convulsions, post-infectious encephalitis and the lethal SSPE. Reduction in the incidence of these secondary illnesses would also follow as a result of a second vaccination programme. Figures of 2.1 and 1.4 per 10 000 cases [59–60] were used in our analysis as an estimate of the incidence of measles encephalitis without and with sequelae respectively. The case fatality rate of these complications was assumed to be 15% [61]. In Israel, under a two dose option, the total number of encephalitis cases would fall from 30.3 to 6.4 cases (Table 1). For SSPE, a figure of 2.32 per 10 000 cases was used for the Israeli Jewish population and 6.57 per 10 000 cases for the non-Jewish populations of Israel [22]. SSPE occurs on average around 7 years after the measles onset and all known cases reported in the 1964–1969 period were under ten years of age [57]. The fall in the number of SSPE cases will be from 35.8 to 17.7 cases under a two-dose (Table 1).

Cost estimates

Estimates of costs per case are listed in the Table 1, which also divides the total cost into the catagories of health service (hospital costs and ambulatory clinic costs including medication) and work losses. The methodology of the calculations is described below for the Israeli Jewish population.

Simple cases (with common complications)

For simple cases of measles, it is assumed that an ambulatory visit will be made related to symptoms of high fever and/or coughing with a second visit to confirm the rash. In addition cases with commonly associated complications as otitis media, mastoiditis, laryngitis and respiratory infections [60–61] will require further ambulatory visits. Therefore 2.2 (15 minute) ambulatory visits which cost (assuming physician wage costs to be $16 per hour) $8.80 are assumed. In addition, a conservative estimate of 3% of cases are estimated to require a physician visit to the home taking 40 minutes including travelling time. This results in a unit cost of $10.66 for a home visit, or $0.32 per simple measles case. Medication (mostly antibiotics and antipyretics) was assumed to be required in 30% of the cases at a unit cost of $8 per course, or $2.40 per measles case. Adult work losses were calculated by multiplying the age and sex specific participation rates in the labor force (being overall 93% for males and 70% for females), adjusted by an estimate of 12% unemployment over the period, by the age and sex specific wage costs (being overall $59.09 per day for males and $35.69 per day for females) [53]. This product was then multiplied by the number of working days missed through illness (4.71 being 5.5/7 of the estimated 6 day period the adult is expected to be ill at home. Child work losses were of course zero, however an addition was made for the work losses incurred by a parent (assumed female) who was conservatively assumed to be at home for 9 days (7.07 working days) to care for her child [60]. The age specific participation rate used was 47.3%, reflecting the younger age group of females when compared to the adult female measles victims. Overall the average work loss per case amounted to $150.79 for Jews and $61.64 for non-Jews, considerably higher than the average health service costs of $11.52 per case.

For cases requiring hospitalization, it was assumed that there would be 2.5 (15 minute) ambulatory or emergency room visits (costing $10 per case), a home visit in 3% of the cases (costing $0.32 per case) and that 70% of cases would require medication (costing $5.60 per case). Total non-hospital health service costs totalled around $416.00 per case. The average length of hospital stay were estimated to be 2.98 days for persons aged sixteen or over and 5.56 days for children under sixteen years, being 88% and 78% of the latest available data from 1979 [58]. The age-specific mix of hospitalized cases (43.4% aged 18 + , 6.4% aged 16–17 and 50.2% aged 0–15) was applied to the above length of stays and multiplied by the average per diem hospital cost of $255 in 1992. This gives the overall average hospital costs for a hospitalized case of measles to be $1241. Work losses were calculated as described in the previous paragraph, based on the assumption that in addition to their 2.98 and 5.56 days in hospital, an adult or child would be at home for a further 5 and 6 days respectively, these totalled $210 per average hospitalized case.

Measles encephalitis

For measles encephalitis cases, the assumed ambulatory, home visits and medication costs were as for an uncomplicated hospital cases ($15.92). However all cases were assumed to be hospitalized for 7 days in an internal medicine ward (costing $255 per day) and for 3 days in an intensive care unit, assumed to cost double the average per diem cost. Therefore each measles encephalitis case costs around $4079 in hospital fees. Work losses were assumed to be double those of an uncomplicated hospitalized measles case and amounted to $421 per case. 75% of the cases were assumed not to have sequelae, requiring a further four follow-up ambulatory visits ($16) and $53 in tests, totalling $4586 per non-sequelae case.

For the cases with neurologic sequelae, work losses three times that of an uncomplicated hospitalized measles case were assumed ($631). In addition, costs of institutional care amounting to a discounted value of $33 000 were estimated. In addition a case fatality rate of 15% was assumed.

SSPE

Cases of SSPE were assumed to require two and a half ambulatory visits, need 90 days hospitalization (averaging $11 775 per case) and incur 60 days of work losses by each working mother (averaging $457 per case).

Mortality costs

Costs of early mortality were based on data between 1979–1986 [63–65] showing the average age at death for non-SSPE cases to be 1.5 years for Jewish children and 37 years for Jewish adults. For non-Jewish Israeli's the average age at death overall was two years and 17 years for adults. An average age of eight years old was assumed for SSPE cases in both the population groups [57].

Instead of valuing a person purely by his discounted expected future earnings (the Human Capital Method), the GNP per Head method assigns an equal value to everyone in society, equivalent to the Gross National Product per head ($10 728 in 1990 [53]). The GNP per Head method of economic valuation of life thereby overcomes the major shortcoming of the human capital method which perhaps unjustly assigns a zero value to childhood years, compulsory military service years, pensioners and housewives who are not in any direct remunerative employment. However, children, military personnel, housewives and pensioners (grandparents etc.) do certainly contribute something to society, be it the production of the next generation, the maintenance of the present workforce or just "tender loving care". Using a 7.5% discount rate, the net present values of Israeli persons were $142 412, $142 069 and $133 882 at ages 1.5, 8 and 37 years old respectively.

Vaccination costs

Costs per vaccination was $0.529 per shot, this figure included costs of the vaccine ($0.42 per dose), swab ($0.0083), disposable syringe ($0.0434) and needle ($0.0062), as well as an additional 18% provision for vaccine wastage, based on recent experience in Israel with second-dose measles vaccine distribution (Dr. F. Mattas, Ministry of Health National Laboratories, personal communication). An additional $0.18 per dose was estimated for the transportation costs of the nurses to the schools to administer the vaccine. A further $0.72 per dose was estimated for the time spent in travelling to the school (0.79 minutes per person), giving the actual injection (2.5 minutes), giving explanations (0.17 minutes per vaccinee as explanations are made to each school-class on a group basis) and administration (2.0 minutes), including follow-up of an estimated 5% non-compliers. Thus the total overall cost is estimated to be $1.43 per vaccinee.

The provision of extra marginal manpower costs is a result of the near strike of public health nurses as a result of the incremental addition of two vaccines in short succession; second dose measles and neonatal hepatitis B. If the nurses claim that they are being diverted from other tasks to carry out the school vaccinations is correct, then it is only right to make provision for these extra opportunity costs incurred on account of the introduction of the two-dose schedule.

A two-dose policy, would require the vaccination of around 98 000 six year olds in Israel in 1988, rising to 158 000 six year olds in 2088. Total additional costs over the period 1988–2008, (using a 7.5% discount rate, for costs incurred other than in the initial year of the project) amount to $1 807 800.

Pneumonia is only expected as a side effect of inoculation in 1 per 100 000 persons, convulsions in 6 per 10 000 persons, brain damage in 1 per million persons [62]. The incidence of these side effects of a second vaccination would be higher if a second vaccination were not to be carried out. In order to simplify calculations, the costs of these side-effects were omitted both from the costs of vaccination and from the benefits of vaccination. Since an early death as a result of a vaccination occurs in less than 1 in 10 million cases, this cost was omitted.

Benefits (or averted costs)

If vaccination is performed there will be a considerable reduction in measles during the period 1988–2008. Table 2 summarizes the estimated reduction in uncomplicated ambulatory and hospitalized cases and the expected reductions in the number of measles encephalitis cases (with and without sequelae), SSPE cases and mortality resulting from non-SSPE cases.

The main monetary benefits of vaccination are the costs averted by caring for a reduced number of cases. Reductions in incidence were multiplied by the relevant unit costs per case, described earlier in this paper. The resultant benefits are listed in Table 3 using a 7.5% discount rate to calculate net present values of future benefits and costs.

Table 2. Expected incidence in Israel (1988–2008)

	Jews	Non-Jews	Total
One dose option			
Simple	72 851	2 8823	10 1674
HOSP	7479	4137	11 616
Encephalitis	15.3	6.1	21.4
Encephalitis + sequelae	6.4	2.5	8.9
SSPE	16.9	18.9	35.8
SSPE-mortality	16.9	18.9	35.8
Non-SSPE mortality	21.7	63.2	84.9
Two dose option			
Simple	15 041	6337	21 378
HOSP	1544	910	2454
Encephalitis	3.2	1.3	4.5
Encephalitis + sequelae	1.3	0.6	1.9
SSPE	8.2	9.5	17.7
SSPE-mortality	8.2	9.5	17.7
Non-SSPE mortality	4.5	13.9	18.4

Table 3. Costs and benefits of a two-dose measles policy in Israel (Period 1988–2008, using a 7.5% discount rate)

	Costs no vaccination	Costs with vaccination	Benefits to society	Benefit to costs
Simple	6 462 227	1 907 140	4 555 086	2.52
Hosp	6 998 014	2 081 939	4 916 075	2.72
Encephalitis	44 760	13 317	31 443	0.02
Encephalitis + seq	155 911	46 418	109 493	0.06
SSPE	202 641	132 925	69 716	0.04
SSPE-mortality	2 379 214	1 560 930	818 284	0.45
Non-SSPE mortality	5 271 111	1 618 735	3 652 376	2.02
Total	$21 513 877	$7 361 405	$14 152 473	7.83
Costs			$1 807 757	

Extension of model to other countries

We adjusted the section of our basic model pertaining to the Jewish population of Israel, in order to take into account the differing age structure [66], reported incidence [18,40,42,67–71], age at mortality [69–71], sex-specific life expectations [66], GNP per capita [72], sex-specific wage levels and labor force participation rates [73] of various developed and developing countries. In the absence of country specific health care costs, these were estimated to reflect the relative level of GNP compared to Israel. In projecting our model for the developing countires of the Philippines and Nigeria, we assumed that there would be no ambulatory visits for measles care and that transportation and cold-chain costs would amount to $3 per dose.

Results

The policy of immunizing all Israeli children aged 6 will cost around $1.81 million and have estimated benefits of $14.15 million, yielding a benefit-to-cost ratio of 7.83/1.

We also calculated the benefits and cost of two further two-dose options, option B (which also includes a one-time mass vaccination campaign of all 7–17 year olds in 1988, resulting in the attainment of a 90% decrease in measles incidence in only 5 years) and option C, including a mass vaccination compaign to all 7–27 year olds, resulting in the attainment of a 90% decrease in incidence in only 4 years. However, the extra costs ($1.58m) of using Option B instead of the initial simple option, exceed the incremental benefits ($1.20m), giving a benefit-to-cost ratio of only 0.76/1. Similarly, the costs of expanding from Option B to Option C ($1.02m) by far exceed the extra benefits ($0.52m), giving a benefit to cost ratio of only 0.51/1.

Therefore, the option of just an additional dose at age 6 with no mass campaign appears to be the most resource-efficient strategy to use in the fight against measles in Israel. However, even if we assume that the secular downward trend from 1968–1985 in incidence rates would have continued into the future, then the benefit to cost ratio to society for the simple two-dose option still exceeds 5.0/1.

When only the benefits resulting from reductions in simple and hospitalized cases are considered, the benefit to cost ratio is still 5.24/1. Benefits from reduced incidence in measles encephalitis and SSPE account for only a small percentage of the total benefits. Hence even if such gains have been grossly overestimated, the effect on the benefit-to-cost ratios will be small.

Sensitivity analyses

Even if simple measles cases required only one ambulatory visit, no home visits or medication, and there was a 50% reduction in hospital length of stay, with an attendant reduction in the number of days lost from work, and a 50% reduction in hospital costs per day, the benefit-to-cost ratio in Israel under the two dose option would still be 5.97/1.

If vaccine costs were to increase from $0.42 to $9.12 per dose, then overall costs would equal benefits under the two dose option in Israel. The break-even point when only benefits to the health services are considered, would occur if vaccine costs were to rise to $2.59 per dose.

In Israel, the basic two-dose vaccination programme would prevent over the next ten years, approximately 28 700 simple cases, 3400 hospitalized cases, 8.6 non-fatal cases of encephalitis, 2.2 cases of SSPE and save 28 lives.

All this can will be gained at no extra cost to society, and indeed a net savings of around $12.3 million will accrue. About $3.1 million of this will be savings to the health services, giving a benefit-to-cost ratio based only on health service benefits of around 2.71/1.

Table 4. Benefit to cost ratios by discount rate and decrease in incidence over 7 years

Decrease in incidence	Discount rate			
	0%	5.0%	7.5%	10.0%
	All benefits			
80%	19.63	8.54	7.11	6.26
90%	21.78	9.43	7.83	6.86
95%	22.88	9.89	8.2	7.18
99%	23.71	10.23	8.48	7.41
	Benefits to health services only			
80%	2.74	2.56	2.46	2.35
90%	3.05	2.83	2.71	2.58
95%	3.21	2.97	2.84	2.70
99%	3.32	3.07	2.93	2.79

The benefit-to-cost ratios were especially sensitive to the choice of discount rate (Table 4), rising to 9.43/1 with a 5% rate and falling to 6.86/1 using a 10% rate. Table 4 also includes the benefit to cost ratios using a 0% discount rate, since we have found that when we present discounted benefit to cost ratios to policy-making committees, they tend to then say "but the benefits occur years into the future", thus effectively psychologically discounting the

Table 5. Benefit to cost ratios by incidence rate and level of gnp per head

True incidence per 100000[a]	Gnp per head			
	$5000	$10 000	$15 000	$20 000
	All benefits			
25	0.99	1.96	2.93	3.90
50	1.97	3.91	5.85	7.79
75	2.96	5.87	8.78	11.69
100	3.95	7.83	11.71	15.59
150	5.92	11.74	17.56	23.38
200	7.89	15.65	23.41	31.17
250	9.87	19.57	29.27	38.96
300	11.84	23.48	35.12	46.76
	Benefits to health services only			
25	0.39	0.77	1.14	1.52
50	0.78	1.54	2.29	3.04
75	1.18	2.30	3.43	4.56
100	1.57	3.07	4.57	6.08
150	2.35	4.61	6.86	9.12
200	3.14	6.14	9.15	12.15
250	3.92	7.68	11.44	15.19
300	4.71	9.22	13.72	18.23

[a]Based on reported incidence rate being 29% of actual (56)

future benefits a second time. By presenting undiscounted benefit to cost ratios, this danger of double discounting is averted. However, comparison between different projects then becomes problematic, if different policy making committees are used for different fields of health care.

Table 4 also shows that even if the incidence rate only fell by 80% in seven years, the benefit to cost ratio (at 7.5% discount rate) would be 7.11/1 to society and 2.46/1 for the health services alone.

The benefit to cost ratios were also sensitive to the true incidence rate (i.e.: reported incidence adjusted by the reporting rate, which was assumed to be the level of 29% found among Israeli Jews [57]) and levels of GNP per capita. However, costs will only exceed benefits if the true incidence rate is below 25/100 000 and the level of GNP per head is $5000 (Table 5). However, when only health services are considered, benefits exceed costs at incidence levels below 60/100 000 when the level of GNP per head is $5000.

Discussion

The incremental benefits of instituting a second measles vaccine to persons aged six in Israel exceed the incremental costs, giving a cost-benefit ratio of 7.83/1. These estimates can be considered to be conservative in nature since the cost of transport and time involved in visiting a hospitalized persons was omitted in the valuation of benefits. Also, it was assumed that parents accompanied children on ambulatory visits outside of their working hours. In addition, benefits accruing after the 20th year of the program as a result of vaccinations in the first twenty years of the program, were also not valued. Finally, no attempt was made to monetarize the benefits in such important but intangible dimensions as reduced pain, worry or grief.

It should be noted that the omission of the costs resulting from side effects of the vaccine and the omission of the larger benefits from a reduction in pneumonia, convulsions and brain damage as a result of a second vaccine programme, act to produce a further downward bias in our cost-benefit ratios.

Extension to other countries

Table 6 lists background data and benefit to cost ratios from the extension of our model to several other countries, which had not instituted a two-dose policy prior to 1990. The high incidence rates and the resultant benefit-to-cost ratios of 86.3/1 to society and 38.6/1 to the health services alone, justify the UK's recent decision to introduce a two-dose schedule. The adoption of such a policy by Spain and Italy in particular, as well as by Canada, Greece, Luxembourg, Iceland and Poland could also be justified on economic grounds alone. For Turkey, the benefit to cost ratio for society is close to unity. However for health services only, the costs of such a programme considerably exceed the benefits.

Table 6. Benefit to cost ratios for second dose measles innoculation

	Reported measles incidence per 100 000 (1983/87)[a]	Measles mortality per million (1987/88)	% of population aged six (1990)	Gnp per head (1990)	Benefit to cost ratios	
					Society	Health services
UK	167.0	0.105	1.27%	$14 902	86.3	38.6
Spain	210.0	0.490	1.34%	$9554	76.1	29.4
Italy	74.0	0.261 ·	1.13%	$15 660	53.0	20.0
Canada	29.0[a]	0.057[a]	1.41%	$19 791	17.9	8.0
Greece	59.0	0.101	1.43%	$5505	11.9	4.5
Luxembourg	16.0	0	1.14%	$16 501	10.9	4.5
Israel	22.5[b]	1.105[c]	1.97%	$10 728	7.83	2.7
Iceland	8.3	0	1.75%	$20 137	4.60	1.90
Poland	59.0	0.317	1.77%	$1790	3.18	1.29
Turkey	33.0	0.462	2.66%	$1460	1.01	0.38
Philippines	99.6	19.27	2.58%	$721	11.4[d]	0.18[d]
Nigeria	3428.6	102.86	2.92%	$251	20.6[d]	2.09[d]

Sources: Refs [55–56,66–72]
[a] (1986/89)
[b] (1983/88)
[c] (1980/88)
[d] assumes, no ambulatory care costs and transport costs of $3 per dose

There can be few doubts as to the external validity of extending our basic cost-benefit model to other developed countries, which have similar health service infrastructure and care practices than Israel. However, there is scope for larger errors to occur in the use of the model for developing countries, which are characterized by having lower hospital bed-to-population ratios, less intensive hospital care, fewer ambulatory services and higher transport and cold-chain costs, especially in rural areas [74]. However, these factors which tend to lower benefit to cost ratios are partially compensated for, by the vaccination having a greater impact in terms of morbidity and mortality reductions [6,49,61,75–78]. Our estimates also do not take into account the benefits resulting from a decrease in post-perinatal child mortality as a result of there being fewer mothers exposed to measles during pregnancy [44], nor from reductions in higher rates of severe complications.

Assuming there were no ambulatory care costs and that transport costs were $3 per dose, we found benefit-to-cost ratios of 11.4/1 for society for the Philippines and 20.6/1 for Nigeria. However, for the Philippines, costs to the health services alone were over five times the expected benefits. In Nigeria however, because of the extremely high incidence rates health service (2.09/1) benefit to cost ratios exceeded unity, the break-even point to the health services occurring where transportation costs are $7.60 per dose.

Immunization coverage of infants in developing countries is even lower than in developed countries, with many unvaccinated children entering the primary schools. Second-dose immunization for school children could also significantly increase the level of herd immunity in developing countries, thus lowering incidence rates in those children under schoolage. Measles outbreaks in schools have been documented in Lesotho, Swaziland, Burundi and Rwanda, among others [21]. Primary school attendance is widespread and mandatory in many developing nations, with enrollment reported, for example, at virtually 100% in the Philippines, and 64–66% in Malawi and Nigeria [7]. School aged vaccination lends itself to mass campaign logistics and can be coupled with other vaccinations such as DPT and polio control activities, so that the additional costs are primarily for the additional vaccine [67].

Although the one-dose policy, aimed at the infant population, is not on its own sufficient to address the problem of measles, further country specific cost-benefit studies in developing countries should be undertaken to see if a supportive economic argument exists for an extension to a two-dose policy.

The two-dose policy has been has already been shown to be successful in many developed countries, but controversy as to whether or not to adopt a two dose policy still rages. While some countries including Canada [40–41] and Australia [79] have decided not to adopt a two dose policy, the UK in 1990, Saudi Arabia and Bahrain in 1991, Papua New Guinea (Dr. S. Rosenthal, EPI, WHO Geneva, Personal communication) and New Zealand in 1992 [79], have recently adopted a two-dose policy.

Limitations in resources for preventive health care remains a serious problem, even in the USA [80–81], our cost-benefit calculations show the two-dose approach to be economically justifiable to society for most of the developed countries as well as the two developing countries, that we considered.

References

1. Hopkins DR, Hinman AR, Koplan JP, Lane JM (1982) The case for global measles eradication. Lancet 1: 1396–1398
2. Walgate R (1988) Talloires, a quiet revolution. World Health: 28–29
3. Clements CJ (1992) World Health Organization experience with Measles Control. Paper presented to World Conference on Poliomyelitis and Measles, New Delhi, India January 7–12, 1992
4. Clements CJ (1980) Progress Toward Achieving the National 1990 Objectives for Immunization. MMWR 37(40): 1–3
5. Albritton RB (1978) Cost-Benefits of Measles Eradication: effects of a federal intervention. Policy Analysis 4: 1–22
6. Axnick NW, Shavell SM, Witte JJ (1969) Benefits due to immunization against measles. Public Health Rep 84: 673–680
7. UNICEF (1991) The state of the world's children 1991. United Nations Children's Fund (UNICEF): Oxford University Press, Oxford
8. Swartz TA (1984) Prevention of measles in Israel: implications of a long term partial immunization program. Public Health Rep 99: 272–277
9. Frank JA, Orenstein WA, Bart KJ, Bart SW, El-Tantawy N, Davis RM, Hinman AR (1985) Major impediments to measles elimination: The modern epidemiology of an ancient disease. Am J Dis Child 139: 881–888
10. Rabo E, Taranger J (1984) Scandinavian model for elimination of measles, mumps, and rubella. Brit Med J 289: 1402–1404

11. Markowitz LF, Orenstein WA (1990) Measles Vaccine. In: Pediatric Vaccinations: Update 1990. Pediatr Clin N America 37: 603–625
12. Hutchins SS, Markowitz LE, Mead P, Mixon D, Sheline J, Greenberg N, Preblud SR, Orenstein WA and Hull HF (1990). A school based measles outbreak: the effect of a selective revaccination policy and risk factors for vaccine failure. Am J Epidemiol 132: 1157–1168
13. Mathias RG, Meekson WG, Arcand TA, Schechter MT (1989) The Role of Secondary Vaccine Failures in Measles Outbreaks. Am J Public Health 79: 475–478
14. Sutter RW (1991) Measles among the Amish: a comparative study of severity in primary and secondary cases in households. J Inf Dis 163: 12–16
15. Epidemiology Department, Ministry of Health of Israel (1966) Measles in Israel. Monthly Epidemiol Bull 22(2): 19–21
16. Kambarami RA, Nathoo KJ, Nkrumah FK, Pirie DJ (1991) Measles epidemic in Harare, Zimbabwe, despite high measles immunization coverage rates. Bull WHO 69: 213–219
17. De Serres G, Boulianne N (1991) Measles outbreak – Ungava Region, Quebec. Can Dis Wkly Rep 17–42: 229–232
18. Canada diseases weekly report (1988) Measles in Montreal, Quebec. Can Dis Wkly Rep 14–52: 233–236
19. Boulianne N, de Serres G, Duval B, Joly JB, Meyer F, Dery P, Alary M, Le Hanaff D, Theriault N (1991) Epidemie majeure de rougeole dans la region de Quebec malgre une couverture vaccinale de 99%. Canad J Public Health 82: 189–190
20. Wong T, Lee-Han H, Bell B, Daley J, Bailey N, Vanderpol M (1991) Measles epidemic in Waterloo region, Ontario, 1990–1991. Can Dis Wkly Rep 17–41: 219–224
21. Markowitz LE, Preblud S, Orenstein WA, Rovira EZ, Adams NC, Haekins CE, Hinman AR (1989) Patterns of transmission in measles outbreaks in the United States, 1985–1986. N Engl J Med 320: 75–81
22. Chen RT, Goldbaum GM, Wassilak SGF, Markowitz LE, Orenstein WA (1989) An explosive point source measles outbreak in a highly vaccinated population: mode of transmission and risk factors for disease. Am J Epidemiol 129: 173–182
23. Centers for Disease Control (1988) Measles – United States. MMWR 38: 601–605
24. Nkowamwane BM, Barts W, Orenstein WA, Baltier M (1987) Measles outbreak in a vaccinated school population: epidemiology, chains of transmission and role of vaccine failure. Am J Public Health 77: 434–438
25. Gustafson TL, Lievens AW, Brunell PA, Moellenberg RG, Buttery CMG, Sehulster LM (1987) Measles outbreak in a fully immunized secondary school population. N Eng J Med 316: 771–774
26. Center for Disease Control (1990) Reported cases of measles, by state, United States, weeks 14–17. MMWR 39: 300
27. Medical letter (1989) Measles Revaccination. 31–797: 69–70
28. Committee on Infectious Diseases, American Academy of Pediatrics (1989) Measles: reassessment of the current immunization policy. Pediatrics 84: 1110–1113
29. Centers for Disease Control (1989) Measles prevention: recommendations of the Immunization Practices Advisory Committee. MMWR 38: 1
30. Edmonson MB, Addiss DG, McPherson JT, Berg JL, Circo SR, Davis JP (1990) Mild measles and secondary vaccine failure during a sustained outbreak in a highly vaccinated population. JAMA 263: 2467–2471
31. Hersh B, Markowitz LE, Hoffman RE, Hoff DR, Doran MJ, Fleishman JC, Preblud SR, Orenstein WA (1991). A measles outbreak at a college with a prematriculation immunization requirement. Am J Public Health 81: 360–364
32. Schlenker TL, Bain C, Baughman AL, Hadler SC (1992) Measles herd immunity: the association of attack rates with immunization rates in preschool children. JAMA 267: 823–826
33. Aaby P, Leeuwenburg J (1990) Patterns of transmission and severity of measles infection: a re-analysis of data from the Machakos area, Kenya. J Infect Dis 161: 171–174
34. National Vaccine Advisory Committee (1991) The measles epidemic: the problems, barriers, and recommendations. JAMA 266: 1547–1552

35. Mast EE, Berg JL, Hanrahan LP, Wassell JT, Davis JP (1990) Risk factors for measles in a previously vaccinated population and cost-effectiveness of revaccination strategies. JAMA 264: 2529–2533
36. Chen RT, Markowitz LE, Albrect P, Stewart JA, Mofenson LM, Preblud SR and Orenstein WA (1990) Measles antibody: reevaluation of protective titers. J Infect Dis 162: 1036–1042
37. Subbarao EK, Amin S (1991) Prevaccination serologic screening for measles in health care workers. J Infect Dis 163: 876–878
38. Davis RM, Whitman ED, Orenstein WA, Preblud SR, Markowitz LE and Hinman AR (1987) A persistent outbreak of measles despite appropriate prevention and control measures. Am J Epidemiol 126: 438–449
39. Centers for Disease Control (1991) Measles outbreak – New York City, 1990–1991. MMWR 40: 305–306
40. Canada disease weekly report (1990) National Advisory Committee on Immunization. Statement on recommended use of measles vaccine in Canada. Can Dis Wkly Rep 16–2: 7–12
41. Epidemiologic Report (1991) Guidelines for measles control in Canada. Can Med Assoc J 144: 997–1000
42. WHO Eurostat Health-For-All Indicators (1991), WHO, European Region, Copenhagen
43. Swartz TA (1991) Epidemiology of measles in the vaccination era: implications for an effective prevention of the disease. Isr J Med Sci 27: 48–50
44. Aaby P, Seim E, Knudsen K, Bukh J, Lisse IM, da Silva MC (1990) Increased postperinatal child mortality among children of mothers exposed to measles during pregnancy. Am J Epidemiol 132: 531–539
45. Aaby P, Knudsen K, Jensen TG, Tharup J, Poulsen A, Sodemann M, da Silva MC, Whittle H (1990) Measles incidence, vaccine efficacy, and mortality in two urban African areas with high vaccination coverage. J Infect Dis 162: 1043–1048
46. Dabis F, Sow A, Waldman RJ, Bikakouri P, Senga J, Madzou G and Jones TS (1988) The epidemiology of measles in a partially vaccinated population in an African city: implications for immunization programs. Am J Epidemiol 127: 171–178
47. Yach D (1990) Letter to the editor, Re: the epidemiology of measles in a partially vaccinated population in an African city: implications for immunization programs. Am J Epidiol 132: 193
48. Koenig MA, Khan MA, Wojtyniak B, Clemens JD, Chakraborty J, Fauveau V, Phillips JF, Akbar J, Barua US (1990) Impact of measles vaccination on childhood mortality in rural Bangladesh. Bull WHO 68: 441–447
49. Ginsberg GM, Tulchinsky TH (1990) Costs and benefits of a second measles innoculation of children in Israel, the West Bank and Gaza. J Epid Comm Health 44: 274–280
50. Tulchinsky TH (1991) Prevention of measles in Israel: short and long-term intervention strategies. Isr J Med Sci 27: 22–29
51. Tulchinsky TH, Abed Y, Ginsberg G, Shaheen, Friedman JB, Shoenbaum ML and Slater PE (1990) Measles in Israel, the West Bank and Gaza: Incidence and the case for a new eradication strategy. Rev Infect Dis 12: 951–958
52. Ministry of Health (1987) A Review of Health Services in Judaea, and Samaria, and Gaza 1985–1986, 16–17. Jerusalem
53. Central Bureau of Statistics (1991) Statistical Abstract of Israel 1991. Jerusalem: Central Bureau of Statistics
54. Neuman L, Shemesh A (1991) The innoculated status of children before admission to first grade schools in 1990. Ministry of Health, Jerusalem, December (In Hebrew)
55. Central Bureau of Statistics (1988) Statistical Tables on Selected Infectious Diseases 1981–1985. Jerusalem: Central Bureau of Statistics
56. Central Bureau of Statistics (forthcoming 1993) Statistical Tables on Selected Infectious Diseases 1986–1990. Central Bureau of Statistics. Special Series. Jerusalem
57. Zilber N, Rannon L, Alter M, Kahana E (1983) Measles, measles vaccination and risk of subacute sclerosing Panencephalitis. Neurology 33: 1558–1564

58. Central Bureau of Statistics (1987) Hospitalization Data 1979. Special series No 803. Jerusalem: Central Bureau of Statistics
59. Centers for Disease Control (1982) Measles Surveillance Report No 11, 1977–1981; Atlanta CDC, September
60. Krugman S, Katz SL (1981) Infectious Diseases of Children. St. Louis: C.V. Mosby, 149–155
61. White CC, Koplan JP, Orenstein WA (1985) Benefits, Risks and Costs of Immunization for Measles, Mumps and Rubella. Am J Public Health 75: 739–744
62. Ada GL (1988). Vaccines and Vaccinations, World Health, July: 5–7
63. Central Bureau of Statistics (1990) Causes of Death 1983–1986. Central Bureau of Statistics Special Series No 869. Jerusalem
64. Central Bureau of Statistics (1985) Causes of Death 1981–1982. Central Bureau of Statistics Special Series No 763. Jerusalem
65. Central Bureau of Statistics (1984) Causes of Death 1979. Central Bureau of Statistics Special Series No 750. Jerusalem
66. UN (1989) UN Demographic Yearbook (1989), New York 1991
67. Tuchinsky TH, Ginsberg GM, Abed Y, Martz J, Bonn J (1993) Measles Control in Developing and developed countries: The case for a two dose policy. Bull WHO (in process).
68. Canada Disease Weekly Report (1989), 15–1: 1–8
69. WHO (1990) World Health Statistics Annual, 1989, WHO Geneva
70. WHO (1991) World Health Statistics Annual, 1990, WHO Geneva
71. WHO (1992) World Health Statistics Annual, 1991, WHO Geneva
72. World Bank (1991) World Development Report, Oxford University Press, New York
73. ILO (1991) Yearbook of Labour Statistics, 1991, International Labour Office, Geneva
74. Creese AL, Henderson RH (1980) Cost-benefit analysis and immunization programmes in developing countries. WHO Bull 58: (3): 491–498
75. Willems JS, Sanders, CR (1981) Cost-effectiveness and cost-benefit analyses of vaccines. J Infect Dis 144: 486–493
76. Koplan JP, White CC (1985) An update on the benefits and costs of measles and rubella immunization. In: Grunberg EM, Lewis C, Goldston (eds) Immunizing against mental disorders: progress in the conquest of measles and rubella. Oxford University Press, Oxford, England
77. Bart KJ, Feng-Ying K (1990) Vaccine-preventable disease and immunization in the developing world: in Pediatric Vaccinations: Update 1990. Pediatr Clin N Am 37: 603–625
78. Pan American Health Organization (1990) Plan to eliminate indigenous transmission of measles in the English-speaking Caribbean countries. Bull Pan Am Health Org 24: 241–247
79. World Health Organization (1991) Measles Surveillance, Weekly Epidemiol Rec 19: 139–142
80. Nassau RD (1992) The costs of preventing measles: Letter to the Editor. JAMA 267: 655–656
81. Fulginti V (1992) The costs of preventing measles: Reply to letter to the Editor. JAMA 267: 656

9

High-titer measles virus vaccines: protection evaluation

M. Garenne[1], **O. Leroy**[2], **J. P. Beau**[3], and **I. Sene**[4]

[1] Harvard School of Public Health, Boston, USA
[2] Pasteur-Mérieux Serums et Vaccins, Marnes-La-Coquette, France
[3] ORSTOM, UR Population et Santé, Dakar, Senegal
[4] Centre Médical de Fatick, Fatick, Senegal

Summary

The immunogenicity and clinical efficacy of high-titer measles vaccines were evaluated in a randomized controlled trial in Niakhar, a rural area of Senegal that is under demographic surveillance. Two high-titer vaccines were studied: a high-titer Edmonston-Zagreb vaccine (EZ-HT) and a high-titer Schwarz vaccine (SW-HT). Both HT vaccines were administered at five months of age, and they were compared to a Schwarz standard vaccine, administered at 10 months of age. At age five months, 56.9% of the children had medium or high levels of maternal antibodies (at least 250 milli-international units), whereas at 10 months of age only 25.5% of children had residual maternal antibodies.

Both high-titer vaccines were immunogenic when given to children with low levels of maternal antibodies; seroconversion rates were 98.6% (95% confidence interval, CI = 95.9 − 99.9) for the EZ-HT and 85.5% (CI = 78.5 − 93.2) for the SW-HT. There was no definitive evidence of seroconversion among children with high levels of maternal antibodies. However, levels of antibodies at age 10 months were significantly higher in the vaccinated groups than in the placebo group. The EZ-HT vaccine produced a higher response than the SW-HT vaccine (P < 0.01). Both the titer at time of vaccination and the age at vaccination had a significant impact on the rate of seroconversion and the percentage of seropositive children at 10 months.

Clinical efficacy (EF) of HT vaccines was highest and significant among children with low levels of maternal antibodies [EF = 100.0%, CI = 50.0–100.0, P = 0.0001 for EZ-HT; and EF = 86.3%, CI = 43.9–96.7, P = 0.0060 for SW-HT], and lowest and not significant among children with high levels of maternal antibodies. Vaccine failures after the HT vaccines were 8.75 times more frequent than after the standard vaccine (CI = 1.02–75.1, P = 0.048). These results indicate that vaccination of children with Schwarz measles vaccine at nine months of age remains the safest and most effective strategy for controlling measles.

Introduction

Measles is a leading cause of mortality and morbidity among children in less developed countries [5,6]. Vaccination is the most effective measure to control measles, and measles vaccination is recognized as one of the most cost-effective public health measures. Recent studies show that measles vaccination has a large impact on child survival in developing countries [2,8].

Age at vaccination has been shown to be a major determinant of vaccine efficacy; children vaccinated too early do not seroconvert and are not protected by the vaccine [12]. The recommended age at vaccination with standard vaccines ranges from 12 to 15 months in most Western countries, but in tropical Africa it has been lowered to nine months in order to increase the protection of infants who can contract measles as early as four months of age [16]. The recommended titer of measles vaccines is at least 1000 50 percent tissue-culture-infective-dose ($TCID_{50}$), as defined by the World Health Organization (WHO). Standard vaccines often contain 5000 to 10 000 $TCID_{50}$, although they have been shown to be efficient even with a titer as low as 20 $TCID_{50}$ [12].

Pioneer work conducted in Iran, Mexico, Brazil and The Gambia found that vaccines grown on human diploid cells (HDC) and having higher titers than those of standard vaccines made possible vaccination earlier in life [10,13,14,17]. Further studies showed that increasing the titer of the vaccine by 100 fold increased its immunogenicity and that the Edmonston-Zagreb strain produced a higher rate of seroconversion that the Schwarz strain [9,15,18]. These results suggested that children could be vaccinated at four to six months of age. Recently, the World Health Organization has recommended the use of the EZ-HT vaccine at six months of age in countries where measles is a significant cause of death [19].

Immunogenicity studies are poorly standardized and their results are seldom matched with clinical protection. Therefore, there results are often difficult to interpret. Among the pitfalls of immunogenicity studies are: (1) the class of antibodies detected by hemagglutination inhibition (HI), plaque neutralization (PN) or enzyme-linked immuno-assay (Elisa); (2) the laboratory methods and their sensitivity; (3) the interval between vaccination and measurement of antibody titers (from 4 weeks to 6 months or more); (4) the definition of seroconversion and seropositivity. In addition, the level of maternal antibodies, which varies significantly from population to population and even within the same population by socioeconomic level and the age at vaccination are both important determinants of immunogenicity of measles vaccines [12].

This study evaluated the protection of high titer measles vaccines in a rural area of tropical Africa. It is unique because it compares the immunogenicity of high titer measles vaccines with their clinical protection, according to the level of maternal antibodies at the time of vaccination.

Material and methods

Study population

The study area was located near Niakhar, in the *département de Fatick* in central Senegal. It consisted of 30 villages, whose total population included about 25 000 people of Sereer origin. A comprehensive demographic and epidemiologic surveillance system, based on weekly visits to households and an annual census, was in place before the study began (July 1987) and has been maintained since that time.

Study design

The study was a randomized controlled trial of the efficacy, safety, and immunogenicity of two high-titer live measles vaccines administered at five months of age: Edmonston-Zagreb (EZ-HT) and Schwarz (SW-HT), in comparison with a Schwarz-standard vaccine adminis-tered at 10 months of age. Children in the Schwarz-standard group received a placebo at five months of age. Children were recruited at birth and vaccinated at the appropriate age. For the study, 24 monthly birth cohorts were recruited: the children born between February 1987 and January 1989. Children who were vaccinated or received the placebo at five months were considered participants. The participants from the first 16 cohorts, who had complete information on serology at five and 10 months, were included in the immunogenicity study. Children of the 24 birth cohorts, who survived until age 20 weeks at least, were included for the final analysis of the clinical efficacy. The details of the vaccine trial and the main results have been reported [3,4].

The study was approved by the Senegalese health authorities (Ministère de la Santé Publique, Dakar), by ORSTOM authorities (Institut Français de Recherche pour le Dévelop-ment en Coopération, Paris), and by the ethical committee of the British Medical Research Council, Fajarah, The Gambia. During the three years of the project, comprehensive vaccina-tions were made available to everyone, and free drugs and medical services were provided to all children and adults of the study population. As a consequence both mortality and case-fatality due to measles were significantly lower than during the preceding three years (1984–1986).

Vaccinations

Oral informed consent was obtained from mothers before administering high titer vaccines. Those who refused received standard vaccines. Three measles vaccines were supplied by the manufacturers: an Edmonston-Zagreb vaccine (EZ-HT: batch 81/3; titer, 5.4 $\log_{10}$ plaque-forming-units [pfu]; Institute of Immunology, Zagreb, Yugoslavia); a Schwarz high-titer vaccine (SW-HT: batch 0980; titer, 5.4 $\log_{10}$ pfu; Institut Mérieux, Lyon, France); and a Schwarz standard vaccine (titer, 3.7 $\log_{10}$ pfu; Institut Mérieux, Lyon, France). The potency of the vaccines was monitored routinely and found to be constant over time. The age at vaccination was strictly standardized: 5.0 months (standard deviation [SD] = 0.35 months) for the HT vaccines and 10.0 months (SD = 0.35 months) for the standard vaccine. Only a few children received the standard vaccine after 10 months of age. By January 1, 1990, 1566 children had been vaccinated; vaccine coverage was 81.6 percent of the resident target population.

Serology

Blood was drawn from all children of the 24 cohorts just before vaccination at 5 and 10 months. In addition, blood was drawn at 3 months of age from volunteer children of three cohorts (05, 06, and 07) and at eight months of age from volunteer children of the first two cohorts

(01, 02). Oral informed consent was obtained from mothers before blood was drawn. A small proportion of mothers (6.3%) refused to allow blood samples taken at the time of vaccination. For the immunogenicity study, the analysis was restricted to 541 children of the first 16 cohorts who came at five and 10 months, whose blood samples were available, and who had no known exposure to the wild measles virus before the second sample was taken. The children were randomized into three groups receiving EZ-HT, SW-HT, or Placebo at 5 months of age. There was no evidence of any bias in the randomization [3].

Blood samples were analyzed at the MRC laboratory in Fajarah, The Gambia. This laboratory uses international standards for antibody assays that have been standardized against the WHO international reference preparation for measles antibody. The test used for measuring the antibody decay or the change in antibody levels after measles vaccination or after exposure to measles was the measles hemagglutination-inhibition test (HI). A micro-method was used. Sera obtained by fingerprick were placed in a serum separator tube (Microtainer, Becton Dickinson), centrifuged at 6000 g for 3 min to separate serum from red cells, and stored at − 16°C until used. The delay between blood sampling and serum analysis averaged 80 days. Serum was decomplemented in the microtainer tube at 56°C for 1/2 h and thereafter adsorbed overnight in another tube with 1/3 volume of packed green-monkey cells. Serum (0.025 mL) was diluted with an equal volume of phosphate-buffered saline containing 1% bovine serum albumin and 0.1% sodium azide in non sterile microtiter plates. Phosphate buffered saline (0.025 mL) containing 4 hemagglutination units of Tween-ether extracted measles hemagglutinin (Behringwerke) was added to an equal volume of each dilution of serum. The mixtures were incubated at 37°C for 3 h; then 0.025 mL of a 0.5% suspension of green monkey cells was added to each of the wells. After standing for 3 h at room temperature, the plates were left overnight at 4°C and were read by eye in the morning. The starting dilution was 1:2, which detected 125 miu (milli international units) of measles HI/mL.

Two thresholds of measles antibodies were considered: 250 miu and 1000 miu. This choice was justified by the different protective value of maternal and vaccine HI antibodies. In case of exposure within the same compound, a level of 250 miu of HI antibodies after vaccination was found 100% protective, a value similar to other studies [1], whereas 1000 miu of maternal antibodies was 98% protective [3].

The immunogenicity of the high-titer vaccine was analyzed in three ways. First, seroconversion was defined as either − ≥ 250 miu at 10 months for those who were negative at five months or − at least fourfold an increase at 10 months for children who had detectable levels of maternal antibodies at five months. Second, seropositivity was defined as ≥ 250 miu at 10 months. Third, the geometric mean titer (GMT) at 10 months was measured for those who were seropositive. These definitions were similar to definitions used in other trials [9,15].

Outbreak investigation

Between August 1987 and July 1990, three measles outbreaks were investigated in the study area, and a total of 601 cases were identified. Most cases occurred during the last year, after the immunogenicity study was completed. Measles cases were validated by clinical examinations performed by a physician and by serology when possible. A clinical score was defined by examining seven clinical signs observed during the physician's examination and summing up the values for each sign: 6 = typical rash or desquamation, 5 = Koplik spot, 2 = atypical rash, 2 = conjunctivitis, 2 = stomatitis, 1 = cough, 1 = rectal temperature above 38.0°C. A clinical score of ≥ 8 was considered as clinical validation. Of all reported cases, 80.1 percent were clinically confirmed; of reported cases examined during the three to 10 days after the onset of the clinical signs, 92.4 percent were clinically confirmed. Serological confirmation was defined as an increase of at least fourfold increase in titer of HI antibodies to measles virus during the acute phase. Of all reported cases, 63.7 percent were serologically confirmed; of reported cases whose first blood sample was taken before day 2 and whose second blood sample was taken after day 28, 100 percent were serologically confirmed. They all had at least a 16-fold increase in titers of HI antibodies. Unconfirmed cases were those not examined early enough by the physician. When restricted to confirmed cases, values of vaccine efficacy were equivalent to the corresponding values for all cases [3]. Therefore, all cases were kept for the final analysis of clinical efficacy.

For calculation of vaccine efficacy, exact person-days at risk were computed for the resident population by use of a DBASE-IV computer program. The starting point was 14 days after vaccination for vaccinated children or age 140 days (20 weeks) for unvaccinated children. The incidence of measles was computed as the ratio of cases to person-years at risk lived by the susceptible population. Vaccine efficacy was computed by the standard formula: 1-(incidence among vaccinated children/incidence among unvaccinated children). The 95 percent confidence intervals (CI) were computed by the standard formulae for relative risks [11].

Results

The decay of maternal antibodies in this population was noticeably early and rapid (Table 1). At age 10–13 weeks, 87.8% of children had at least 250 miu; at 18–25 weeks (the age range for vaccination with high titer vaccines), this proportion was 56.8% and by 42–49 weeks (the age range for vaccination with standard titer vaccines), this proportion had fallen to 25.4%. The GMT declined markedly between three and six months and by age 7 months it was down to a low value of 133 miu (Fig. 1). These figures are difficult to compare to others since very few studies provide detailed information in the same system of international units. However they seem to be similar to values found in Kenya [16].

Among unvaccinated children, the titer at 10 months was virtually independent from the titer at five months (see Table 2). The decay of maternal antibodies was faster among children with the highest titer at five months. By age 10 months, most children had lost their antibodies, and it was therefore impossible to define a "half-life" without introducing the titer at five months. However, some children seem to keep moderate levels of antibodies well beyond their first birthday [3].

Table 1. Decay of maternal antibodies according to age among children who were never vaccinated, who never had measles, and had no known exposure to measles, Niakhar, Senegal, 1987–1990

Age group (weeks)	Mean age (months)	GMT (miu)	No of samples	Percent distribution of titers (in miu)			
				< 250	250–999	≥ 1000	Total
10–13	2.9	490	(33)	12.1	57.6	30.3	100.0
14–17	3.5	366	(82)	19.5	58.5	22.0	100.0
18–25	5.0	207	(871)	43.2	45.5	11.4	100.0
26–33	6.5	112	(19)	73.7	21.1	5.3	100.0
34–41	8.8	133	(82)	70.7	14.6	14.6	100.0
42–49	10.0	121	(334)	74.6	15.3	10.2	100.0

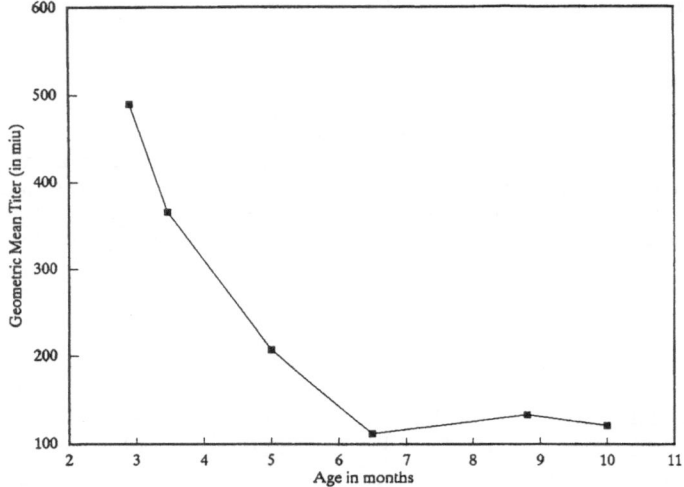

Fig. 1. Decay of maternal antibodies to measles virus, Niakhar, Senegal

Table 2. Immunogenicity of high titer measles vaccines, according to the level of maternal antibodies at time of vaccination (five months), Niakhar, Senegal, 1987–1990

Titer of maternal antibodies at five months	EZ-HT (N = 169)		SW-HT (N = 168)		Controls (Placebo) (N = 204)	
	Mean	95% CI	Mean	95% CI	Mean	95% CI
% Seroconversion (5–10 months)						
< 250	98.6*	[95.9–99.9]	85.5*	[78.0–93.8]	28.6	[20.4–40.1]
250–999	61.6*	[51.4–73.9]	26.4*	[17.9–38.8]	2.0	[0.5– 7.8]
≥ 1000	5.3	[0.8–35.5]	5.0	[0.7–33.8]	5.3	[0.8–35.5]
% Seropositive at 10 months (≥ 250 miu)						
< 250	98.6*	[95.9–99.9]	88.2*	[81.2–95.7]	31.0	[22.5–42.6]
250–999	97.3*	[93.6–99.9]	66.7*	[56.6–78.5]	20.8	[14.2–30.4]
≥ 1000	84.2*	[69.3–99.9]	45.0	[27.7–73.1]	21.1	[8.8–50.3]
GMT at 10 months (for those seropositive)						
< 250	3217*	[2173–4763]	2434*	[1630–3636]	619	[325–1178]
250–999	1276*	[1006–1619]	930*	[697–1242]	359	[232– 556]
≥ 1000	878	[469–1643]	630	[273–1453]	707	[202–2476]

*P < 0.05, with respect to placebo group
GMT in milli-international units

Immunogenicity

Data on immunogenicity were available only for the high titer vaccines and were compared to the control group receiving the placebo at five months (Table 2). Among children with a low level of maternal antibodies (< 250 miu) who received the EZ-HT vaccine, 98.6% (CI = 95.9–99.9, P < 0.0001) had seroconverted by 10 months, and 98.6% (CI = 95.9–99.9, P < 0.0001) were seropositive with a GMT of 3217 miu (CI = 2173–4763, P < 0.0001). Comparable figures were 85.5% (CI = 78.0–93.8, P < 0.0001), 88.2% (CI = 81.2–95.7, P < 0.0001) and 2434 miu (CI = 1630–3636, P = 0.0002) among children who received the SW-HT vaccine, as compared with 28.6% (CI = 20.4–40.1), 31.0% (CI = 22.5–42.6), and 619 miu (CI = 325–1178) in the placebo group. Both vaccines were immunogenic and the immunogenicity of the EZ-HT vaccine was significantly higher than that of the SW-HT vaccine (P = 0.0020 for seroconversion; P = 0.0060 for seropositivity). However, there was no significant differences in the GMT among seropositive children between the two groups (P = 0.2347).

Among children with medium levels of maternal antibodies (250–999 miu), who received the EZ-HT vaccine, 61.6% (CI = 51.4–73.9, P < 0.0001) had seroconverted by 10 months, and 97.3% (CI = 93.6–99.9, P < 0.0001) were

Table 3. Multivariate analysis of the correlates of the immunogenicity of high-titer measles vaccines, Niakhar, Senegal, 1987–1990

Variable	Odds ratio	Coefficient	Standard Error	T	P (2-tail)
Model 1: Dependent variable = seroconversion					
Constant		−4.527	1.890	−2.395	0.017
Vaccine: EZ-HT	75.02	2.318	0.410	10.538	1.0E–13*
SW-HT	17.22	2.846	0.355	8.013	3.0E–13*
Maternal antibodies (unit = 2-fold)	0.25	−1.395	0.141	−9.907	1.0E–13*
Age at vaccination (weeks)	1.20	0.178	0.085	2.102	0.0360*
CHI^2(5) : 351.1, P = 1.0E–13					
Model 2: Dependent variable = seropositivity					
Constant		−7.286	1.793	−4.063	0.0001
Vaccine: EZ-HT	116.33	4.756	0.478	9.958	1.0E–13*
SW-HT	11.30	2.425	0.268	9.052	1.0E–13*
Maternal antibodies (unit = 2-fold)	0.68	−0.389	0.082	−4.761	2.0E–06*
Age at vaccination (weeks)	1.36	0.309	0.081	3.836	0.0001*
CHI^2(5) : 280.5, P = 1.0E–13					
Model 3: Dependent variable = GMT at 10 months for seropositive children					
Constant		4.048	1.575	2.570	0.0104*
Vaccine: EZ-HT		1.938	0.319	6.067	2.0E–09*
SW-HT		1.541	0.329	4.680	3.0E–06*
Maternal antibodies (unit = 2-fold)		−0.468	0.076	−6.131	1.0E–09*
Age at vaccination (weeks)		0.017	0.070	0.239	0.4172
F(3,284) = 18.372, P = 1.0E–13					

*P < 0.05

seropositive with a GMT of 1276 miu (CI = 1006–1619, P < 0.0001). Corresponding figures were 26.4% (CI = 17.9–38.8, P = 0.0002), 66.7% (CI = 56.6–78.5, P < 0.0001), and 930 miu (CI = 697–1242, P = 0.0002) among those who received the SW-HT vaccine as compared with 2.0% (CI = 0.5–7.8), 20.8% (CI = 14.2–30.4), and 359 miu (CI = 232–556) in the placebo group. In this case again, both vaccines were immunogenic, and the immunogenicity of the EZ-HT vaccine was significantly higher than that of the SW-HT vaccine (P < 0.0001 for seroconversion; P < 0.0001 for seropositivity). However, there was no significant differences in GMT among seropositive children between the two groups (P = 0.1067).

Among vaccinated children with high levels of maternal antibodies (≥ 1000 miu), there was no evidence of seroconversion. However, the proportion of seropositive children was significantly higher in the EZ-HT group than in the placebo group: 84.2% (CI = 69.3–99.9, P = 0.0012) but not in the SW-HT group: 45.0% (CI = 27.7–73.0, P = 0.0679). Among the seropositive children, the GMT of measles antibodies in the two vaccinated groups was not significantly different from that of the placebo group.

Correlates of seroconversion, seropositivity and GMT were analyzed with a linear logistic regression model (LOGIT). Correlates of Log-GMT was analyzed with a linear regression model (Table 3). Results confirmed the previous analysis. Both vaccines were immunogenic, and the EZ-HT vaccine was more immunogenic than the SW-HT vaccine; it induced a higher percentage of seroconversion (P = 0.0068), and a higher percentage of seropositivity (P < 0.0001), but not a higher GMT among seropositive children (P = 0.3868). The level of maternal antibodies reduced significantly the three parameters of immunogenicity. Age at vaccination, which had a narrow range (18 to 26 weeks), was also found to be a significant covariate of seroconversion and seropositivity, but not of GMT. These results were stable even after controlling for time between obtaining the two blood samples, sex, and season, variables that were not significant.

Clinical efficacy

The clinical efficacy of the three vaccination strategies was studied in relation to the level of antibodies at time of vaccination and compared with results for unvaccinated controls. There were five vaccine failures in the EZ-HT group; two of these children had medium levels of maternal antibodies, one had a high level, and for two the level of antibodies at time of vaccination was unknown. There were five vaccine failures in the SW-HT group; two had low levels of maternal antibodies, two had medium levels, and one had a high level. There was only one vaccine failure in the group that received the standard vaccine; this child had low levels of maternal antibodies at time of vaccination (Table 4).

Clinical efficacy of the EZ-HT vaccine was 100% (P = 0.0001) among children with low levels of maternal antibodies, 83.3% (CI = 31.5–95.9, P = 0.0136) among those with medium levels, and 70.4% (not significant)

Table 4. Clinical efficacy of three measles vaccination strategies, according to the vaccine and the titer of maternal antibodies at time of vaccination, Niakhar, Senegal, 1987–1990

Titer of antibodies at time of vaccination	Person-years at risk	No of measles cases	Incidence/ 1000 person-years	Vaccine efficacy		P (2-tail)
				Mean	95% confidence interval	
EZ-HT at 5 months						
< 250	421.6	0	0.00	100.0	--100.0	0.0001*
250–999	294.8	2	6.78	83.3	31.5– 95.9	0.0136*
≥ 1000	83.2	1	12.03	70.4	0.0– 95.9	0.2288
SW-HT at 5 months						
< 250	359.6	2	5.56	86.3	43.9– 96.7	0.0060*
250–999	288.6	2	6.93	82.9	30.0– 95.8	0.0146*
≥ 1000	74.4	1	13.45	66.9	0.0– 95.4	0.2740
Standard at 10 months						
< 250	607.8	1	1.65	95.9	70.7– 99.4	0.0016*
250–999	126.1	0	0.00	100.0	--100.0	0.0197*
≥ 1000	78.5	0	0.00	100.0	--100.0	0.1123
Controls (not vaccinated)						
Total	1329.4	54	40.62	(ref)		

*P < 0.05

among those with high levels (CI = 0.0–95.9, P = 0.2288). Corresponding values for the SW-HT vaccine were 86.3% (CI = 43.9–96.7, P = 0.0060) at low levels of maternal antibodies, 82.9% (CI = 30.0–95.8, P = 0.0146) at medium levels, and 66.9% (not significant) at high levels (CI = 0.0–95.4, P = 0.2740). Corresponding values for the standard vaccine were 95.9% (CI = 70.7–99.4, P = 0.0016) at low levels of maternal antibodies, 100% (P = 0.0197) at medium levels, and 100% (not significant) at high levels; P = 0.1123).

The correlates of clinical efficacy were also investigated in a multivariate analysis. The model was a proportional hazard model, where the dependent variable was the probability of contracting measles within 36 months after vaccination. This model allowed direct comparison of the vaccination strategy (high-titer vaccines at five months vs standard vaccines at 10 months), as well as comparison of the two strains (Edmonston-Zagreb and Schwarz). Two models were built: one comparing the groups directly, without any other control, and another one that controlled for the level of maternal antibodies at vaccination (five or 10 months of age) (Table 5).

In both models measles vaccination was highly efficacious, reducing measles incidence by 99.7% (CI = 82.9–99.9%, P = 0.0048) and 98.3% (CI = 0.0–99.7%, P = 0.0560) respectively. In the second model, the level of maternal antibodies was a strong determinant of vaccine failure, the rate of failure being multiplied by 1.93 for each fourfold increase in level of maternal

Table 5. Multivariate analysis of the clinical efficacy of three strategies for administration of measles vaccines, (proportional hazards model on the probability to contract measles within 36 months after vaccination), Niakhar, Senegal, 1987–1990

Covariate	Relative risk	95% Confidence interval	Estimate	Standard error	T	P (2-tail)
Model I: All children (N = 2978, cases = 62)						
Measles vaccine	0.003	[0.001–0.171]	−5.795	2.054	2.821	0.0048*
Strategy (HT-5/St-10)	8.75	[1.02 −75.1]	+2.169	1.097	1.978	0.0480*
Strain (EZ/SW)	0.59	[0.17 −2.05]	−0.521	0.633	0.822	0.4112
CHI^2(3) = 107.8, P = E-13						
Model II: Known maternal antibodies at vaccination (N = 1490, cases = 17)						
Measles vaccine	0.017	[0.003–1.11]	−4.061	2.124	1.912	0.0560*
Strategy (HT-5/St-10)	5.13	[0.59 −44.3]	1.636	1.000	1.487	0.1372
Maternal antibodies (unit = 4 log)	1.93	[0.99 −3.77]	0.658	0.342	1.924	0.0546*
Strain (EZ/SW)	0.54	[0.13 −2.25]	−0.618	0.730	0.846	0.3976
CHI^2(4) = 23.7, P = 0.0001						

antibodies (CI = 0.99–3.77, P = 0.0546). In both models, the coefficient of the strain was consistent with previous findings of immunogenicity (0.59 and 0.54), but none of the difference was statistically significant. In the first model, the strategy of giving high titer measles vaccines at 5 months multiplied the rate of vaccine failure by 8.75 (CI = 1.02–75.1, P = 0.0480). When controlling for the level of maternal antibodies, the risk ratio of vaccine failure was reduced to 5.13 and no longer significant (CI = 0.59–44.3, P = 0.1372).

Vaccine failures were also compared to the seroconversion status. Among the five vaccine failures in the EZ-HT group, one did not seroconvert and was seronegative at time of exposure, although it was seropositive at 10 months of age; the seroconversion status of the other four children was unknown. In the SW-HT group, the status of the five vaccine failures was known: none had seroconverted, and none were seropositive at 10 months. The study was not designed to relate seroconversion to clinical efficacy of the standard vaccine.

Discussion

The results did not show a close relationship between clinical protection and immunogenicity, although the relative ranking of the immunogenicity of HT vaccines matched the evaluations of their relative protective power. Seroconversion rates tended to underestimate clinical protection, whereas rates of seropositivity at 10 months of age tended to overestimate clinical protection for the EZ-HT vaccine, but not for the SW-HT vaccine among children with high levels of maternal antibodies. Discrepancies may be due partly to random fluctuations and partly to inaccuracies in the HI method. The standard error of this method was analyzed be comparison of two titrations of 106 blood samples. There was a standard error of 0.90 dilution on the average, and a third of the samples deviated by more than one dilution. The inaccuracy of the HI method also explains why a relatively large number of placebo recipients met the criteria of seroconversion. Although this type of inaccuracy is normal, it would have been virtually impossible to have an appropriate estimation of clinical protection only on the basis of immuno-genicity data.

Precise comparisons of the immunogenicity of HT vaccines observed in this study with results of other studies were difficult because of differences in vaccines, laboratory methods, age at vaccination, profiles of antibodies in the population, time between vaccination and drawing of the second blood sample, thresholds and definitions used for seroconversion and seropositivity. The most similar study was probably the second trial conducted in Mexico [9]. The investigators used identical vaccines (same batch), the age at vaccination was slightly older (six months), the threshold for seropositivity was slightly lower (200 miu), and the delay before obtaining the second blood sample was slightly shorter (18 weeks), although the titrations of antibody were done by plaque neutralization. In Mexico, the EZ-HT group had a seropositivity rate of 94% (96% in Senegal), and the SW-HT group a seropositivity rate of 79% (74% in Senegal). However, the GMT were much lower in Mexico and did not reflct the ranking found in Senegal: the EZ-HT group had a GMT of 1078 miu (1638 in Senegal) and the SW-HT group a GMT of 1224 (678 in Senegal).

Very few data were available to compare with the immunogenicity of the standard vaccine. However, the GMT of 31 children vaccinated with the Standard vaccine at 10 months (4090 miu) was 2.5 times higher than the GMT of children vaccinated with the HT vaccines.

The rationale for using high-titer vaccines was the hypothesis that they would "pass the barrier of maternal antibodies". However, both in the Mexican study and in this study, the level of maternal antibodies was a major barrier to immunogenicity of high-titer vaccines. Seroconversion rates, the proportion of seropositive children and the GMT among seropositives were dramatically lower when levels of maternal antibodies exceeded 250 miu. Seroconversion rates and GMT were even not significant at levels of ≥ 1000 miu. The level of maternal antibodies was also a major factor in clinical protection. The relatively good performance of the high-titer vaccines was primarily due to the choice of a population and an age group among which the proportion of children with at least 1000 miu was relatively low.

Multivariate analysis showed that age at vaccination was a significant factor in both seroconversion and clinical protection, even after controlling for the level of maternal antibodies. Part of this effect may be due to inaccuracies in the HI titration method. More importantly, age may be a proxy for other classes of antibodies that may prevent successful vaccination. A third possible explanation is age-specific immunocompetence. This last point has been poorly documented in the literature and requires further research.

No data on the optimal age at vaccination for high-titer vaccines have been published. Since the level of maternal antibodies is a major determinant of successful vaccination, one would expect the optimal age to be when the proportion of children with ≥ 250 miu becomes minimal. In the Senegalese population, this plateau seems to be reached at age seven months. This optimal age appears rather close to the optimal age for the Schwarz standard vaccine of nine month.

The benefit of changing the strategy from using standard vaccine at nine months to vaccinating with HT vaccines at six months seems to be minimal. Assuming a constant incidence of measles of 10% per person per year, susceptibility to measles starting at four months, and clinical efficacy values observed in this trial, a cohort of 1000 children would experience 373 cases of measles by the age of five years if no one was vaccinated, 87 cases if all infants were vaccinated with SW-HT at six months, 60 cases if all infants were vaccinated with EZ-HT at six months, and 57 cases if all infants were vaccinated with the standard vaccine at nine months. Therefore, even in terms of the prevention of clinical measles, the HT strategy would not be worthwhile between birth and five years of age. The situation would be even worse if all ages up to 30 years were taken into consideration.

Furthermore, the cost of using the HT vaccines is considerable. First, the HT vaccines are more costly to produce than standard vaccines, a factor that lowers their cost-effectiveness. Second, the HT vaccines were found to increase significantly the risk of death for children within three years after vaccination [4]. There was no correlation between immunologic response and risk of death. In particular, there was no death among the children who had a strong immune response, i.e. antibody titers of $\geq 16\,000$ miu in any of the two HT groups, and no death among the children who did not seroconvert in the EZ-HT group. There was some evidence of rising antibody levels among children between three and five months after vaccination, as found in other studies [14]. This observation suggests the possibility of a persistent infection, that may explain the excess risk of death.

Vaccination with standard measles vaccines at nine months of age still appears to be a sound strategy for tropical Africa. If well implemented with a high coverage rate, it will produce herd immunity and thereby reduce the incidence before the recommended age at vaccination (nine months). Together with the appropriate management of measles cases, this strategy may lead to satisfactory control of mortality due to measles. If the goal is measles eradication, alternative strategies need to be investigated, either with new vaccines or with several doses given at various ages.

Acknowledgment

The study was funded by the Task Force for Child Survival, Atlanta, GA, USA. The authors thank the Senegalese health authorities who supported the project, in particular Dr Colonel

Mame Thierno Sy and Dr Fodé Diouf, Dr Whittle who did the laboratory work, Dr Mary Adams and Ms Meg Tyler who provided comments and critiques on the manuscript.

References

1. Chen RT, Markowitz LA, Albrecht P, Stewart JA, Mofenson LM, Preblud SR, Orenstein WA (1990) Measles antibody: reevaluation of protective titers. J Infect Dis 162: 1036–1042
2. Garenne M, Cantrelle P (1986) Rougeole et mortalité au Sénégal. Etude de l'impact de la vaccination effectuée à Khombole 1965–1968 sur la survie des enfants. In: Estimation de la mortalité du jeune enfant (0–5 ans) pour guider les actions de santé dans les pays en dévelopement. Séminaire INSERM 145: 515–532
3. Garenne M, Leroy O, Beau JP, et al (1991) Efficacy, safety and immunogenicity of two high-titer measles vaccines: Final report. ORSTOM, Dakar, June 1991, 229 p
4. Garenne M, Leroy O, Beau JP, et al (1991) Child mortality after high-titer measles vaccination: a prospective study in Senegal. Lancet 338: 903–907
5. Hayden GF, Sato PA, Wright PF, et al (1989) Progress in worldwide control and elimination of disease through immunization. J Pediatr 114: 520–527
6. Henderson RH, Keja J, Hayden G, et al (1988) Immunizing the children of the world: progress and prospects. Bull WHO 66: 535–543
7. Hersh BS, Markowitz LE, Hoffman RE, et al (1991) A measles outbreak at a college with a prematriculation immunization requirement. Am J Public Health 81(3): 360–364
8. Koenig MA, Khan MA, Wojtyniak B, et al (1990) The impact of measles vaccination on childhood mortality in Matlab, Bangladesh. Population Council, programs division, Working papers 3: 18
9. Markowitz LE, Sepulveda J, Diaz-Ortega JL, et al (1990) Immunization of six month-old infants with different doses of Edmonston-Zagreb and Schwarz measles vaccines. N Engl J Med 332: 580–587
10. Nirshansy H, Shasyi A, Bahrani S et al (1977) Comparative field trial of five measles vaccines produced in human diploid cells MRC-5. J Biolog Standard 5: 1–18
11. Orenstein WA, Bernier RH, Hinman AR (1988) Assessing vaccine efficacy in the field: further observations. Epidemiol Rev 10: 212–241
12. Preblud SR, Katz SL (1988) Measles vaccine. In: Plotkin SA, Mortimer EA (eds) Vaccines. WB Saunders, Philadelphia pp 182–222
13. Sabin AB, Arechiga FA, Fernandez de Castro J, et al (1983) Successful immunization of children with and without maternal antibody by aerosolized measles vaccine. I. Different results with undiluted human diploid cell and chick embryo fibroblast vaccines. JAMA 249: 2651–2662
14. Sabin AB, Arechiga FA, Fernandez de Castro J et al (1984) Successful immunization of children with and without maternal antibody by aerosolized measles vaccine. II. Vaccine comparisons and evidence for multiple antibody response. JAMA 251: 2363–2371
15. Tidjani O, Grunitsky B, Guérin N et al (1989) Serological effects of Edmonston-Zagreb, Schwarz and AIK-C measles vaccine strains given at ages 4–5 or 8–10 months. Lancet ii: 1357–1360
16. Van Ginneken JK, Muller AS (1984) Maternal and child health in rural Kenya, Croom Helm, London
17. Whittle HC, Rowland MGM, Mann GF, Lamb WH, Lewis RA (1984) Immunization of 4–6 month old Gambian infants with Edmonston-Zagreb measles vaccine. Lancet ii: 834–837
18. Whittle HC, Mann G, Eccles M, et al (1988) Effect of dose and strain of vaccine on success of measles vaccination of infants aged 4–5 months. Lancet i: 963–966
19. WHO/EPI (1990) Measles immunization before 9 months of age. Wkly Epidem Rec 2: 8–9

10

Recent approaches in the development of measles vaccines

O. Leroy

Pasteur Mérieux Sérums et Vaccins, Marnes-La-Coquette, France

Summary

Different approaches are followed in order to develop measles vaccines which will induce a protective cellbound and humoral immunity in the face of maternal antibodies. The use of massive doses of different strains of live attenuated measles vaccines at 6 months of age have reached deadlock and the improvement of the currently used vaccines is of limited interest. The recent advances in genetic and molecular techniques pave the way for the development of recombinant and synthetic vaccines, which is closely linked to the understanding of the molecular and structural determinant of the protective immune response. The synthetic vaccines are still at an experimental level, and it is possible to design immunogenic peptides, containing both T and B cell recognition sites. The major limitation of the synthetic vaccine approach is the unresolved question on the mechanism of the protective immunity related to the structure of the peptides – length, construction, way of presentation – and their recognition by the largest possible number of individuals. The development of adjuvants is complementary to synthetic vaccines. Different adjuvants are developed in order to enhance the immunogenicity of sub-unit vaccines. Liposomes and the immunostimulating complex (ISCOM) are the most promising for measles sub-unit vaccines. Many efforts are devoted to the development of recombinant measles vaccines. The construction of recombinant measles vaccines is based on the insertion of the genes coding for the fusion and haemagglutinin proteins of the measles virus envelope in poxvirus vaccines used as vectors. Vaccinia, fowlpox and canarypox viruses are the most commonly used vectors. There are strong experimental demonstrations of the expression of the recombinant antigens at sufficient levels, inducing humoral and protective immunity.

Introduction

Vaccines are the most cost-effective strategy for measles control. For almost thirty years and since the availability of the first vaccines, immunization has resulted in a dramatic reduction in both morbidity and mortality. Measles is still however one of the leading killers of children in the world with an estimated number of 1 million deaths in 1990 [1]. The vaccines currently used are live attenuated vaccines which induce a B and T cell-mediated immune response and also a protective immune response at the level of mucosal membranes, closely mimicking natural infection. When the vaccines are used properly, they confer long-lasting immunity which prevents measles and they are capable of providing sufficient antibody prevalence rates for herd immunity [2,3]. Different factors may affect the quality of the immune response to live attenuated vaccines, factors related directly to the vaccines and the vaccinated individuals, but also in relation to the heterogeneity of the human populations and the existence of clusters of higher risk of virus transmission [4]. The major and well-known obstacle to seroconversion is the presence, at the time of vaccination, of passively acquired measles antibodies, such as maternal antibodies or immunoglobulins. Intercurrent infections, even as mild as "colds," can also interfere with the immune response to live attenuated vaccines [5,6]. Nevertheless, several studies in developing countries found that sick children could be successfully immunized [7–9]. Although most of these vaccines derive from the original Edmonston strain [10], seroconversion can differ, mainly in young infants, from one strain to another [11,12], and can also depend on the route of administration [13,14], the dose of vaccine [15–20] and on improper handling of the vaccine [21,22]. The age at vaccination interferes with the immune response and, in younger infants, the persistence of maternal transplacentally derived antibodies has been found to be responsible for measles vaccine failure [23]. The optimal age for the first vaccination must therefore be established in relation to the level and the persistence of maternal antibodies [4]. In populations where the risk of severe measles at younger age is high, the measles vaccine must be delivered as soon as possible after the loss of maternal protection. This is particularly important in low income populations where children become susceptible earlier. The variation in persistence of passive protection can be related to the level of measles antibodies in mothers, to genetic or environmental differences influencing the transport of IgG across the placenta and to the rate of loss of maternally acquired antibodies in children [4].

A measles vaccine, which could be administered as early as two months of age or even before, would be a major advance in measles control. It would avert the occurrence of measles in the high case-fatality rate age group and, moreover, it would certainly lead to a better compliance with measles vaccination if included in the first sessions of the current calendar of the Expanded Programme for Immunization, limiting the drop-out after the third dose of Diphtheria-Tetanus-Pertussis vaccine.

High-titer live attenuated measles vaccines

Recent research has focused on the means to overcome the maternal antibody obstacle. Efforts have been made towards the possibility of using massive doses of supposedly more potent strains of live attenuated measles vaccines at six months of age [4,11,24]. The first promising serological results of the use of the high-titer Edmonston-Zagreb strain measles vaccine (EZ-HT) at 4–6 months of age were based essentially on the evaluation of the short-term immunogenicity [7,18,25–27]. Nevertheless, the EZ-HT did not overcome, as expected, the barrier of maternal antibodies: in the Mexico and Senegal studies, the level of maternal antibodies at time of vaccination interfered with the seroresponse [18,28]. Moreover, because few studies have been properly designed at look at protective efficacy, there is currently no clear evidence that this strategy would prevent more cases of measles than vaccination with standard vaccine at nine months of age: on the contrary, the EZ-HT vaccine given at 5 months of age was found in the Senegal study to have a much higher rate of vaccine failure than the standard Schwarz strain measles vaccine given at 10 months of age [29,30]. Lastly, a major concern on the long-term safety has been raised with the use of high-titer vaccines: those children, particularly the girls, vaccinated with the EZ-HT vaccine were found to have an increased risk of dying, several months after immunization. These findings were possible in the Senegal study because of the well-established demographic surveillance system and the routinely conducted monitoring by verbal autopsy of the causes of death [30–34]. This study has confirmed the as yet unpublished observations made by a researcher in Guinea-Bissau. The review of several other EZ-HT vaccine studies leads to the same conclusions, particularly in one study conducted in Haiti. The biological explanation of this excess delayed mortality is not yet clear, even if there are some indications of a cell-mediated immune dysfunction. After two meetings held in February 1991 and June 1992, the World Health Organization took the decision to suspend the use of the high-titer vaccines at younger age [35,36].

Even if the assessment of high-titer live attenuated measles vaccines has reached deadlock, the way is always open for the evaluation of more immunogenic strains of live attenuated vaccine used at standard titer, such as the AIK-C strain.

New strategies for the development of measles vaccines

Because of the limited field for the improvement of live attenuated measles vaccines, it is even more important to develop new approaches in measles vaccines. Here is a "shopping list" of the characteristics of the ideal vaccine: heat-stable, inexpensive, totally safe, inducing life-long protection and herd immunity, a single dose administered as soon as possible after birth, combined or associated with other antigens and given by a painless route [37].

The recent improvements in peptide synthesis and molecular and genetic techniques have led to different strategies. Most are based on the presentation as antigens to the immune system, of well-defined viral subunits, produced from the whole virus, recombinant proteins or synthetic peptides. In order to enhance the immunogenicity of the viral subunit vaccines, efforts are being devoted to the development of improved adjuvants such as liposomes or the immunostimulating complex (ISCOM). Another interesting avenue of research is the targeting and controlled release of antigens by microcapsule delivery system [38].

The advance in recombinant DNA technology has also made it possible to engineer recombinant viruses or bacteria as live vectors for the expression of protective antigens.

The necessary condition for the choice of antigens is their ability to induce both cellular and humoral immune responses which will confer clinical protection without deleterious effects and whatever the pre-existing immune status. The pre-requisite information on measles virus is the structural and molecular understanding of the protective immune response. There are in fact creative interactions between the development of synthetic peptides or recombinant vaccines and the knowledge of the intricate mechanisms of the protective immune response.

The measles virus is a pleiomorphic enveloped particle. The virus genome consists of a single strand of RNA of negative sense; a polymerase-associated protein (P) is necessary to the transcription into message of positive sense before the translation for the production of virus-coded proteins. The association of the viral RNA, the large polymerase protein (L), the P protein and the molecules of a nucleoprotein (N) build up a long nucleocapsid complex. The internal nucleocapsid folds itself into an outer lipoprotein envelope. The lipidic structure of the envelope derives from the host cell membrane. The envelope contains three structural proteins: two virus-coded glycosylate proteins – the haemagglutinin (H) protein and the fusion (F) protein – and the matrix (M) protein. The H protein is the cell receptor binding protein, responsible for the adsorption of the virus to the host cells. The F protein is responsible for haemolysis, for virus fusion with the host cell and for cell-to-cell spread of the virus causing the giant cells. The M protein has a stabilizer function for the envelope. In addition to the six structural proteins, there is a non-structural virus protein, the C protein [39]. Schematically, during the viral replication cycle, the virus binds with the cell receptors by the H protein. Then, after cleavage by cellular proteases of an F_0 precursor protein in two molecules F_1 and F_2, the virus envelope fuses with the cell membrane and the viral nucleocapsid is released into the cell cytoplasm where the viral proteins and RNA are synthesized. The H, F and M synthesized proteins link to the cytoplasmic membrane which buds, including a newly produced nucleocapsid. In the later stages of infection, the envelope glycoprotein accumulation results in the fusion of the neighbouring cells and produces syncitium. The formation of syncitia elicits the cell-to-cell transmission of the infection. The cleavage of the F precursor protein is essential for infectivity and in some cases determines pathogenicity. Early inactivated vaccines were shown to elicit abnormal immune response, which resulted in so-called atypical measles in vaccinees after exposure to the wild virus. The denaturation of the F protein by the inactivation process (formalin or Tween ether), was held responsible for the loss of the functional antigenicity of the F protein.

The vaccinees were non-protected despite the production of high levels of anti-H antibody [40–44]. There is some evidence that disturbed humoral response can last even after booster vaccination with live attenuated vaccine [45].

Thus, the vaccine acquired immunity will intervene at three different steps of viral infection. Firstly, the vaccine will induce the production of secretory IgA to prevent infection and replication in the epithelium cells and proximate lymph nodes. Secondly, it will produce antibodies which will neutralize the circulating virus. Thirdly, it will avert the replication and the cell-to-cell spread of the virus, with the intervention of cytotoxic cells leading to the cytolysis of infected cells.

Measles induces IgG, IgM and IgA antibodies which can be detected in serum (IgG and IgM) and in nasal secretions (IgA) [46–49]. The humoral response following natural infection or after immunization with live attenuated vaccines is generally investigated using standardized antibody assays such as complement-fixation [50], haemagglutination inhibition, ELISA and neutralization tests (for review see [51]). Monospecific antibodies against the different structural proteins of the measles virus can be detected by immunochemical techniques and by dosage of their biological activity, e.g. agglutination, haemolysis or fusion [41,52–54]. The cell mediated immune (CMI) responses play an important role in the recovery from measles and have been studied mostly on immunocompromised patients. Patients who present a cell mediated immunodeficiency but without abnormality of the humoral response develop severe giant-cell pneumonia [55,56], whereas patients presenting an antibody deficiency syndrome with proper CMI responses recover normally [57]. A subset of cytotoxic T lymphocytes (CTL) is involved in the recovery from the disease: CD8$^+$ CTL restricted to the class I major histocompatibility complex (MCH) [58]. The measles virus is able to replicate in B and T lymphocytes and monocytes and is the first virus which has been involved in cellbound immunodeficiency. After the acute phase of the disease, von Pirquet described a loss of reactivity to tuberculin skin tests and the reactivation of tuberculosis [59]. The evidence of measles-induced immunodeficiency is observed *in vitro*; there is a decrease in the production of lymphokines [60], of the natural killer cell activity [61] and of the proliferative responses to mitogens [62–64].

Due to the non-comparable pathogenicity of the measles virus in other animal species, except the *macacus* monkey, it is difficult to assess experimentally the participation of the different structural proteins in the protective immune response. Different experimental alternatives can circumvent these difficulties. The measles virus and the canine distemper virus (CMV) belong to the morbillivirus group and dogs vaccinated with a measles vaccine are protected against canine distemper. This heterotypic vaccination approach has been considered for the evaluation of the immune properties of the different antigens of the measles virus. The immunization of dogs with a purified F CMV protein gives conclusive indications that the F protein is involved in the protection [65]. Several strains of measles virus have been adapted to the cerebral tissues of small rodents. These neurotropic strains can be used for challenge in measles immunized animals.

Monoclonal antibodies against both F and H proteins neutralize the virus *in vitro* and protect animals from infection [66,67]. The H,F and N proteins have been shown to protect rats from measles infection [53,66,68]. The involvement of the other structural M and P proteins in the protective response has not yet been proven. The protective immunity related to F and N proteins is conferred by both the CTL and antiviral antibody synthesis, while the response to the H protein is associated with the production of high titers of neutralizing antibodies [53]. Because of the important involvement of the F and H measles-virus glycoproteins to elicit a protective immune response, they must be considered for use in recombinant or synthetic vaccines.

Synthetic vaccines

The potential interest of synthetic vaccines is the possibility of production of viral selected epitopes. This assertion has been verified at least once. The synthetic vaccine against foot and mouth disease has been sucessfully used to protect cattle. The comparison of sera from protected and unprotected animals demonstrates the higher affinity of the antibodies for both the peptide and the whole virus in protected animals. These findings could be a useful but non-exclusive tool for the assessment of the efficacy of synthetic vaccines [69].

Despite the apparent simplicity of this approach, in the case of measles virus, it comes up against the difficulties of the identification of the determinant epitopes for the immune response. As demonstrated by Milich, there is a direct implication for vaccine development in the fact that the peptide must contain the two distinct B and T cell recognition sites [70]. The complete amino acid sequences of the F and H proteins are now available [71,72]. Peptide synthesis techniques have enabled the preparation of every possible overlapping peptide. The B cell epitopes are localized and identified by systematically reacting synthesized peptides with antibodies and especially with monoclonal antibodies. The putative T cell epitopes are predicted using computerized algorithms. After their synthesis, the peptides are assessed as T cell recognition sites of the antigen.

Partidos et al, identified amino acid residues of the fusion protein as a putative T cell epitope [73]. This sequence was synthesized and, after inoculation in different inbred strains of mice, the peptide was determined as a B cell epitope. This finding was confirmed in humans, based on the response to lymphoproliferation tests. In both humans and mice, this peptide binds to a wide range of class II MCH alleles. Furthermore, Partidos et al constructed a chimeric peptide which contains both T and B cell recognition sites. This synthetic peptide is shown to be immunogenic since it induces in mice the production of antibodies which bind to the F protein and to the measles virus. The authors highlight the importance of the orientation of the B and T cell epitopes in the chimeric peptide for the induction of an appropriate and specific immune response [74]. There is some indication that the length and flanking sequence of the peptide interfere with the T helper cell activation [75].

Because of the genetic restriction, the selected peptides must be recognized by the largest possible number of individuals. This could be attempting the impossible because of the genetic heterogeneity of the human population. The T cell epitopes must be associated with protective immunity and the widest possible range of MHC-CTL before they can be considered as components of a candidate vaccine. In the light of these results, the concept of a measles synthetic vaccine receives strong indications that it is possible to design immunogenic peptides. The use of synthetic peptides as components of a measles vaccine remain an attractive alternative with regard to the advantage of having a chemically-defined and contamination-free vaccine. But, whatever the difficulties of this development, the way the peptide is presented to the organism still seems to be essential in inducing adequate efficacy. Thus the complementary approach for synthetic vaccines is the development of carriers and adjuvants [76].

New adjuvants

Immunostimulants such as aluminium salts (alum) are widely used. Nevertheless, there remain unresolved problems concerning protein adsorption, vaccine storage and aluminium associated pathology. Different adjuvants have been developed as substitutes for alum and the strongest experimental one is Freund's adjuvant. However, the soluble H and F measles-virus glycoproteins mixed in complete Freund's adjuvant did not induce a satisfactory humoral response [54]. Two other adjuvants enhance the immune response to viral-subunit measles vaccines: liposomes and the immunostimulating complex.

Liposomes are artificial phospholipid bilayer vesicles formed by the dispersion of phospholipids in aqueous solution. The production of liposomes is easy and uniform. They can be considered as carriers with adjuvant properties. They can contain a large amount of antigens and, possibly, of adjuvant. The effect of the liposomes passes through their phagocytose by macrophages which play a role of antigen-presenting cells. The incorporation of a viral membrane antigen (VMA) into liposomes has been realized by Mougin et al, in order to assess the humoral response to the H and F glycoprotein after injection in rats. The liposome-form VMA led to a humoral IgG response as that of the virus [77]. The adjuvanticity of liposomes was also demonstrated in naive mice by comparison of the T cell response induced by a free-form purified H glycoprotein with that induced by a liposome-form H glycoprotein. A 20-fold greater amount of free-form H protein is needed to induce a T cell response similar to that of the liposome-form H protein [78]. The composition, size and stability of the liposomes influence their capacity to enhance the immune response [78].

The immunostimulating complex (ISCOM) is a cage-like particle in which viral proteins or peptides are presented. It is spontaneously formed when phosphatidyl choline and cholesterol are mixed in the presence of saponin (Quil A extracted from the bark of *Quillaja saponaria* Molina). The adjuvant properties of saponin were observed initially in France in the 1930s and widely used in veterinary vaccines, e.g. foot-and-mouth vaccine [79]. Morein et al tested the immunizing potency of different ISCOM-form antigens in animals. The measles-ISCOM was prepared by incorporation of the H and F protein micelles. The measles-ISCOM or the protein micelles were injected intramuscularly in different rats; the neutralizing and haemolysis-inhibiting humoral responses were much higher in ISCOM-immunized rats [80], without side effects [81]. An ISCOM model presenting only the F protein has been tested in mice. The mice were protected against experimental measles; they produced haemolysis-inhibiting and fusion-inhibiting antibodies and presented a stimulation of the T cell [82]. These first studies were performed with glycoproteins extracted from the virus. Pedersen et al have incorporated synthetic peptides representing different regions of the H protein, into an ISCOM and have used it sucessfully for immunization of rabbits [83].

In a recent review of several ISCOM systems, Osterhaus points out the different advantages of the ISCOM approach. ISCOM induce a B cell response, long-lasting, even in the presence of pre-existing antibodies. ISCOM stimulate T cell immunity, inducing $CD4^+$ CTL, even in the presence of pre-existing antibodies, and $CD8^+$ CTL, which are demonstrated to be essential in the recovery from measles. They confer a protective immunity in many experimental systems. No toxicity has yet been described and ISCOM vaccines are licensed for veterinary use, which proves that they are adaptable for large-scale production [84]. There are still unresolved points in the saponin or ISCOM approach. The mechanism of adjuvanticity is not yet resolved and little is known of the structure activity and the biological properties of the saponins in relation to their adjuvant activity [79]. Further studies are necessary in order to assess the balance between the quantity and the quality of the immune response.

Recombinant vaccines

The recent improvements in biochemical characterization and genetic engineering techniques have been applied to the development of recombinant vaccines. There is a wide range of new strains of viral or bacterial vectors. Most of them derive from existing live attenuated vaccines. The insertion of foreign genes into these vaccines used as vectors has made it possible to induce an immune response against both intrinsic and foreign antigens. In relation to their replication competence, two different classes of virus vectors can be used. Among the replicative viruses, herpesviruses, adenoviruses and poxviruses are able to express heterologous genes without loss of their infectivity. The replication-defective virus vectors need complementation by specifically transformed cells (SV40, adeno-associated viruses) or by helper-virus super-infection (retroviruses) [85].

For measles vaccine strategy, many efforts are devoted to the development of transgenic replication-competent poxviruses vectors. Among the poxviruses, the vaccinia, canarypox and fowlpox viruses have been developed in immunological studies of the different structural measles proteins [66,68,86–88] and the research of immunogens for new measles vaccine candidates [89–91]. Other viruses, such as baculovirus and adenovirus have been used for the understanding of the role and the immunogenic pathway of different measles proteins [71,92,93].

The recombinant genome construction is based on the cleavage of the viral genome and its ligation with the foreign DNA. The hybrid DNA is then introduced into the cells by transfection and the viral replication is made possible. In the case of poxviruses with a non-infectious viral DNA, it is necessary to use specific promoters for transcription [94].

Vaccinia virus is a live attenuated strain of the smallpox virus. It has been used sucessfully to eradicate smallpox. The vaccinia virus has a large, double-stranded DNA genome and replicates into infected cells. Because of the cytoplasmic replication of the vaccinia in the host cells, it is easier to introduce foreign genes into the genome and to assess the expression of the subsequent recombinant DNA. The vaccinia is not restricted to humans and can infect a wide range of animal cells. The viral DNA contains regions of non-essential genes where large or multiple foreign genes can be inserted. Despite its worldwide use, some safety issues have arisen. These difficulties have been overcome by strain attenuation improvement and the development of a genetically engineered highly attenuated strain, e.g. the NYVAC strain derived from the Copenhagen vaccinia virus vaccine strain [95]. The construction of recombinant vaccinia virus expressing foreign genes is now routinely performed in many laboratories. The first step is the construction of a chimeric gene containing the foreign protein coding sequence which is linked to a vaccinia promoter and then assembled in a plasmid vector. The plasmid is flanked by DNA sequences which are homologous to a non-essential region of the vaccinia genome and represent the insertion site. The recombinant plasmid and the vaccinia virus are transfected and infected respectively into host cells. Within the infected cells, homologous recombination between vaccinia sequence of the plasmid and the virus results in site-specific insertion of the foreign gene into the viral genome. Finally, the recombinant vaccine can be isolated and purified before being used to express the inserted gene [96].

The construction of vaccinia virus (VV) recombinants expressing different measles-antigens have been tested as candidate vaccines in several experimental studies. The gene coding for either the H or F measles proteins has been inserted in a VV. The VVH and VVF recombinants protect mice against a lethal challenge [66] and also protect dogs against canine distemper virus challenge [91]. The role of the nucleocapsid protein in measles infection of the central nervous system has been demonstrated after immunization of rats with a VVN recombinant vector; the immunized rats developed a specific CD4$^+$ cell-mediated immune response to the N protein at a level sufficient to protect them from encephalitis after challenge with a neurotropic measles virus [68]. The comparison of a number of VV-measles recombinant viruses containing different combinations of gene coding for H, F and N proteins shows that it is possible to express more than one measles virus gene and suggests that the association of measles antigens in a single recombinant vaccine could overcome host-related restriction of the immune response to particular antigens [97]. Recombinant measles-vaccinia viruses have been demonstrated to express genes of interest at a level sufficient to induce strong humoral immunity and cell-mediated immune response which protect against subsequent challenge. These findings are supported by studies with other antigens, e.g. the glycoprotein G of rabies virus. A VV-rabies recombinant vaccine has been used in Europe with conclusive results for the use of such a vaccine for the eradication of rabies [98].

The development of measles-vaccinia viruses is still at an experimental level but has led to major improvements in the understanding of the immunology of measles.

Canarypox virus (CPv) and *fowlpox virus* (FPv) are members of the *Poxviridae* family, *Avipox virus* genera, specific to avian species [99]. They are being investigated as alternatives to vaccinia viruses. The construction of recombinant avipox viruses is similar to that of vaccinia recombinant vaccines; the chimeric foreign gene is constructed under the control of vaccinia virus promoters.

It has been demonstrated that the infection of non-avian species cells with recombinant avipox viruses results in abortive replication but there is nevertheless an authentic expression of the foreign genes [88,100–102]. Two different avipox recombinant rabies vaccines have been constructed with CPv or FPv and their immunogenic and protective properties have been assessed in laboratory animals. Inoculation with rabies FPv vaccine induces neutralizing antibody in rodents, cattle, horses and swine [100,101]. The recombinant rabies CPv and FPv vaccines have protected mice, cats and dogs against challenge [101,102]. Furthermore, the safety and immunogenicity of the CPv expressing rabies glycoprotein has been evaluated in humans. Twenty-five adult volunteers were immunized following a two-dose schedule four weeks apart, with vaccines at increasing titers in comparison with the human diploid cell vaccine [103]. The authors report that the neutralizing antibody response may be dependent on the vaccine-titer or the regimen of immunization. The side-effects of the CPv were similar to those of the rabies human diploid cell vaccine.

A recombinant canarypox virus expressing the measles virus fusion and haemagglutinin glycoproteins is currently under investigation [90]. This vaccine has been constructed following the standard procedure described below. Firstly, two CPv recombinant vaccines expressing F or H proteins were developed resulting respectively in F-CPv and H-CPv recombinant vaccine. The F-CPv recombinant vaccine was then used as a recipient for the H-CPv recombinant vaccine, leading to a single CPv expressing both proteins. The replication was investigated in a number of non-avian cell systems [90]; there is no evidence of either production of progeny virus or adaptation of the virus to the non-avian cells. Moreover, replication is blocked at an early stage of the viral DNA synthesis. The expression of F and H protein was confirmed by immunoprecipitation analysis. Interestingly, the F_1 and F_2 cleavage products of the F precursor were expressed. The fusion activity of the F + H-CPv recombinant was demonstrated with Vero cell fusion assays. When the Vero cells were inoculated with F-CPv or H-CPv individually, no cell fusion was observed but cell fusion activity was shown at a comparable level after inoculation with the F-H CPv or after co-infection with both F-CPv and H-CPV. Furthermore, the heterotypic immunization of dogs with the F + H-CPv vaccine results in protection against a lethal canine distemper virus challenge. The neutralizing antibody production was equivalent to that observed after the F-H vaccinia virus recombinant vaccine. These latter findings are of great interest because they could imply that the productive replication of the vector is not essential to the induction of protective immunity [90]. The abortive replication of the avipox vector in non-avian species would avert dissemination within the vaccinated individuals or transmission to nonvaccinated individuals and would circumvent the safety issues of the vaccinia virus vaccine.

The results of the experimental studies on the recombinant measles poxviruses vectors give us strong indications that the appropriate recombinant antigens are expressed at sufficient levels, conferring humoral and protective immunity. Canarypox measles vaccine may be a better approach to human vaccination than vaccinia virus, due to the demonstration of its immunogenic properties in spite of the absence of replication in non-avian species. Another potential advantage of the recombinant vaccine is that it will certainly be less expensive to produce than the sub-unit or inactivated vaccines.

In conclusion, a wide range of new strategies in measles vaccine development is currently under investigation, addressing the inability of live attenuated vaccines to overcome the barrier of maternal antibodies in young infants. Efforts are devoted to the design of a vaccine which would induce a protective cellbound and humoral immunity in the face of prior vaccination or maternal antibodies, without any deleterious effects. In the past few years, rapid progress has been made in the structural and molecular understanding of measles virus interactions with the immune system and this is closely interlinked with vaccine development.

Acknowledgements

I would like to thank ML. de Wazières and J. Symonds for their expert technical support in the preparation of the manuscript.

References

1. (1992) Measles control in the 1990s: plan of action for global measles control. WHO/EPI/ GEN/92.3
2. Miller C (1987) Live measles vaccine: a 21-year follow-up. BMJ 295: 22–24
3. Markowitz LE, Preblud S, Fine PEM, Orenstein WA (1990) Duration of live measles vaccine-induced immunity. Pediatr Infect Dis J 9: 101–110
4. Black FL (1989) Measles active and passive immunity in a worldwide perspective. Prog Med Virol 36: 1–33
5. Krober MS, Stracener CE, Bass JW (1991) Decreased measles antibody response after measles-mumps-rubella vaccine in infants with colds. JAMA 265: 2095–6
6. Edmonson MB, Davis JP (1992) Measles vaccination (MV) during the respiratory virus season as a risk factor for vaccine failure, 32nd Interscience conference on Antimicrobial Agents and Chemotherapy (ICAAC) Anaheim, California, USA
7. Halsey NA, Boulos R, Mode F, André J, Bowman L, Yaeger RG, Toureau S, Rohde J, Boulos C (1985) Response to measles vaccine in Haitian infants 6 to 12 months old. New Engl J Med 313: 544–549
8. Ndikuyeze A, Munoz A, Stewart J, Modlin J, Heymann D, Herrmann KL, Polk BF (1988) Immunogenicity and safety of measles vaccine in ill African children. Int J Epidemiol 17: 448–455
9. Harris MF (1977) The safety of measles vaccine in severe illness. S Afr Med J: 38
10. Enders JF, Peebles TC (1954) Propagation in tissue cultures of cytopathogenic agents from patients with measles. Proc Soc Exp Biol Med 86: 277–286
11. Sabin AB, Arechiga AF, Fernandez de Castro J, Albrecht P, Sever JL, Shekarchi I (1984) Successful immunization of infants with and without maternal antibody by aerosolized measles vaccine. II. Vaccine comparisons and evidence for multiple antibody response. JAMA 251: 2363–2371
12. Mirchamsy H, Bahrami S, Shafyi A, Kamali M, Razavi J, Nazari, Razavi J, Ahouri P, Fatemi S, M Amin-Salehi M (1977) A comparative field trial of five measles vaccines produced in human diploîd cell, MRC5. J Biol Stand 5: 1–18
13. Sabin A, Albrecht P, Takeda AK, Ribeiro EM, Veronesi R (1985) High effectiveness of aerosolized chick embryo fibroblast measles vaccine in seven-month-old and older infants. J Infect Dis 152: 1231–1237
14. Khanum S, Garelick H, Uddin N, Mann G, Tomkins A (1987) Comparison of Edmonston-Zagreb and Schwarz strains of measles vaccine given by aerosol or sub-cutaneous injection. Lancet: 150–153
15. Zourbas J (1971) Essais cliniques et immunologiques comparatifs de 4 lots de vaccin antirougeoleux Schwarz de titres differents. La Femme et l'Enfant 4: 5–17
16. Makino S (1983) Development and characteristics of live AIK-C measles virus vaccine: a brief report. Rev Infect Dis 5: 504–505
17. Whittle H, Eccles M, Jupp L, Hanlon L, Mann G, O'Neill K, Hanlon P, Marsh V (1988) Effects of dose and strain of vaccine on success of measles vaccination of infants aged 4–5 months. Lancet: 963–966
18. Markowitz LE, Sepulveda J, Diaz-Ortega JL, Valdespino JL, Albrecht P, Zell ER, Stewart J, Zarate ML, Bernier RH (1990) Immunization of six-month-old infants with different doses of Edmonston-Zagreb and Schwarz measles vaccines. New Engl J Med 322: 580–587
19. Tidjani O, Guérin N, Lecam N, Grunitsky B, Levy-Bruhl D, Xueref C, Tatagan K (1989) Serological effects of Edmonston-Zagreb, Schwarz and AIK-C measles vaccine strains given at ages 4–5 or 8–10 months. Lancet: 1357–60
20. Kiepiela P, Coovadia HM, Loening WE, Coward P, Botha G, Hugo J, Becker PJ (1991) Lack of efficacy of the standard potency Edmonston-Zagreb live, attenuated measles vaccine in African infants. Bull WHO 69: 221–7
21. Wassilak S, Orenstein W, Strickland P, Butler C, Bart K (1985) Continuing measles transmission in students despite a school-based outbreak control program. Am J Epidemiol 122: 208–217

22. Adu FD, Akinwolere OAO, Tomori O, Uche LN (1992) Low seroconversion rates to measles vaccine among children in Nigeria. Bull WHO 70: 457–460
23. Albrecht P, Ennis FA, Saltzman EJ, Krugmann S (1977) Persistence of maternal antibody in infants beyond 12 months: mechanism of measles vaccine failure. J Pediatr 91: 715–718
24. Sabin A (1983) Immunization against measles by aerosol. Rev Infect Dis 5: 514–523
25. Aaby P, Hansen JL, Tharup J, Sodemann M, Knudsen K, Jensen TG, Kristiansen H, Poulsen A, Jakobsen M, Da Silva MC, Whittle H (1988) Trial of high-dose Edmonston-Zagreb measles vaccine in Guinea-Bissau: protective efficacy. Lancet: 809–814
26. Job JS, Halsey NA, Boulos R, Holt E, Farrell D, Albrecht P, Brutus JR, Adrien M, Andre J, Chan E, Kissinger P, Boulos C, Cité Soleil/JHU Project Team (1991) Successful immunization of infants at 6 months of age with high dose Edmonston-Zagreb measles vaccine. Pediatr Infect Dis J 10: 303–11
27. Whittle H, O'Neill K, Marsh V, Hanlon P, Hanlon L, Jupp E, Aaby P (1988) Trial of high-dose Edmonston-Zagreb measles vaccine in the Gambia: antibody response and side-effects. Lancet: 811–814
28. Gareme M, Leroy O, Beau JP, Sene I (1992) High-titer measles vaccines: protection evaluation. Forthcoming
29. Garenne M, Leroy O, Beau JP, Sene I Efficacy of measles vaccines when controlling for exposure. Forthcoming
30. Garenne M, Leroy O, Beau JP, Sene I, Whittle H, Sow AR (1991) Efficacy, safety and immunogenicity of two high-titer measles vaccines: a study in Niakhar, Senegal. Final report. ORSTOM, Dakar
31. Garenne M, Leroy O, Samb B, Beau JP, Whittle H, Sene I, Sow AR (1991) Survival after measles vaccination with high titer vaccines given at five months of age. In WHO consultation on mortality rates following administration of measles vaccine before nine months, Geneva. Working paper #2
32. Garenne M, Fontaine O (1986) Assessing probable causes of deaths using a standardized questionnaire. In IUSSP (ed.) A study in rural Senegal. Seminar on morbidity and mortality, Sienna: 123–142
33. Garenne M, Leroy O, Beau J, Sene I (1991) Child mortality after high-titre measles vaccines in Senegal — the complete data set — reply. Lancet 338: 1518–1519
34. Garenne M, Leroy O, Beau JP, Sene I (1991) Child mortality after high-titre measles vaccines — prospective study in Senegal. Lancet 338: 903–907
35. Weiss R (1992) Measles battle loses potent weapon. Science 258: 546–547
36. (1992) EPI. Safety of high titer measles vaccines. Week Epi Rec 67: 357–361
37. Minor P (1992) La voie périlleuse vers un vaccin rougeoleux amélioré. CVI Forum 2: 8–10
38. Mestecky J, Eldridge JH (1991) Targeting and controlled release of antigens for the effective induction of secretory antibody responses. Curr Opin Immunol 3: 492–495
39. Barrett T (1987) The molecular biology of the morbillivirus (measles) group. Biochem Soc Symp 53: 25–37
40. Norrby E (1975) Occurence of antibodies against envelope components after immunization with formalin inactivated and live measles vaccine. J Biol Stand 3: 375–380
41. Norrby E, Enders-Ruckle G (1975) Differences in the appearance of antibodies to structural components of measles virus after immunization with inactivated and live virus. J Infect Dis 132: 262–269
42. Norrby E, Lagercrantz R (1976) Measles vaccination. VII. The occurence of antibodies against virus envelope components after immunization with inactivated vaccine. Effects of revaccination with live measles vaccine. Acta Paed Scand 65: 171–176
43. Norrby E (1985) Measles vaccination, today and tomorrow. Ann Inst Pasteur/Virol 136 E: 561–570
44. Sato TA, Kohama T, Sugiura A (1989) Protective role of human antibody to the fusion protein of measles virus. Microbiol Immunol 33: 601–607

45. Mäkelä MJ, Marusyk RG, Norrby E, Tyrrell DLJ, Salmi AA (1989) Antibodies to measles virus surface polypeptides in an immunized student population before and after booster vaccination. Vaccine 7: 541–545

46. Bellanti JA, Sanga RL, Klutinis B, Brandt B, Artenstein MS (1969) Antibody responses in serum and nasal secretions of children immunized with inactivated and attenuated measles virus vaccines. N Eng J Med 280: 628–633

47. Heffner RR, Schluederberg AE (1967) Specificity of the primary and secondary antibody responses to myxoviruses. J Immunol 98: 668–672

48. Pedersen IR, Mordhorst CH, Ewald T, Von Magnus H (1986) Long-term antibody response after measles vaccination in an isolated Arctic society in Greenland. Vaccine 4: 173–178

49. Graves M, Griffin DE, Johnson RT, Hirsch RL, Lindo de Soriano I, Roedenbeck S, Vaisberg A (1984) Development of antibody to measles virus polypeptides during complicated and uncomplicated measles virus infections. J Virol 49: 409–412

50. Bech V (1959) Studies on the development of complement fixing antibodies in measles patients. Observations during a measles epidemic in Greenland. J Immunol 83: 267–275

51. Brunell PA (1990) EPI, Measles control in the 1990s: Measles serology. WHO/EPI/GEN/90.4

52. Liebert U, ter Meulen V (1987) Virus aspects of measles virus-induced encephalomyeletis in Lewis and Bn rats. J Gen Virol 68: 1715–1722

53. Brinckmann UG, Bankamp B, Reich A, ter Meulen V, Liebert UG (1991) Efficacy of individual measles virus structural proteins in the protection of rats from measles encephalitis. J Gen Virol 72: 2491–2500

54. Varsanyi TM, Utter G, Norrby E (1984) Purification, morphology and antigenic characterization of measles virus envelope components. J Gen Virol 65: 355

55. Nadel S, McGann K, Hodinka RL, Rutstein R, Chatten J (1991) Measles giant cell pneumonia in a child with human immunodeficiency virus infection. Pediatr Infect Dis J 10: 542–4

56. Shelhamer JH, Toews GB, Masur H, Suffredini AF, Pizzo PA, Walsh TJ, Henderson DK (1992) Respiratory Disease in the Immunosuppressed Patient. Ann Int Med 117: 415–431

57. Burnet FM (1968) Measles as an index of immunological function. Lancet 2: 610–613

58. van Binnendijk RS, Poelen MCM, de Vries P, Uytedehaag FGCM, Osterhaus ADME (1990) A role for CD8+ class I MHC-restricted CTLs in recovery from measles: implications for the development of inactivated measles vaccines. In Modern approaches to new vaccines including the prevention of AIDS. Cold Spring Harbor Laboratory Press: 299–303

59. Von Pirquet C (1908) Das verhalten der kutanen tuberkulin reaktion wärhend der masern. Deutsche Med Wochenschr 34: 1297–1300

60. Joffe MI, Rabson AR (1981) Defective helper factor (LMF) production in patients with acute measles infection. Clin Immunol Immunopathol 20: 215–223

61. Ward BJ, Johnson RT, Vaisberg A, Jauregui E, Griffin DE (1990) Spontaneous Proliferation of Peripheral Mononuclear Cells in Natural Measles Virus Infection: Identification of Dividing Cells and Correlation with Mitogen Responsiveness. Clin Immunol Immunopathol 55: 315–326

62. Hirsch RL, Griffin DE, Johnson RT, Cooper SG, Lindo de Soriano I, Roedenbeck S, Vaisberg A (1984) Cellular immune response during uncomplicated measles virus infections of man. Clin Immunol Immunophathol 31: 1–12

63. Griffin DE, Ward BJ, Jauregui E, Johnson RT, Vaisberg A (1989) Immune activation in measles. New Engl J Med, 320: 1667–1672

64. Griffin DE (1991) Immunologic abnormalities accompanying acute and chronic viral infections. Rev Infect Dis 13 (Suppl 1): 129–133

65. Norrby E, Utter G, Orvell C, Appel M (1986) Protection against canine distemper virus in dogs after immunization with isolated fusion protein. J Virol 58: 536

66. Drillien R, Spehner D, Kirn A, Giraudon P, Buckland R, Wild F, Lecocq JP (1988) Protection of mice from fatal measles encephalitis by vaccination with vaccinia virus

recombinants encoding either the hemagglutinin or the fusion protein. Proc Natl Acad Sci USA 85: 1252–1256

67. Malvoisin E, Wild F (1990) Contribution of measles virus fusion protein in protective immunity: anti-F monoclonal antibodies neutralize virus infectivity and protect mice against challenge, J Virol 64: 5160–5162

68. Bankamp B, Brinckmann UG, Reich A, Niewiesk S, ter Meulen V, Liebert UG (1991) Measles virus nucleocapsid protein protects rats from encephalitis. J Virol 65: 1695–1700

69. Steward MW, Stanley SM, Dimarchi R, Mulcahy G, Doel TR (1991) High-affinity antibody induced by immunization with a synthetic peptide is associated with protection of cattle against foot-and-mouth disease. Immunology 72: 99–103

70. Milich DR (1989) Synthetic T and B cell recognition sites: implications for vaccine development. Adv Immunol 45: 195–264

71. Alkhatib G, Briedis DJ (1986) The predicted primary structure of the measles virus hemagglutinin, Virology 150: 479–484

72. Richardson C, Hull D, Greer P, Hasel P, Berkovich A, England D, Bellini W, Rima B, Lazzarini R (1986) The nucleotide sequence of the mRNA encoding the fusion protein of the measles virus (Edmonston strain): a comparison of fusion proteins from several paramyxoviruses. Virology 155: 508–523

73. Partidos CD, Steward MW (1990) Prediction and identification of a T cell epitope in the fusion protein of measles virus immunodominant in mice and humans. J Gen Virol 71: 2099–2105

74. Partidos CD, Stanley CM, Steward MW (1991) Immune responses in mice following immunization with chimeric synthetic peptides representing B and T cell epitopes of measles virus proteins. J Gen Virol 72: 1293–1299

75. Partidos CD, Steward MW (1992) The effects of a flanking sequence on the immune response to a B-cell and a T-cell epitope from the fusion protein of measles virus. J Gen Virol 73: 1987–1994

76. Arnon R, Horwitz RJ (1992) Synthetic peptides as vaccines. Curr Opin Immunol 4: 449–453

77. Mougin B, Bakouche O, Gerlier D (1988) Humoral immune response elicited in rats by measles viral membrane antigens presented in liposomes and ISCOMs. Vaccine 6: 445–449

78. Garnier F, Forquet F, Bertolino P, Gerlier D (1991) Enhancement of in vivo and in vitro T cell response against measles virus haemagglutinin after its incorporation into liposomes: effect of the phospholipid composition. Vaccine 9: 340–5

79. Bomford R, Stapleton M, Winsor S, Beesley JE, Jessup EA, Price KR, Fenwick GR (1992) Adjuvanticity and ISCOM formation by structurally diverse saponins. Vaccine 10: 572–577

80. Morein B, Sundquist B, Höglund S, Dalsgaard K, Osterhaus A (1984) Iscom, a novel structure for antigenic presentation of membrane proteins from enveloped viruses. Nature 308: 457–460

81. Speijers GJA, Danse LHJC, Beuvery EC, Strik JJTWA, Vos JG (1988) Local reactions of the saponin quil A and a quil A containing Iscom measles vaccine after intramuscular injection of rats: a comparison with the effect of DPT-Polio vaccine. Fundamental and applied Toxicology 10: 425–430

82. De Vries P, Van Binnendijk RS, Van der Marel P, Van Wezel AL, Voorma HO, Sundquist B, Uytdehaag FGCM, Osterhaus ADME (1988) Measles virus fusion protein presented in an immune-stimulating complex (Iscom) induces haemolysis-inhibiting and fusion-inhibiting antibodies, virus-specific T cells and protection in mice. J Gen Virol 69: 549

83. Pedersen IR, Bog-Hansen TC, Dalsgaard K, Heegaard PMH (1992) Iscom immunization with synthetic peptides representing measles virus hemagglutinin. Virus Research 24: 145–159

84. Osterhaus ADME, De Vries P (1992) Vaccination against acute respiratory virus infections and measles in man. Immunobiology 184: 180–192

85. Esposito JJ, Murphy FA (1989) Infectious recombinant vectored virus vaccines. In Advances in veterinary science and comparative medicine. Academic Press Inc: 195–247

86. Wild TF, Malvoisin E, Buckland R (1991) Measles virus: both the haemagglutinin and fusion glycoproteins are requied for fusion, J Gen Virol 72: 439–442

87. Spehner D, Kirn A, Drillien R (1991) Assembly of nucleocapsidlike structures in animal cells infected with a vaccinia virus recombinant encoding the measles virus nucleoprotein, J Virol 65: 6296–300

88. Wild TF, Giraudon P, Spehner D, Drillien R, Lecocq JP (1990) Fowlpox virus recombinant encoding the measles virus fusion protein: protection of mice against fatal measles encephalitis. Vaccine 8: 441–2

89. Tartaglia J, Pincus S, Paoletti E (1990) Poxvirus-based vectors as vaccine candidates. Crit Rev Immunol 10: 13–30

90. Taylor J, Weinberg R, Tartaglia J, Richardson C, Alkhatib G, Briedis D, Appel M, Norton E, Paoletti E (1992) Nonreplicating viral vectors as potential vaccines – recombinant canarypox virus expressing measles virus fusion (F) and hemagglutinin (HA) glycoproteins. Virology 187: 321–328

91. Taylor J, Pincus S, Tartaglia, Richardson C, Alkhatib G, Briedis D, Appel M, Norton E, Paoletti E (1991) Vaccinia virus recombinants expressing either the measles virus fusion or hemagglutinin glycoprotein protect dogs against canine distemper virus challenge. J Virol 65: 4263–4274

92. Vialard J, Lalumière M, Vernet T, Briedis D, Alkhatib G, Henning D, Levin D, Richardson C (1990) Synthesis of the membrane fusion and hemagglutinin proteins of measles virus, using a novel baculoviurs vector containing the beta-galactosidase gene. J Virol 64: 37–50

93. Takehara K, Hashimoto H, Ri T, Mori T, Yoshimura M (1992) Characterization of baculovirus-expressed hemagglutinin and fusion glycoproteins of the attenuated measles virus strain AIK-C, Virus Research 26: 167–175

94. Pastoret PP, Brochier B (1990) Le virus de la vaccine et ses proches parents. Ann Méd Vét 134: 207–220

95. Tartaglia J, Perkus ME, Taylor J, Norton EK, Audonnet J-C, Cox WI, Davis SW, van der Hoeven J, Meignier B, Rivière M, Languet B, Paoletti E (1992) NYVAC: a highly attenuated strain of vaccinia virus. Virology 188: 217–232

96. Smith GL, Moss B (1984) Vaccinia virus expression vectors: construction, properties and applications. Biotechniques: 306–312

97. Wild TF, Bernard A, Spehner D, Drillien R (1992) Construction of vaccinia virus recombinants expressing several measles virus proteins and analysis of their efficacy in vaccination of mice. J Gen Virol 73: 359–367

98. Brochier B, Kieny MP, Costy F, Coppens P, Bauduin B, Lecocq JP, Languet B, Chappuis G, Desmettre P, Afiademanyo K, Libois R, Pastoret P-P (1991) Large-scale eradication of rabies using recombinant vaccinia-rabies vaccine. Nature 354: 520–522

99. Moss B (1990) Poxviridae and their replication. In BN Fields, DM Knipe et al (eds.), Virology, Raven Press Ltd, New York: 2097–2111

100. Taylor J, Paoletti E (1988) Fowlpox as a vector in non-species. Vaccine 6: 466–468

101. Taylor J, Weinberg R Languet B, Desmettre P, Paoletti E (1988) Recombinant fowlpox virus inducing protective immunity in non-aviar species. Vaccine 6: 497–503

102. Taylor J, Trimarchi C, Weinberg R, Languet B, Rivière M, Desmettre P, Paoletti E (1991) Efficacy studies on a canarypox-rabies recombinant virus. Vaccine 9: 190–193

103. Cadoz M, Strady A, Meignier B, Taylor J, Tartaglia J, Paoletti E, Plotkin S (1992) Immunisation with canarypox virus expressing rabies glycoprotein. Lancet 339: 1429–1449

11

Molecular biological basis of measles virus strain differences

B. K. Rima

Division of Molecular Biology, School of Biology and Biochemistry, The Queen's University of Belfast, Belfast, Northern Ireland

Summary

Biological differences between measles strains have been described in relation to plaque type, fusiogenicity, haemagglutination, neurovirulence in animal models, temperature-sensitivity, cold adaptation, ability to induce interferon, presence of defective interfering particles as well as, most importantly, attenuation in human infection. The molecular (nucleotide sequence) basis of few, if any, of these changes has been elucidated. Recently described differences in monoclonal antibody (Mcab) binding properties of the nucleocapsid proteins of different strains have been linked to sequence variation in B cell epitopes. Similarly, B cell epitopes in the haemagglutinin have now been defined by sequencing Mcab escape mutants. These appear to be localized in a region of moderate strain variability. We have also identified a mutation linked to the salt dependency for haemagglutination of some strains. A large number of mutations have been described for viruses isolated from SSPE and other persistent infections. Mutations in the M gene (biased hypermutation and premature stop codons) and in the F gene (truncation of cytoplasmic tails) have been reviewed and their significance in SSPE is discussed. The recent discovery of a number of different lineages of MV has indicated co-circulation and global distribution of some strains. All vaccines belong to one lineage. The other virus strains are classified into various lineages. Differences in glycosylation sites have been observed and these may be of biological importance. However, it is as yet unknown whether the different lineages have biological significance beyond escape from immunological pressure.

Introduction

Measles virus (MV) is an important human pathogen which is responsible for a large number of deaths primarily in the developing world. Only one serotype of the virus has been described and no differences with respect to serological reactivity have been observed. As a matter of fact MV cross-reacts serologically with all the members of the genus Morbillivirus including rinderpest virus, peste-des-petits-ruminants virus, canine and phocine distemper virus. The monotypic nature of the virus, the lack of a natural animal reservoir, and the life-long immunity which occurs after infection with the virus, make it theoretically possible to eradicate measles. However, many practical difficulties remain to be overcome before this goal can be achieved.

In order to aid surveillance of the virus and to monitor the response of the virus to wide-spread control programmes, it is important to define the molecular characteristics of co-circulating wild-type strains of the virus and of currently used vaccines. Since its propagation in tissue culture cells in 1954, it has been understood to be an example of a monotypic virus. However, variation between MV strains exists and this chapter deals with its molecular basis.

First it may be important to review briefly the structure and strategy of genome expression of the virus. The virion consists of six structural proteins

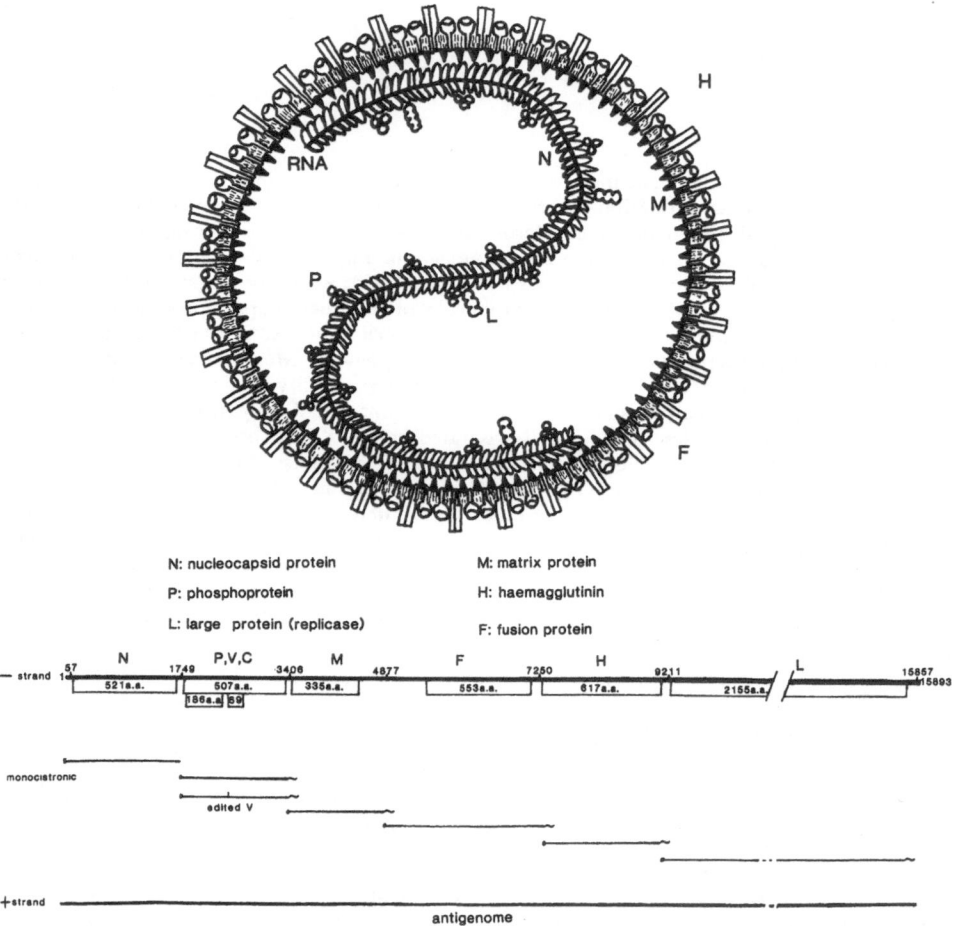

Fig. 1. Measles virus, virion structure, transcription and replication

(See Fig. 1) (reviewed in 22). Three proteins are associated with the internal nucleocapsid structure and these are referred to as the nucleocapsid or N protein (Mr 60K), the phospo- or P protein (Mr 54K but migrating in SDS-PAGE at a position corresponding to a 70K protein) and the large (L) protein (Mr 220K) which forms probably the main component of the RNA polymerase/replicase/transcriptase complex. Three proteins are associated with the envelope of the virus: the matrix or membrane (M) protein with a Mr of 37K, the fusion or F protein which is a heterodimer of two proteins (F_1 and F_2) derived from a F_0 precursor protein by proteolytic cleavage, and the haemagglutinin (H) protein which is supposed to be the main attachment protein that interacts with the cellular receptor. Two non-structural proteins have also been described. The "C" protein is derived from an overlapping reading frame (ORF) in the P mRNA [5] and the V protein is generated in common with many other paramyxoviruses from an edited mRNA derived from the P gene by insertion of one G residue. In this manner, an ORF is accessed which provides a 69 aa C terminal, cysteine-rich extension to the N terminal 230 residues of the P protein [11]. The gene order on the negative stranded non-segmented genome of the virus (Fig. 1) has been established [23] and the nucleotide sequence of the Edmonston reference strain consisting of 15 893 nucleotides has been determined by collaboration of several groups (reviewed in 3, 13).

Biological differences between MV strains have been described in relation to plaque type [20] and fusiogenicity [9]. These characters could be associated with changes in the fusion or haemagglutinin protein of MV. It is now recognized that both glycoproteins are required for cell to cell or virion to cell fusion, since vectors expressing either one of these proteins do not lead to cell fusion [29]. No molecular changes have been documented and correlated with changes in these biological characteristics. The region of the F molecule, that is the so called fusion related external domain (which comprises the hydrophobic N terminus of the F1 portion of the protein) has not been implicated directly in fusiogenicity since mutants that were resistant to fusion inhibiting oligopeptide analogues of the F_1 terminal amino acid sequence contained mutations not in this region [15] but at distant sites in the primary sequence (although not necessarily distant in secondary or tertiary structure). Fusiogenicity and plaque morphology have also been linked to neurovirulence of the morbilliviruses in some animal model systems [9]. Other variations in MV relate to haemagglutination properties; some strains (salt-dependent variants) require high concentrations of salt in the medium before the haemagglutinate monkey red blood cells [27]. We have tentatively linked this property to one amino acid change in the protein but this requires formal proof by the demonstration of altered haemagglutination properties from H molecules expressed from cloned and mutagenised vectors (S. Flanagan and B. K. Rima, unpublished observations). A few temperature-sensitive and cold-adapted mutants of MV have been described. Some of the ts mutants are unstable and complementation analysis has not been able to link them to the various genes of the virus. None of the ts- or cold-adapted mutants have been characterised in molecular terms, although this would be of great importance since some of the strains have remarkable phenotypes when tested in experimental animal infection. At least one ts mutant was associated with an increased frequency of hydrocephalus in infected hamsters [6]. The ability to induce interferon also varies from strain to strain but this characteristic can be rapidly altered by passage of the virus in interferon inducing cells or non-inducing cells. Passage of a virus stock adapted to BSC-1 cells into Vero cells led to an immediate modification in the amount of interferon induced upon infection [18]. It is also possible that these host cell changes influenced the amounts and types of defective interfering (DI) particles present in the infection and that these are responsible for interferon induction. This property has been linked to the presence of DI particles and it is known that the host cell type can influence the replication of certain DI particles. The presence of DI particles in virus stocks used in vaccine production [8] may be of importance not only for consistent growth of the virus as well as for its ability to induce interferon. Most importantly, attenuation of infection in the human host has been demonstrated for egg grown virus preparations and these have been used successfully to generate live attenuated viral vaccines. Again the molecular basis of this is not known and it is important to investigate this further. Variants with different abilities to react with monoclonal antibodies have been described [26] and in some cases it has been possible to correlate these with sequence variations in different lineages of MV that have been shown to co-circulate in mankind (see below). Also nucleotide sequence analysis of virus strains or viral RNA molecules isolated from cases of subacute sclerosing panencephalitis (SSPE) or measles inclusion body encephalitis (MIBE) has demonstrated a pattern of mutations in these viruses that leads to reduced expression of one or more of the three envelope proteins of the virus. This chapter summarizes what is known about differences between MV strains in molecular terms. However, it must be emphasised again that none of the more interesting variant phenotypes of MV have been correlated with molecular differences.

Virus mutations in subacute sclerosing panencephalitis

A number of groups have been investigating the importance of mutations of MV in the pathogenesis of diseases such as SSPE and MIBE. The latter is prevalent in patients who are immunocompromised and therefore not able either to control and clear MV either in the primary infection, or in the subsequent persistent infection which has been suggested as a potential explanation for the life-long immunity after MV infection. A number of different mutations and quantitative changes in virus gene expression have been observed in viruses that were either cultured from SSPE brain tissue, or in nucleotide sequences directly cloned from RNA derived from the brains. The latter has been successful in cases where it was impossible to grow the defective virus associated with such infections. Many of the changes observed seem to have as common denominator a reduction in the expression of the three virus membrane associated proteins; M, F and H. This is achieved by different mechanisms and in the majority of the cases investigated the expression of the M protein appears most affected [12]. This correlates well with earlier findings that in many but not all SSPE sera antibodies to the M protein are underrepresented [14]. One of the first mechanisms discovered for the reduction of expression of the M protein in some SSPE cases was that the normal pattern of transcription of the virus was disturbed in some SSPE brains and in persistent infections derived from them. A reduction in the amounts of monocistronic mRNA encoding the P and M protein and a corresponding increase in the dicistronic RNA read-through transcripts which contained the P and M ORFs in tandem was observed. It has been shown that the M gene cannot be translated from such mRNAs and this mechanism would thus lead to a selective inhibition of the expression of the protein, without necessarily affecting the expression of the promoter distal F, H and L genes (See Fig. 1). The mechanisms by which the transcription mechinery selectively ignores the normal intergenic sequence between the P and M genes is unclear in one SSPE case as no apparent mutations are observed [2]. In the case of one persistent infection it could be related to mutation of the intergenic 3'GAA 5' sequence to 3'GGA 5' [30]. It is noteworthy that the corresponding positive strand U to C change in this case is probably caused by the "unwindase" activity described below and that the ORF of the M gene is severly affected by this process in this strain. Another mechanism for the suppression of envelope antigen expression observed in SSPE brain RNAs and also in slow infections of MV in rat brain is the apparent increase in the gradient of gene expression associated with the normal transcription of MV [10, 24]. Figure 1 shows that the transcription complex starts off at the 3' end of the negative strand genome and that a gradient of expression of each of the genes is created by the fact that each intergenic junction there is a finite chance that the complex leaves the template and aborts transcription of the distal genes. This leads to a situation in which the number of copies of the mRNA for the first (N) gene is much larger than that for the second and third gene and so forth. In SSPE brain it has been observed that the chance of the transcription complex leaving the template is increased so that a steeper gradient occurs in which the numbers of copies of the distal M, F, H and L genes are much smaller per copy of the N mRNA than under normal lytic virus growth conditions [10]. In such a situation, defective growth of the virus could be caused by the relative lack of the envelope components which enable virus budding and replication. The copy number of total viral genomes in such infection could also be further reduced by the lack of replicate enzyme in the cell as a result of the expected low abundance of the L protein mRNAs resulting from the steeper expression gradient. In other cases of SSPE, premature stop codons have been observed in the M gene leading to the expression of very short truncated molecules and in other cases strains expressing mutant M protein which greatly decreased stability have been isolated [1]. The specific nucleotide changes in the M ORF responsible for this phenotype have not yet been identified. Independent of the mechanism, low steady state levels of M protein in infected cells would prevent effective budding of the virus.

Nucleotide sequence analysis of a number of SSPE viruses but specifically of the one single nucleotide sequence of a MIBE virus RNA indicated that M genes are also susceptible to an apparently strange form of "biased" hypermutation [4,10]. In many positions, U residues in wild type sequences are replaced with C residues. For example in the MIBE M gene positive strand sequence out of 235 potential U residues 126 have been mutated to C residues. An explanation for this phenomenon has been put forward [4]. An enzyme activity has been discovered which is able to convert a high proportion of adenine residues to inosine in double stranded RNA. This can pair with C residues in subsequent cycles of replication. Due to the uneven chance of fixation of these nucleotide changes as a result of the fact that in most cases one of the strands will be a mRNA strand rather than a positive antigenomic RNA molecule, the process leads to a bias for A to G mutation in the negative stranded genome and thus to U to C changes in the mRNA strands. One single example of A to G hypermutation in the positive strand strands has been identified in the H gene of an SSPE virus [12] but in all other cases analysed high proportions of U residues have been found mutated into C residues in the positive sequences. This enzyme is called an "unwindase" and its prime function is supposed to be the unwinding of double stranded RNA in the cell so that this can be removed by RNase. The enzyme is present in a very wide varient of cell types but very prevalent in neurones [21]. The latter could provide an explanation for the high frequency with which biased hypermutation is found in MV RNA molecules extracted from SSPE or MIBE brains. The finding that in some brains both wild type and sequences which have undergone biased hypermutation are present (Baczko et al., in preparation) strengthens the hypothesis that this mutation event occurs late during virus persistence, after infection of specific neuronal cell types. The hypermutation events that have been observed seem to be mostly confined to the M gene region and it is possible that this region is preferentially involved in the formation of double stranded intermediates due to some specific difference in the transcription of this gene. Alternatively, the genome with defective M genes are still capable of anatomical spread within the infected brain as a transcription inhibitory function is associated with the M proteins of vesicular stomatitis virus and Sendai virus. Thus, if the M protein of MV had the same function, biased hypermutation in the M gene could inactivate a functional control on MV replication.

Mutations that inactivate the F and H protein respectively have also been observed in SSPE viruses, but these are rarer than those affecting the M protein. In one case the H gene is inactivated largely as a result of biased hypermutation and the resulting attachment protein is defective in haemadsorption activity [12].

In several virus strains an mutation in the F gene has been identified which is associated with the deletion of one A residue in the part of the gene that encodes the cytoplasmic tail of the F protein [12,16]. It is of interest to note that the sequence preceding this area in which the A deletion takes place resemble the nucleotide sequence at sites involved in editing the mRNA of the P gene to generate V like protein in morbilliviruses [11]. At least one of the strains (Ya) is apparently capable of lytic growth in the tissue culture [16]. Recent studies in M. Billeter's laboratory have indicated (personal communication) that with respect to this mutation both wild type and sequences with the A residue deleted can be found in a single brain. The alteration of the cytoplasmic tail of the F protein and the resulting deletion of the C terminal pentadekapeptide which is conserved throughout the morbilliviruses, is of interest and could provide an example where the deletion of specific nucleotides would generate diversity between virus genomes.

We thus see that the expression of viral envelope antigens is often suppressed in strains replication in SSPE brains. However, it must be remembered that it is not yet clear whether the viral mutations are causative of the disease or are an epiphenomenon. The remarkable difference in susceptibility of the two sexes to the disease (boys have a two times higher chance of getting SSPE) and the rural prevalence suggest that other factors may be involved in SSPE. The low frequency of MIBE in immunocompromised hosts also suggests that other factors than solely viral mutation are involved.

Lineages of measles virus

Measles has been considered an antigenically stable and essentially monotypic virus and the small antigenic differences detected with monoclonal antibodies have not yet been linked to biological differences [26]. However, immunological tests for cell-mediated immunity mechanisms have been applied in only very few laboratories and to few virus strains, so that it is impossible to estimate the extent of variation. The life-long protection induced by infection with the wild-type virus appears to confirm that clinically significant strain differences have not been observed. With the recent introduction of new and relatively simple RNA-PCR technology a new tool in measles virus surveillance has become available and this technique has been applied to a wide variety of isolates. Comparisons between paramyxoviruses and within the morbillivirus genus indicate that the C terminus of the N protein is the most variable region of the genome with exception of the non-coding region between the ORFs for the M and F protein. However, differences in the latter region do not correlate to a protein antigen that has been identified as yet. It is also noteworthy that the H gene is far less variable than the part of the N gene encoding the C terminus of the protein [27]. This part of the protein has been shown to be localised on the outside of the nucleocapsid structure and to contain a number of important B cell epitopes [7]. Furthermore, it has been recognized that in the early humoral immune response in acute infection, antibodies to internal antigens are prevalent.

Thus, it is possible that the C terminus of the N protein changes under immunological pressure and does not appear to be under much functional and structural constraint. We have used this fact to analyse nucleotide sequences encoding this domain in a number of MV strains including now a large number of wild type viruses [28]. The data on MV variability, which were published earlier were heavily biased towards inclusion of virus strains associated with neurological diseases [12]. The most recent data indicate that

Table 1

Strain	Approx. year of isolation	Country	Character
Edmonston	1954	U.S.A.	wild type
Schwarz			vaccine strain from EdmB
Moraten			vaccine strain from EdmB
P9			SDA derivative from EdmB
Leningrad		Russia	vaccine from Leningrad isol.
CAM	1959	Japan	vaccine from Tanabe strain
Hallé	late 1960's	U.S.A.	SSPE?
Hu2	1974	N. Ireland	Schwarz vaccine derived case
R118	1984	Gabon	wild type
Y22	1983	Cameroun	wild type
JM	late 1970's	U.S.A.	wild type
S(B)	mid 1980's	Austria	SSPE
BAB	1990	Germany	wild type imm. compr. patient
BIL	1991	Netherlands	wild type
Woodfolk	early 1970's	U.S.A.	wild type
CM	late 1970's	U.S.A.	wild type
S(C)	1981	U.S.A	MIBE
Braxator	1972	Germany	measles encephalitis (MIBE?)
IP3A	early 1970's	U.S.A.	SSPE
MF	early 1970's	Europe	SSPE
Biken	1975	Japan	SSPE
S(A)	mid 1980's	Germany	SSPE
S(K)	mid 1980's	Germany	SSPE
Yamagata	late 1980's	Japan	SSPE
MVO	1974	U.K.	wild type
MVP	1974	U.K.	wild type
S33	1983	N. Ireland	SSPE
S81	1986	N. Ireland	SSPE
CL & SE	1985	U.K.	wild type
TT	1985	U.K.	Kawasaki disease case London
CH1	1989	U.S.A.	wild type
Snd	1989	U.S.A.	wild type
Ch2	1989	U.S.A.	wild type
Still to be grouped			
Niagata	early 1970's	Japan	SSPE
ZH	late 1970's	Iran	SSPE

amongst MV strains a number of different co-circulating lineages can be distinguished. These lineages are characterised by the presence of a large number of shared mutations away from a consensus or root sequence. Table 1 shows our current understanding of the relationships between MV strains. This is drawn up on the basis of rooted tree diagrams for sequence comparisons available for the N, P, M, H and F genes of several virus strains. First, it must be noted that trees drawn on the basis of any of the genes are the same to trees drawn on the basis of the N gene, so that it is unlikely from the present data that recombination between virus strains has occurred. Secondly, the classification of strains in these various lineages shows that some of these are circulating in wide geographical areas over a long time span and thirdly, the strain differences within a lineages are very small and witness the stability of the virus, particularly when one takes into account that the gene area chosen for the analysis is the most variable part of the genome.

We can distinguish six different lineages of measles virus. The first and oldest lineage of which we have no recent wild type isolates is represented by the Edmonston vaccine strain sequence. The total sequence of 15 893 nucleotides of this strain has been determined and it must be stated that this is the only complete sequence that we have for MV. For all other strains only partial sequences (several whole genes or parts of genes) are available. This lineage includes all the other vaccine viruses that have been analysed. A strange inclusion within this group is the Hallé virus originally isolated from lymph glands of a case of SSPE. We are not certain whether this is indeed a true SSPE derived virus (it does not show any of the characteristics of a wild type lineages), whether it represents a vaccine associated case or whether it is a laboratory contaminant. It is interesting to note that all vaccines currently in use belong to this lineage and that even though in 1954 it represent a wild type virus, there appear to be no surviving wild-type viruses in the field that belong to this lineage. However, their relationship to some of the African strains (lineage 2) merits further analysis.

A third lineage is represented by four strains of MV. The first (JM) was isolated in the late seventies in the USA and this seems to be related to the virus isolated from a SSPE case from Austria, a German isolate from 1990 and a Dutch isolate from 1991. This lineage thus has been found on both sides of the Atlantic Ocean and appears to be currently circulating on the European continent. It has not yet been isolated in the United Kingdom.

A fourth lineage includes wild type strains (CM and Woodfolk) and a MIBE strain (case C) and a MV encephalitis case (Braxator). The latest isolate of this lineage has been made in 1981 and it is thus not certain whether it is in circulation.

A fifth lineage consists of a number of older SSPE isolates from various countries in the world. No wild type isolates have been found in this group. We do not believe that this is of special significance to the aetiology of SSPE, but reflects a lack of MV wild-type isolates.

A sixth lineage includes a number of U.K. wild type and SSPE isolates and has recently been shown to include viruses isolated in Birmingham in 1985 and London from a case Kawasaki disease [25]. It appears that a significant number of mutations present in this lineage are also present in the recent U.S. isolates from 1991. But the relationship between these latter strains and currently circulating lineages is not yet clear.

It is difficult to establish interrelationships between separate lineages. In the absence of original root sequences or much older isolates we have analysed diversification of strains from a consensus sequence which is very similar to a root sequence derived from computer analysis of the strains. Because the sequences now include a number of different lineages and since none of the lineages is overrepresented in the sequence data base the root and consensus sequences should be similar. It must be stressed however that the relationships between lineages cannot easily be investigated within the short time frame and with the number of strains available. The speed with which the "molecular clocks" tick in MV may be established by comparison of sequence data from year to year in the same lineage. However, extrapolation backwards may be difficult since it is unknown how the wide spread vaccination programmes would affect the rate of evolution of the virus.

Immunogenic regions of measles virus proteins

The relative importance of the two arms of the immune system in clearance of MV is not well established. Generally, the fact that children with dysgammglobulinemea are able to clear the infection suggests that humoral immunity is probably not as important as cell mediated immune systems. For the latter, it has been shown that class II restricted lymphocytes do play an important role but recent evidence indicates that class I restricted cytotoxic cells also play a role in control of MV infection.

B cell epitopes have been found and localised in the primary sequence of the N protein of MV [7]. At least one is present in a conserved region in the N terminal half (residues 122–150) of the protein [7], one consists of the C terminal 7 residues and these vary only in the vaccine lineage, whereas in all the other lineages it is conserved and a further one is localised between amino acids 457 and 476 in a hypervariable region of the protein. No B cell epitopes have been localised on the P, M, F, C, V and L proteins as yet. An antigenic region of the H protein has been studies using synthetic peptides [17] and this area (between residues 368 and 377), although not variable to a great degree between various lineages has also been identified by the study of monoclonal antibody resistant mutants. One T cell epitope has been described in the F protein of MV and this again appears to consist of an invariable region of the protein [19]. The analysis of hypervariable parts of the genome as potential epitopes due to immunological pressure has not been rewarding due to the stability of the MV genome. Thus, overall, the delineation of B and T cell epitopes and the antigenic properties of the virus is still in its infancy.

Conclusions

Few of the biological differences between MV strains have been linked to molecular changes, so that a large amount of work remains to be done before we will have a good understanding of the functional importance of various parts of the genome and their gene products in pathogenesis. Variations between different lineages of MV have been observed but their biological significance is unclear at present. Future comparisons of vaccine and wild type strains from the same lineage are required to understand virus attenuation in this system. The development of a cDNA rescue system and the generation of in vitro recombinant viruses will remain one of the most important research priorities for proper development of new generations of virus vaccines. Also we do need to increase our understanding of the interactions of MV with the immune system of the infected host and of the role that these may play in pathogenesis.

Acknowledgements

I would like to thank Knut Baczko and Volker ter Meulen for unpublished sequence data and Martin Curran, Paul Yeo, Lizzie Godfrey and Malcolm Taylor for generation of sequence data in our laboratory. The work reported here was supported by the Medical Research Council, The Wellcome Trust and The Multiple Sclerosis Society of Great Britain and Northern Ireland. I would like to thank Denise McKay for typing this manuscript.

References

1. Ayata M, Hirano A, Wong TC (1989) Structural defect linked to nonrandom mutations in the matrix gene of Biken strain subacute sclerosing panencephalitis virus defined by cDNA cloning and expression of chimeric genes. J Virol 63: 1162–1173
2. Baczko K, Carter MJ, Billeter M, ter Meulen V (1984) Measles virus gene expression in subacute sclerosing panencephalitis. Vir Res 1: 589–595
3. Barrett T, Subbarao SM, Belsham GJ, Mahy BWJ (1991) The Molecular Biology of the Morbilliviruses. In: Kingsbury DW (ed) The Paramyxoviruses. Plenum Press, New York pp 83–102
4. Bass BL, Weintraub H, Cattaneo R, Billeter MA (1989) Biased hypermutation of viral RNA genomes could be due to unwinding/modification of double stranded RNA. Cell 56: 33
5. Bellini WJ, Englund G, Rozenblatt S, Arnheiter H, Richardson CD (1985) Measles virus P gene codes for two proteins. J Virol 53: 908–919
6. Breschkin AM, Haspel MV, Rapp F (1976) Neurovirulence and induction of hydrocephalus with parental, mutant and revertant strains of measles virus. J Virol 18: 809–811
7. Buckland R, Giraudon P, Wild F (1989) Expression of measles virus nucleoprotein in *Escherichia coli*: use of deletion mutants to locate the antigenic sites. J Gen Virol 70: 435–441
8. Calain P, Roux L (1988) Generation of measles virus defective interfering particles and their presence in a preparation of attenuated live-virus vaccine. J Gen Virol 70: 435–441
9. Carrigan D (1986) Round cell variant of measles virus: neurovirulence and pathogenesis of acute encephalitis in newborn hamsters. Virology 148: 349–359

10. Cattaneo R, Rebmann G, Baczko K, ter Meulen V, Billeter MA (1987) Altered ratios of measles virus transcripts in diseased human brains. Virology 160: 523–526
11. Cattaneo R, Kaelin K, Baczko K, Billeter MA (1989) Measles virus editing provides an additional cysteine rich protein. Cell 56: 759–764
12. Cattaneo R, Schmid A, Spielhofer T, Kaelin K, Baczko K, ter Meulen V, Pardowitz J, Flanagan S, Rima BK, Udem SA, Billeter MA (1989) Mutated and hypermutated genes of persistent measles viruses which caused lethal human brain diseases. Virology 173: 415–425
13. Galinski MS (1991) Annotated nucleotide and protein sequences for selected Paramyxoviridae. In: Kingsbury DW (ed) The Paramyxoviruses. Plenum Press, New York and London, pp 537–568
14. Hall WW, Lamb RA, Choppin PW (1979) Measles and substrate sclerosing pan-encephalitis virus proteins: lack of antibodies to the M protein in patients with subacute sclerosing panencephalitis. Proc Natl Acad Sci USA 76: 2047–2051
15. Hull JD, Krah DL, Choppin PW (1987) Resistance of measles virus mutant to fusion inhibitory oligopeptides is not associated with mutations in the fusion peptide. Virology 159: 368–372
16. Komase K, Haga T, Yoshikawa Y, Sato TA, Yamanouchi K (1990) Molecular analysis of structural proteins of the Yamagata-1 strain of defective subacute sclerosing pan-encephalitis virus. IV. Nucleotide sequence of the fusion gene. Virus Genes 4: 173–182
17. Mäkelä MJ, Lund GA, Salmi AA (1989) Antigenicity of measles virus haemagglutinin studied by synthetic peptides. J Gen Virol 70: 603–614
18. McKimm J, Rapp F (1977) Variation in ability òf measles virus plaque progeny to induce infection. Proc Natl Acad Sci USA 74: 3056–3059
19. Partidos CD, Stanley CM, Steward MW (1991) Immune responses in mice following immunization with chimeric synthetic peptides representing B and T cell epitopes of measles virus proteins. J Gen Virol 72: 1293–1299
20. Rapp F (1964) Plaque differentiation and replication of virulent and attenuated strains of measles virus. J Bact 88: 1448–1458
21. Rataul SM, Hirano A, Wong TC (1992) Irreversible modification of measles virus RNA in vitro by nuclear-unwinding activity in human neuroblastoma cells. J Virol 66: 1769–1773
22. Rima BK (1983) The proteins of morbilliviruses. J Gen Virol 64: 1205–1219
23. Rima BK, Baczko K, Clarke DK, Curran MD, Martin SJ, Billeter MA, ter Meulen V (1986) Characterisation of clones for the sixth (L) gene and a transcriptional map for morbilliviruses. J Gen Virol 67: 1971–1978
24. Schneider-Schaulies S, Liebert UG, Baczko K, Cattaneo R, Billeter M, ter Meulen V (1989) Restriction of measles virus gene expression in acute and subacute encephalitis of Lewis rats. Virology 171: 525–534
25. Schulz TF, Hoad JG, Whitby D, Tizard EJ, Dillon MJ, Weiss RA (1992) A measles isolate from a child with Kawasaki disease: sequence comparison with contemporaneous isolates from "classical" cases. J Gen Virol 73: 1581–1586
26. Sheshberadaran H, Chen SN, Norrby E (1983) Monoclonal antibodies against five structural components of measles virus: I characterisation of antigenic determinants on nine strains of measles virus. Virology 128: 341–353
27. Shirodaria PV, Dermott SE, Gould EA (1976) Some characteristics of salt-dependent haemagglutinating measles virus. J Gen Virol 33: 107–115
28. Taylor MJ, Godfrey E, Baczko K, ter Meulen V, Wild TF, Rima BK (1991) Identification of several different lineages of measles virus. J Gen Virol 72: 443–447
29. Wild TF, Malvoisin E, Buckland R (1991) Measles virus: both the hemagglutinin and fusion glycoproteins are required for fusion. J Gen Virol 72: 439–442
30. Yoshikawa Y, Turuoka H, Matsumoto M, Haga T, Shioda T, Shibuta H, Sato TA, Yamanouchi K (1990) Molecular analysis of structural protein genes of the Yamagata-1 strain of defective subacute sclerosing panencephalitis: II nucleotide sequence of cDNA corresponding to the P plus M dicistronic mRNA. Virus Genes 4: 151–162

12

Measles virus antigenic variations and the role of individual antigens in immunization

T. F. Wild

Unité d' Immunologie et Stratégie Vaccinale, Institut Pasteur, Lyon, France

Summary

Measles remains one of the major childhood diseases. Advances in genetic manipulation make possible new approaches to vaccination. However, a number of problems need to be studied before such a vaccine could be considered for field trials. (1) Does measles virus exhibit antigenic variation and if so what is its significance in protection. (2) Which antigen(s) should be incorporated in future vaccines. In the present study we compared field isolates and vaccine virus strains using monoclonal antibodies. No antigenic variation was observed on either of the two glycoproteins, the haemagglutinin (HA) and the fusion (F) proteins. In contrast, differences were observed on the nucleoprotein (NP) and one of the epitopes distinguished wild-type and vaccine strains.

To investigate the role of the different measles virus proteins in immunization, the measles virus proteins were expressed using vaccinia virus (VV) as a vector. Immunization with the HA or F recombinants, but not the NP protected mice against a lethal challenge. Protection with the F recombinant depended on the haplotype of the animal, but a low response to this antigen could be increased by the co-expression with the NP protein. Immunization of mice with the VV-recombinants expressing the F, NP or M (matrix) proteins, but not the HA, partially protected animals from a lethal challenge with canine distemper virus. In the case of the NP and M the protection was not mediated by antibodies.

Introduction

Measles virus (MV) is a member of the morbillivirus family which also includes rinderpest virus (RPV) and canine distemper virus (CDV). On the basis of immunological cross-reactivity, it has been suggested that RPV is the archevirus of the group, from which CDV evolved first, followed by MV [23]. The host range of MV is limited to primates. Only the family Hominoidiae maintains the virus although many monkey species are sometimes naturally infected. The host range is dictated by the ability of the virus to become attached and to penetrate into the potential host cell. The haemagglutinin

Table 1. Amino acid homology between the different
morbillivirus proteins

Protein	% Homology		
	MV/CDV	MV/RPV	CDV/RPV
HA	36	60	36
F	66	79	66
NP	66		
M	76	78	77
P	44		

The sequence data were obtained from references
2, 3, 4, 5, 7, 8, 11, 15, 20, 26, 31, 32

(HA) envelope protein of MV is responsible for the attachment, whereas the fusion (F) protein aids the passage of the virus through the cell membrane.

In nature, the three major morbilliviruses have different hosts. Nucleotide sequence analyses of the two envelope proteins from this group of viruses predict that the HA has the least homology whereas the F protein is much higher (Table 1).

These differences may reflect the nature of the function of the two viral proteins. The HA needs to recognize a species specific cell receptor, whereas the interaction of the F protein with the cell membrane may not be 'species specific, but the stringent conditions for the physical interaction may restrict variation. The internal structural virus proteins have high homologies amongst the three morbilliviruses (Table 1). Several studies using monoclonal antibodies have compared the immunological relationship between these viruses. In general, these studies confirm the order of magnitude of cross-reactivity of the individual proteins of the viruses compared to the predicted homology from the sequence data [2–5,7,8,11,15,20,26,31,32].

The diversification within the morbilliviruses probably occurred when variants overcame the species barrier and the susceptible host population was sufficient to maintain the new variant within the population. This situation did not arise until the development of the Middle-Eastern river valley civilization about 4,300 years ago [6]. Measles as we know it today must only have existed from this point. On this time scale, it is important to know if once measles virus became "fixed" in man, further evolution occurred.

The well documented observations in the Faroe islands confirm that MV immunity from natural infections persists for at least 65 years [24] suggesting that, if measles virus is evolving, then it does not do so radically. This may not be true in the case of attenuated MV vaccines. After their introduction into the United States in 1967, the number of measles cases fell a thousandfold [6]. However, with the recrudescence of measles in the population, it was necessary to consider whether MV was evolving within a vaccinated population or whether the epidemics were due to either the waning immunity in vaccinees or an accumulation in the population of vaccine failures.

Most laboratory studies on MV have used the original Edmonston strain. Apart from viruses isolated from SSPE cases few wild-type strains had been studied. In an effort to examine the possibility of antigenic changes, it was necessary to obtain fresh isolates with a minimum history of laboratory passage. To do so we isolated MV strains from Africa and obtained other fresh isolates from Japan (K. Yamanouchi) and France (P. Lebon). All these isolates were examined using a panel of monoclonal antibodies [16,17].

Antigenic characterization of the HA protein

The anti-HA monoclonal antibodies had been characterized by competition binding studies to map to a single antigenic site consisting of four overlapping epitopes [18]. Monoclonal antibodies to three of these sites neutralized virus and to two, passively protected animals against a lethal challenge. All the antibodies recognized conformational epitopes.

Studies in several laboratories have shown antigenic differences in MV antigens when comparing laboratory and SSPE strains [29]. However, in our experience the results often depend on the stringency of the test applied. Immunoprecipitation in the presence of SDS often reveals differences in antigens which are indistinguishable by other techniques (Mougin & Wild, unpublished observations). In the present study, we considered an epitope present if it was positive by both immunofluorescence and immunoblot. Using these criteria, no evidence of antigenic variation was found with MV strains isolated on three continents between 1981 and 1987 (Table 2).

Antigenic characterization of the F protein

Anti-F monoclonal antibodies prepared against purified MV have been shown to react with a large number of epitopes on the F protein. Most of

Table 2. Antigenic analysis of MV field isolates with anti-HA monoclonal antibodies

HA epitope	Source of virus/year		
	France 1981	Africa 1983/84	Japan 1987
A	+	+	+
B	+	+	+
C	+	+	+
D	+	+	+

The French isolates were supplied by P. Lebon and the Japanese by K. Yamanouchi. The African isolates were from epidemics in the Cameroun, 1983 and the Gabon, 1984

Table 3. Antigenic analysis of MV field isolates with anti-F monoclonal antibodies

Monoclonal antibody	Source of virus/year		
	France 1981	Africa 1983/84	Japan 1987
16EE8	+	+	+
19CG6	+	+	+
19FF4	+	+	+
9DB10	+	+	+
16GD10	+	+	+
19FF10	+	+	+

The anti-F monoclonal antibodies were a gift from E. Norrby

these antibodies also reacted with CDV [27], so it was not surprising to find that when these antibodies were tested with the different MV isolates they were all positive (Table 3). The importance of anti-F antibodies in protection became apparent in the initial attempts to use inactivated MV vaccines. Although these preparations stimulated high anti-HA and neutralizing antibodies, the children were not protected [14]. Subsequent studies have shown that the antibody response to the F protein was different from that in MV infections [22]. The murine anti-F monoclonal antibody prepared against MV was similar to the antibodies in human serum after vaccination

Table 4. Characterization of anti-F monoclonal antibodies from mice immunized with VV-F

Monoclonal antibody	Epitope	FI	Neutralization	
			in vitro	in vivo
186B19	1	N.D.	1×10^4	2×10^3
186A	1	5×10^3	3×10^4	$> 10^6$
319	1	$> 10^4$	3×10^4	10^4
186B22	1	5×10^3	5×10^4	$> 10^5$
M77–4	2	4×10^4	8×10^2	10^2

FI fusion inhibition
Neutralization in vitro was determined as the highest dilution of the ascites fluid which completely neutralized 50 p.f.u. of MV. For the in vivo determination, suckling mice were inoculated i.c. with the Yamagata-1 strain of MV and $50\,\mu l$ of ascites fluid were inoculated i.p. into the mice. Seven days later, the brain was removed and the virus titrated on vero cells. The titres are expressed as the decrease in virus yield compared to mice which received no antibody

with inactivated virus i.e. the antibodies neither neutralized virus infectivity nor inhibited viral fusion. In contrast, our studies using vaccinia-MV-F recombinants showed that low levels of neutralizing antibodies were induced in mice [12]. We therefore postulated that perhaps anti-F monoclonal antibodies with biological activities similar to those in humans would only be obtained by immunizing mice with a vectored antigen which synthesized the protein endogenously. To do so, we immunized mice with a vaccinia recombinant expressing the MV-F protein. Anti-F monoclonal antibodies obtained using this method of immunization not only neutralized MV infectivity, but also inhibited fusion activity and protected in vivo (Table 4). By selecting escape mutants, we identified a minimum of two epitopes. In contrast to the previously isolated anti-F monoclonals, there was no cross-reaction with CDV. Such monoclonals may be better tools to search for possible antigenic variation.

Antigenic characterization of the NP

The NP's structural role within the virus is probably to protect the genome from nucleases. Unlike the glycoprotein, the NP is not subjected to immune pressure. Mutations which occur should be outside regions which have more defined roles such as protein- RNA or protein-protein interactions. Comparative studies of the NP of MV and CDV have shown that there is a high degree of homology except at the carboxy terminus [26].

Studies with MV anti-NP monoclonal antibodies revealed the presence of three antigenic sites [19]. By expressing fragments of the NP protein in bacteria we were able to map the three sites on the NP protein [9]. Site I which was between amino acids 122–150 reacted with all the wild-type MV-strains examined. Monoclonal antibodies produced against the laboratory strain of virus (Hallé) which identified site II (amino acids 457–476) reacted with a number of MV isolates from Japan, Africa and Europe. However, the presence or absence of this marker varied even within an epidemic. This would suggest that MV strains co-circulate, a phenomenon which has been shown for other paramyxoviruses [33].

Epitope III was defined at the carboxy terminus (amino acids 518–525). Monoclonal antibody defining this site produced against the laboratory strain reacted with all the vaccine strains examined, but non of the wild-type isolates or SSPE strains. To study the differences in antigenicity found in these virus strains, we analyzed a number of vaccine and wild-type strains by PCR and nucleotide sequence analysis [28]. Table 5 shows the analysis of two African strains which differed in their reactivity at site II. The strain (R118) exhibiting the same antigenicity as the vaccine strain for this site had a single conservative amino acid change, whereas the strain not reacting with this antibody had two non-conservative changes. Epitope III had the same amino acid sequences in all the wild-type strains examined and differed from the vaccine strains by two non-conservative substitutions. It was interesting

Table 5. Amino acid sequences of epitopes II and III of the nucleoprotein of different measles virus strains and their reactivity with monoclonal antibody (Mab)

Virus strain	Epitope II sequence	Mab 25
Edmonston	S D A R A A H L P T G T P L D I	+
Y22	* * * * R * * P * * * * * * * *	−
R118	* * * * * T * * * * * * * * * *	+

Virus strain	Epitope III sequence	Mab 105
Edmonston	T P I V Y N D R N L L D	+
Y22	* * R * * * * * D * * *	−
R118	* * R * * * * * D * * *	−
SSPE-33	* * R * * * * * D * * *	−

to note that in these studies a number of SSPE strains were also analyzed and in each case epitope III had the wild-type sequence.

Our studies on four of the MV proteins show that, with the exception of the NP, the antigenicity, as measured by murine monoclonal antibodies does not vary. The significance of the variation in the NP protein is difficult to assess as, unlike the glycoproteins, there is no readily available biological functional measure.

Evaluation of the MV antigens in the immune response

The role that specific viral antigens may play in immunoprotection can be studied in a number of ways: (i) passive immunoprophylaxis using mono-clonal or monospecific antibodies (ii) immunization with purified antigens (iii) vectored antigens using viruses such as vaccinia or adenovirus.

In humans, administration of immunoglobulin blocks MV infections. Using murine models, we have shown that passive administration of either anti-F or anti-HA, but not anti-NP MV monoclonal antibodies effectively blocks infection [18,21]. Further, our observations that an antibody response to the F protein resembling that of humans was only obtained in mice when the F antigen was expressed from vaccinia virus (VV) have led us to use this system to investigate the response of MV proteins.

Our initial approach was to evaluate the contribution of the individual MV antigens in the immune response. In parallel studies, we also became aware that certain biological activities of the virus, such as fusion, were the result of more than one protein [35]. Thus, having dissected the immune response, it was then necessary to reconstruct it by expressing combinations of antigens from the same vector in order to study possible synergic effects.

Table 6. Pathogenicity of vaccinia virus in suckling mice

Virus	LD50
wildtype	7×10^2
KIL –	10^4
tk –	6×10^5
tk –, KIL –	$6–10 \times 10^5$

Balb/c suckling mice were inoculated with 10-fold dilutions of the different VVs and the mortality recorded during 3 weeks. The LD50 is expressed as the number of p.f.u. of VV required to kill half of the mice inoculated

Vaccinia (VV)-MV recombinants

The MV genes corresponding to the HA, F, M or NP were expressed using the vaccinia 7.5 K early-late promotor and inserted either in the tk locus or at a host range locus KIL in the Copenhagen strain of vaccinia [12,36]. Further constructions were made in which two or three MV genes were expressed in the same recombinant.

Inactivation of the tk gene of the WR strain of VV reduced its neurovirulence in mice [10]. The Copenhagen strain is more attenuated than the WR strain, so to test if the former becomes further attenuated after introduction of MV genes into the KIL and tk sites, suckling mice were inoculated with dilutions of the VV recombinants (Table 6). Inactivation of the KIL and tk genes reduced the LD50 by 10 and 1000-fold respectively.

The expression of the measles virus proteins by the different recombinants was examined in several cell lines using a variety of techniques. In each case, with the exception of the M protein, the size, antigenicity and distribution of the proteins was identical to that found in MV-infected cells.

Immunization

Balb/c mice were immunized with VV recombinants encoding either HA, F or NP or combinations of two or three of these MV antigens. Sera from these mice specifically immunoprecipitated the corresponding MV protein. Immunization of mice with VV recombinants encoding up to three antigens (HA, F and NP) gave serum antibodies recognizing all three MV components [36]. The sera from the VV-HA vaccinated mice had high levels of MV neutralizing antibody, as compared to those immunized with VV-F. Vaccination with VV recombinants expressing different combinations of MV antigens (HA:HA-NP:HA-F:HA-F-NP) did not increase the virus neutralization titre. Thus, at the quantitative level no synergic effects were observed between the antigens in inducing neutralizing antibody.

Table 7. Protection of mice immunized with VV-MV recombinant viruses

Dose (p.f.u./mouse)	VVwt	Recombinant*		VV-NP
		VV-HA	VV-F	
Balb/c				
3×10^7	0/4	6/6	4/5	6/20
3×10^6		5/5	6/6	
3×10^5		5/5	4/4	
CBA				
3×10^7	0/6	6/6	1/6	0/10
3×10^6		6/6	1/6	
3×10^5		6/6	1/6	

Balb/c and CBA mice were immunized with the recombinant viruses and
challenged with MV 21 days later
*Animals protected/total inoculated

To measure protection in the vaccinated mice, we challenged the animals by intracerebral inoculation with a MV isolate from an SSPE case [34]. In unvaccinated animals, this resulted in an acute encephalitis with mortality between the 7th and 12th day. Balb/c mice immunized with either VV-HA or VV-F, but not VV-NP, were protected from the lethal MV challenge (Table 7).

VV has been used as a vector for expressing a number of antigens. Certain strains of VV are more efficient than others in inducing an antibody response to the vectored antigen. The reasons for these differences have not been investigated, but may reflect differences in attenuation or the virus encoded proteins which may affect the processing of the virus antigens. As insertion of MV genes into the tk and KIL loci reduced the virulence of VV, it may in turn reduce the efficiency of immunization. To study this, we inoculated mice with VV-F recombinants in which the F gene was inserted into either of the two loci, i.e. the F-tk construction is a hundred-fold more attenuated in mice than the KIL insertion. Mice were challenged with MV 21 days after vaccination with either of the VV-constructions. In both cases 10^4 p.f.u. of the recombinant virus protected approximately half of the mice, whereas 10^5 p.f.u. protected all the mice against MV infection. These experiments show that the immunogenic potential of the vectored antigens is not hindered by the level of attenuation of the VV strain. This will be important in the development of new and safer strains of vaccinia.

Affect of host genetics of immunization

Studies with several viruses have shown that the immune response to an antigen is controlled by the genetics of the host. In the case of cell mediated

Table 8. Synergic effect of immunization of CBA mice with VV-recombinants expressing the MV F and NP proteins

Recombinant	Protected/Total
VV-NP	0/10
VV-F	7/26
VV-F.NP	18/23

immunity (CMI), only a limited number of viral epitopes are presented. Against certain genetic backgrounds, the viral antigens may fail to induce certain types of CMI, e.g. H2k mice are incapable of inducing a CTL class I response to vesicular stomatitis virus [25].

To extend our studies to mice of a different haplotype, we immunized CBA (H2k) mice with recombinants VV-HA or VV-F and challenged them with MV 21 days later (Table 7). Although vaccination of mice with VV-HA protected animals against infection, VV-F did not. Non-responsiveness to certain antigens is sometimes a quantitative factor and not absolute [30]. We have noted that increasing the amount of VV-F inoculated into the CBA mice, although not affecting the eventual outcome, did increase the survival period after the challenge [36]. Further, immunization of CBA mice with a second inoculation of VV-F induced MV neutralizing antibodies. Thus, the restriction against the F protein in CBA mice is not absolute.

Studies on both rabies and influenza viruses have shown that pre-immunization with NP increases the immune response to the other glyco-proteins [13]. It would therefore be possible that the co-expression of the NP with the F protein in CBA mice might increase the response to the glycoprotein. To test this hypothesis, CBA mice were immunized with a VV recombinant expressing both MV antigens and later challenged with MV (Table 8). The co-expression of the NP and F proteins increased resistance to MV to more than 75%. The mechanism involved has still to be elucidated.

In search of further models

The experiments on protection reported so far have all used a model in which animals were inoculated with a defective cell-associated MV isolate. We have been unable to develop a similar model using cell-free virus strains. However, in parallel studies, we adapted CDV to mice and developed a virus strain which gave high mortality in adult animals. In contrast to MV, the CDV strain gave cell-free virus.

In the veterinary field, dogs can be vaccinated against CDV with MV vaccines. This protects the animals against a lethal infection with CDV. An analysis of the antibody response in the immunized animals suggests that the

Table 9. Protection of mice immunized
with VV-MV and VV-CDV-F recombinants
against a challenge with CDV

Immunizing virus	Protected/Total
VVwt	0/9
VV-HA	0/7
VV-F	5/7
VV-NP	5/9
VV-M	6/10
VV-CDV-F	10/10

Mice were immunized with the VV recombinants and challenged 21 days later by i.c.
inoculation with 4×10^3 p.f.u./mouse of
CDV

protection in part is mediated by an antibody response to the F protein [1].
In many cases, the protection appears incomplete, only limiting the development of symptoms in the presence of virus replication. The MV/CDV model
presents a novel system to investigate immunoprotection between two closely
related viruses.

To study the contribution of the individual MV proteins in the protection
against CDV, Balb/c mice were vaccinated with the different VV-MV
recombinants and as a positive control with a VV-recombinant expressing
the F protein of CDV (VV-CDV-F). Studies of the serum antibodies of the
mice immunized with the VV-MV recombinants showed that only the VV-F
vaccinated mice reacted with MV antigens and no CDV neutralising
antibodies were found. When the mice were challenged with CDV, those
vaccinated with VV-HA were not protected, whereas 50% or more of the
mice immunized with the F, NP or M recombinants were. The VV-CDV-F
recombinant completely protected the mice (Table 9). Analysis of the sera
of surviving mice revealed the presence of CDV neutralizing antibodies even
in the VV-CDV-F vaccinated animals. In situ hybridization studies on the
brain of the mice 7 days after challenge showed an active replication of CDV
in the VV-MV vaccinated animals but not in the VV-CDV-F vaccinated
mice.

These observations show that several of the MV proteins can participate
in protecting mice from a lethal infection. Although we do not know the
mechanism involved in the case of the NP and M proteins, it probably does
not involve antibody. The role of the CMI in this model will be interesting
to elucidate as it may dictate some of the basic rules for MV vaccination.

Conclusion

Studies in immunology have dissected the immune response into several
pathways. Certain may only be stimulated under specific conditions such as

in a viral infection. Thus, although endogenous and exogenous antigens can stimulate class II T cells, only the former induce a class I response. Further, studies on a number of viruses have shown that this response is often restricted to a single virus protein or may be absent. At present, we do not know which of the immune responses are essential for an efficient vaccination against MV. If neutralizing antibodies have a role to play in the initial contract, the failure of inactivated MV vaccine suggests that the correct response is more important to the F protein. Our studies have shown that such antibodies could only be induced by a live virus. Our antigenic studies on the two glycoproteins failed to show evidence that antigenic drift may be a source of vaccine failure.

Using a heterologous (MV/CDV) mouse model, we were able to demonstrate a role in protection for the two internal virus proteins, M and NP. Immunization with these antigens did not block virus replication but limited development of symptoms. These results, together with our observations that NP can aid the F response in CBA mice, suggest that the two internal proteins play a role in CMI.

The NP was the only antigen in which antigenic variation (B cell epitopes) was observed. However, as this antigen's role is probably in CMI, such observations may be irrelevant except as a marker for vaccine and wild-type virus strain differentiation. It will be more relevant to define the T cell epitopes and study their variation in different isolates.

Acknowledgements

The studies reported here represent a summary of the combined efforts of a number of people over the last few years. Those involved include R. Buckland, P. Giraudon, P. Beauverger, A. Bernard, E. Malvoisin, B. Rima and R. Drillien.

References

1. Appel MJG, Shek WR, Sheshberadaran H, Norrby E (1984) Measles virus and inactivated canine distemper virus induce incomplete immunity to canine distemper. Arch Virol 82: 73–82
2. Barrett T, Shrimpton SB, Russel SEH (1985) Nucleotide sequence of the entire protein encoding region of canine distemper virus polymerase-associated (P) protein mRNA. Virus Res 3: 367–372
3. Barrett T, Clarke DK, Evans SA, Rima BK (1987) The nucleotide sequence of the gene encoding the F protein of canine distemper virus: a comparison of the deduced amino acid sequence with other paramyxoviruses. Virus Res 8: 373–386
4. Bellini WJ, Englund G, Rozenblatt S, Arnheiter H, Richardson CD (1985) Measles virus P gene codes for two proteins. J Virol 53: 908–919
5. Bellini WJ, Englund G, Richardson CD, Rozenblatt S, Lazzarini RA (1986) Matrix genes of measles virus and canine distemper virus: cloning, nucleotide sequences and deduced amino acid sequences. J Virol 58: 408–416
6. Black FL (1991) Epidemiology of paramyxoviridae. In: Kingsbury DW (ed) The Paramyxoviruses. Plenum Press, New York, pp 509–536

7. Buckland R, Gerald C, Barker R, Wild TF (1987) Fusion glycoprotein of measles virus: nucleotide sequence of the gene and comparison with other paramyxoviruses. J Gen Virol 68:1695–1703

8. Buckland R, Gerald C, Barker R, Wild TF (1988) Cloning and sequencing of the nucleoprotein gene of measles virus (Hallé strain). Nucleic Acids Res 16:11821

9. Buckland R, Giraudon P, Wild TF (1989) Expression of Measles virus nucleoprotein in Escherichia coli: use of deletion mutants to locate the antigenic sites. J Gen Virol 70:435–441

10. Buller RML, Smith GL, Cremer K, Notkins AL, Moss B (1985) Decreased virulence of recombinant vaccinia virus expression vectors is associated with a thymidine kinase-negative phenotype. Nature 317:813–815

11. Curran MD, Clarke DK, Rima BK (1991) The nucleotide sequence of the gene encoding the attachment protein H of canine distemper virus. J Gen Virol 72:443–447

12. Drillien R. Spehner D, Kirn A, Giraudon P, Buckland R, Wild TF, Lecocq JP (1988) Protection of mice from fatal measles encephalitis by vaccination with vaccinia virus recombinants encoding either the hemagglutinin or the fusion protein. Proc Natl Acad Sci USA 85:1252–1256

13. Ertl HCJ, Dietzschold B, Gore M, Otuosjr L, Larson JK, Wunner WH, Koprowski H (1989) Induction of rabies virus-specific T-helper cells by synthetic peptides that carry dominant T-helper cell epitopes of the viral ribonucleoprotein. J Virol 63:2885–2892

14. Fulginiti VA, Eller JJ, Downie AW, Kempe CH (1967) Altered reactivity to measles virus. Atypical measles in children previously immunized with inactivated measles virus vaccines. J Am Med Ass 202:1075–1080

15. Gerald C, Buckland R, Barker R, Freeman G, Wild TF (1986) Measles virus haemagglutinin gene: cloning, complete sequence analysis and expression in COS cells. J Gen Virol 67:2695–2703

16. Giraudon P, Wild TF (1981a) Monoclonal antibodies against measles virus. J Gen Virol 54:325–332

17. Giraudon P, Wild TF (1981b) Differentiation of measles virus strains and a strain of canine distemper virus by monoclonal antibodies. J Gen Virol 57:179–183

18. Giraudon P, Wild TF (1985) Correlation between epitopes on hemagglutinin of measles virus and biological activities: passive protection by monoclonal antibodies is related to their hemagglutination inhibiting activity. Virology 144:46–58

19. Giraudon P, Jacquier MF, Wild TF (1988) Antigenic analysis of African measles virus field isolates: identification and localisation of one conserved and two variable epitope sites on the NP protein. Virus Res 18:137–152

20. Limo M, Yilma T (1990) Molecular cloning of the rinderpest virus matrix gene: comparative sequence analysis with other paramyxoviruses. Virology 175:323–327

21. Malvoisin E, Wild TF (1990) Contribution of measles virus fusion protein in protective immunity: anti-F monoclonal antibodies neutralize virus infectivity and protect mice against challenge. J Virol 64:5160–5162

22. Norrby E, Enders-Ruckle G, Ter Meulen V (1975) Differences in the appearance of antibodies to structural components of measles virus after immunization with inactivated and live virus. J Inf Dis 132:262–269

23. Norrby E, Sheshberadaran H, McCullough KC, Carpenter WC, Örvell C (1985) Is rinderpest virus the archevirus of the morbillivirus genus? Intervirology 23:228–232

24. Panum PL (1940) Observations made during the epidemic of measles on the Faroe islands in the year 1846. American Publishing Association, New York

25. Reiss CS, Evans GA, Margulies DH, Seidman JG, Burakoff SJ (1983) Allospecific and virus-specific cytolytic T lymphocytes are restricted to the N or C1 domain of H-2 antigens expressed on L cells after DNA-mediated gene transfer. Proc Natl Acad Sci USA 80:2709–2712

26. Rozenblatt S, Eizenberg O, Ben-Levy R, Lavie V, Bellini WJ (1985) Sequence homology within the morbilliviruses. J Virol 53:684–690

27. Sheshberadaran H, Norrby E, McCullough KC, Carpenter WC, Örvell C (1986) The

antigenic relationship between measles, canine distemper and rinderpest viruses studied with monoclonal antibodies. J Gen Virol 67: 1381–1392

28. Taylor MJ, Godfrey E, Baczko K, ter Meulen V, Wild TF, Rima B (1991) Identification of several different lineages of measles virus. J Gen Virol 72: 439–442

29. Ter Meulen V, Loffler S, Carter MJ, Stephenson JR (1981) Antigenic characterization of measles and SSPE virus haemagglutinin by monoclonal antibodies. J Gen Virol 57: 357–364

30. Tite JP, Russell SM, Dougan G, O'Callaghan D, Jones I, Brownlee G, Liew FY (1988) Antiviral immunity induced by recombinant nucleoprotein of influenza A virus. I. Characterization and cross-reactivity of T cell responses. J Immunol 141: 3980–3987

31. Tsukiyama K, Sugiyama M, Yoshikawa Y, Yamanouchi K (1987) Molecular cloning and sequence analysis of the rinderpest virus mRNA encoding the hemagglutinin protein. Virology 160: 48–54

32. Tsukiyama K, Yoshikawa Y, Yamanouchi K (1988) Fusion glycoprotein (F) of rinderpest virus: entire nucleotide sequence of the F mRNA, and several features of the F protein. Virology 164: 523–530

33. Van Wyke Coelingh KL, Winter C, Murphy BR (1985) Antigenic variation in the hemagglutinin-neuraminidase protein of human parainfluenza type 3 virus. Virology 143: 569–583

34. Wild TF, Giraudon P, Bernard A, Huppert J (1979) Isolation and characterization of a defective measles virus from a subacute sclerosing panencephalitis patient. J Med Virol 4: 103–114

35. Wild TF, Malvoisin E, Buckland R (1991) Measles virus: both the haemagglutinin and fusion glycoproteins are required for fusion. J Gen Virol 72: 439–442

36. Wild TF, Bernard A, Spehner D, Drillien R (1992) Construction of vaccinia virus recombinants expressing several measles virus proteins and analysis of their efficacy in vaccination of mice. J Gen Virol 73: 359–367

13

Cell-mediated immune response to measles virus

A. A. Salmi

Department of Virology, University of Turku, Turku, Finland

Summary

Measles virus (MV) causes a strong but temporary depression of cell-mediated immunity (CMI) in infected individuals Despite of the immunosuppression, both proliferative and cytotoxic T cell responses appear after MV infection. CMI response develops slowly, reaches only a relatively low level, and wanes rapidly afterwards. Memory T cells are maintained a long time after measles. The major structural MV proteins are recognized by T cells immune individuals but the role of nonstructural MV proteins has not been clarified. A few T cell epitopes have been described on the internal protein N and the envelope proteins F and H but their importance in the protective immune response has not been shown. The major class of cytotoxic T cells are CD8 positive, MHC-class I restricted, but CD4 positive, class II restricted cytotoxic T cells have also been described.

Introduction

The measurement of antibodies is used to evaluate immunity and predict protection against a viral disease. Neutralizing antibodies are necessary for the prevention of secondary virus infection but cell-mediated immune (CMI) response also plays an important role both in primary and secondary viral infections.

Antibody response to most viral antigens is T cell dependent. Therefore, a good protective antibody response depends on a proper helper T cell (T_h) response. As the antibody response to most viruses is good in the majority of individuals, it can be assumed that a proper helper T cell response is also operative in all such infections. Virus-specific T_h cells have been poorly characterized in most virus infections.

Cytotoxic T cells (T_c/CTL) develop early during virus infections. They play a role in recovery from some, if not all, virus infections. Cytotoxic T cells are also important for controlling persistent virus infections. This has been demonstrated, for example with lymphocytic choriomeningitis virus in mice which clear presistent infection after injections of virus-specific cytotoxic cells.

Although antibodies to measles virus (MV) are necessary for protection against reinfection, CMI is necessary for the recovery of patients from measles infection. This is seen in "experiments of nature": patients with agammaglobulinemia recover normally from measles but patients with impaired CMI have often a fatal form of the disease. Despite the central role of CMI in measles immunity and pathogenesis, little information has accumulated on the subject and molecular details still wait to be revealed. The present review aims at giving a summary of the development of CMI studies in measles published mainly in the last 10 years.

Effect of measles infection on CMI in vivo

Any virus infection may disturb the immunological balance of the infected individual. The severe immunosuppression by HIV infection has been thoroughly studied but mechanisms of the milder disturbances caused by other viruses are poorly known. MV has a dual effect on CMI of the infected individual. It depresses temporarily the pre-existing T-cell reactivity to non-MV antigens and simultaneously stimulates immunological cells specific for MV. The spontaneous proliferation of peripheral blood mononuclear cells (PBMC) observed during measles may represent the developing MV-specific response [49].

The suppressing effect of measles infection was already observed early this century when CMI response to tuberculin was found to be reduced during measles infection [47]. This observation has been confirmed and extended in numerous studies. Live attenuated measles virus vaccine causes immunosuppression which is less severe than that found during measles infection [4]. Immunosuppression during natural measles may totally abolish a strong skin test reactivity for a few days. Full recovery from the immunological anergy may take weeks and unresponsiveness persists longer in patients with complications [44]. Suppression of cell-mediated immunity may also disturb the process of viral elimination and lead to virus persistence [8]. On the other hand, suppression of the cell-mediated immunity may also give temporary relief in some autoimmune or allergic diseases [2,40]. General immunosuppression during measles explains at least partially the fact that bacterial infections are common sequlae of this disease. Measles may also have an adverse effect on chronic bacterial infections, such as tuberculosis.

Effect of MV infection on CMI in vitro

Immunosuppression by MV and its mechanism have also been studied in vitro. When PBMC are infected with MV and stimulated with specific antigens or general mitogens, DNA synthesis as measured by tritiated thymidine incorporation is strongly reduced compared to uninfected, stimulated controls. Although all MV strains have this property, the magnitude of suppression varies between different strains [20].

All the subpopulations of PBMC can be infected in vitro but the bulk of infectious virus is released from activated T cells. Productive infection of PBMC requires stimulation of the cells with antigen or nonspecific mitogens, such as phytohemagglutinin or concanavalin A. However, productive infection is not required for the immunosuppression [48].

As T cells are infected by MV and produce large quantities of progeny virus after stimulation, they should have a role in immunosuppression. Monocytes seem also to take part in the immunosuppression although their role has been controversial [14,,26,38,43]. Death of productively infected T cells may be a partial explanation [38] and release of inhibitory molecules from infected cells into the supernatants has also been described [38,39].

Mechanisms of immunosuppression at the molecular level are poorly known. As activation signals can be transmitted normally through the membranes of MV-infected T cells [45], the block of cellular response has to be later during the activation cycle. It has been shown that the cell cycle is arrested in the G1 phase by MV infection of T cells which may explain the suppression [28]. Inhibition of the lymphoproliferative reaction may in part also be due to changes in synthesis of soluble mediators [50]. Despite these explanations for the MV-induced immunosuppression the detailed mechanisms still require further investigations.

Proliferative response of PBMC in measles patients and immune individuals

Although MV is a strongly suppressive virus, measles-specific CMI is still induced in patients. CMI to MV were earlier studied with tests measuring inhibition of PBMC migration in vitro or skin reaction in humans or foot-pad swelling in experimental animals. The results of the early studies were often conflicting and difficult to interpret. More recent studies measuring the proliferative response after antigen stimulation and specific cytotoxic tests have revealed more definitive information about CMI to MV.

In the proliferation test, also called blast transformation test, measles virus antigen is used to stimulate PBMC and the response is measured by incorporation of radioactively labelled tritiated thymidine. T lymphocytes are the major responding cells in this assay and the helper phenotype is the major contributor to the results [42]. As helper T cells are central in the development and maintenance of the specific immunity, proliferative response in vitro reflects the status of the cell-mediated arm of the specific immunity.

Early results of proliferative studies were also controversial. Strong response to MV in all patients were found in some studies while others reported no response at all. Some of the conflicting results can be explained by poorly defined stimulating antigens which might have also been contaminated, for example by mycoplasma strains. It was later shown that the purity of the antigen preparation has an effect on results [19]. The physical form of the antigen is also important for optimal results in the proliferation test [31].

Using defined antigens we showed that MV-specific lymphoproliferative response appears in measles patients 1–3 weeks after the onset of rash and reaches the maximum level within two months [17]. In another study response was seen in only a small percentage of patients 3 to 15 weeks after the onset of rash [12]. The response was significantly decreased already in samples taken 4 to 12 weeks later suggesting that the majority of the responding cells disappear soon after measles from the circulation or that immunological anergy develops.

Development of lymphoproliferative response after vaccination has not been thoroughly studied with defined MV antigens. The results of the published studies indicate, however, that successful vaccination induces a proliferative response, but its level is even lower than after natural measles [9]. Poor MV-specific response has also been observed in children revaccinated with an attenuated measles vaccine [25].

Memory lymphocytes specific to MV antigen can be maintained for a long time after infection as shown by proliferation assays of normal blood donors with a history of uneventful measles infection [16,29]. However, the level of the response is much lower than found in similar tests for other respiratory virus infections. In fact, only a minority of seropositive individuals react in these assays and only a small percentage of them have a strong positive reaction [12]. As no active cellular suppression can be demonstrated [12], the results indicate that there is no prolonged antigen stimulation of T cells after the acute phase of measles or that MV-specific T cells sequester to other immunological organs.

Proliferative response of PBMC in subacute sclerosing panencephalitis and in chronic diseases without known etiology

The pathogenesis of a rare, but fatal late complication of measles, subacute sclerosing panencephalitis (SSPE), is not thoroughly understood. It was suggested that diminished CMI to MV is the fundamental defect predisposing to SSPE [5]. This idea was tested in a number of studies, but no conclusive evidence proving the hypothesis were produced. When we used defined MV antigens to study four SSPE patients, three had weak reactions similar to normal controls and only one had an unusually strong reaction to MV antigens [19]. When this patient was tested regularly for a longer period of time, the strong reaction type was maintained but large fluctuations at the level of reactivity was observed [21]. Based on these and other published studies it can be concluded that no specific defect in CMI to MV is found in SSPE patients although the response may be more heterogenous and variable than in normal controls.

Measles etiology of multiple sclerosis (MS) has been suspected for a long time. A large number of studies on immune response to MV in MS has been published in the last 30 years. Although there is a general agreement that MV antibodies are significantly elevated in these patients, their CMI response to MV as measured by proliferation assay seems to be more variable. Some reports have found elevated, some decreased and the rest similar responses as in controls. Real and significant differences between MS and controls may be still possible but the selection of the patient groups has been difficult due to exacerbations and remissions in MS.

Although the results of MV-specific proliferation assay do not differ in MS patients as a group compared to controls, some patients react strongly. Studies on 28 twin sets concordant or diconcordant for MS showed that seven affected twins had an elevated response to MV, whereas the unaffected twins had a low response [13]. These findings suggest that at least in some MS patients the number of clonally expanded measles-specific T cell is increased. The significance of this observation is not known.

Polypeptide specificity of the stimulating antigens in the proliferation assay

The polypeptide-specificity of the proliferating T lymphocytes has been analyzed with isolated MV components. MV polypeptides were eluted from polyacrylamide gel and used to stimulate lymphocytes from both low- and high-responder subjects [37]. Lymphocytes from only high-responders proliferated and the magnitude of the reaction was at about the same level when stimulated with MV hemagglutinin (H)-, nucleocapsid (N)-, fusion (F)-, or matrix(M)-polypeptide. Later studies on T cell lines and clones generated from MS patients have given similar results [36] but the number of T cell clones was too small for a conclusion on the abundance of different specificities. The results demonstrated, however, that memory T cells specific for the major structural proteins circulate in the peripheral blood of immune individuals.

PBMC from the majority of blood donors with a history of measles infection proliferate only weakly in the presence of measles virus antigens [18]. However, even a small number of MV- specific T cells can be expanded in vitro and T cell lines and clones established. In a recent study, we generated 66 T cell clones from two DR1/DR2 heterozygous healthy adult donors with antibodies to MV and tested them for reactivity to purified H and N proteins of MV [18]. More than 50% of the clones were specific for these two proteins. Most clones specific for the N protein were HLA DR1 restricted whereas clones specific for the H protein were more frequently recognized in association with the DR2 antigen.

Several studies have shown that H and F proteins of MV are potent T cell antigens and that their immunogenicity can be improved by insertion of the proteins into liposomes [1,10,31]. When the role of individual MV structural proteins in protection against encephalitis was studied in rats [3],

CMI response to MV N-, M-, F-, H-, and phospho(P)-proteins were observed. Animals immunized with N, F or H proteins were protected against encephalitis. CD8 + cells were not protective but N- and F- specific CD4 + cells prevented development of encephalitis in the absence of neutralizing antibodies.

Stimulating epitopes on antigens in the proliferation assay

In recent years, synthetic peptides have become important tools in analyzing antigenicity of viral proteins and localizing B and T cell epitopes. Thus far, only a few studies have employed synthetic peptides for studies on CMI in measles. Several different prediction methods were used to scan the sequence of MV F protein for putative T cell epitopes [34]. Eight of the ten tested individuals with a prior history of measles responded to stimulation with a peptide with residues 288 to 302. We have also used prediction methods in designing peptides for T cell epitopes of MV hemagglutinin but none of the synthesized peptides including those with predicted amphipathicity did stimulate human T cell clones [18, our unpublished data].

Recently it has been shown [33] that a chimeric peptide containing a predicted T cell epitope from the F protein of MV functions in vivo as a helper determinant for a B cell response to the peptide 188 to 199, earlier shown to be immunogenic and to induce neutralizing antibodies in experimental animals [30]. Mice with a range of different H-2 haplotypes responded to the chimeric peptide which suggests that the helper determinant from the F protein is promiscuous and may be used in a subcomponent vaccine [33].

Helper T cell response to MV N protein was studied in mice [11]. The major epitopes were shown to be between amino acids 67–98 of MV N protein both in H2d and H2k mice. In addition, an epitope was identified between amino acids 457–525 in H2d mice. Although these results suggest that a short area of the MV N protein may be generally immunodominant in mice, final conclusions can not be made before a larger number of mice with different genetic backgrounds have been tested. Neither can the results from mice be directly applied to human population which is genetically more diverse than a few mouse strains combined.

Cytotoxic T cell in measles

The specific effector cell of the cell-mediated arm of the immunity are MHC-restricted cytotoxic T cells (T_c). As the test requires autologous or HLA-matched MV-infected cells as targets, accumulation of data on T_c immunity in humans has been slow. Other less specific tests for cell mediated cytotoxicity have also been described including NK cells and antibody mediated cellular cytotoxicity.

Cellular cytotoxicity to MV infected cells without HLA restriction have been described. PBMC both from immune and non-immune individuals contain cytotoxic cells which exert their function within 4 hours in vitro [6]. This type of cytotoxicity seems not to be antibody-dependent and may, in fact, represent natural killer (NK) cell activity. NK cell activity increases during acute viral infections and can be important for recovery from measles [15]. Antibody-dependent cellular cytotoxicity against MV-infected cells has also been described in vitro [35] but it is not clear if such activity has any role in vivo.

Specific cytotoxic T lymphocytes (CTL) are important for clearance of virus-infected cells in acute or chronic infections. It has been difficult to demonstrate MV-specific HLA-restricted CTLs directly in PBMC. The best evidence for the existence of circulating MV-specific CTLs is from a study of nine adults with acute measles infection [41]. Only four individuals had HLA-restricted CTLs during the acute phase of measles and the reactions were weak. Attempts to generate MV-specific HLA-restricted CTLs from immune individuals failed in this study. This suggests that the pool size of the CTL memory cells has to be low. Despite this, cytotoxic T cell lines and clones can be generated in vitro from PBMC of immune individuals [22,46].

In general, CTL against viruses are CD8 positive and recognize their target antigen in context with the MHC class I molecules (HLA-A, B, -C in humans) whereas T helper cells are CD4 positive and recognize antigens in association with the class II molecules (HLA-DR, -DP, -DQ in humans). However, MV-specific cytotoxic T cell clones generated from a MS patient were shown to be class II restricted and CD4 positive [22]. Similar findings were later reported with normal blood donors [23]. All the major structural proteins including the internal N and M proteins as well as the surface proteins H and F were shown to be targets of the CTL response [7,23].

The most convincing study on class II restricted CTL employed transfection of murine fibroblasts with HLA-DR antigen and cytoplasmic N and M proteins of MV [24]. In these experiments, class II restricted cytotoxic T cell lines or clones lysed target cells efficiently. These results show that endogenously synthesized MV proteins can be presented by class II antigens. Similar observations have recently been made with exogenous influenza A matrix protein which can be presented to class II-restricted T cells also by the endogenous pathway [32].

All the experiments showing class II restricted cytotoxic T cells have used lines or clones generated by stimulation with purified MV antigens in vitro. This may have lead to a selection of an atypical minority of CTL. The concept of CD4 + class II restricted CTLs as the major cytotoxic T cell class in measles infection has recently been challenged [46]. It was shown that CD4 + T cell lines are generated when PBMC are stimulated with inactivated MV antigen, whereas CD8 + clones are obtained when MV-infected B cell lines are used. The results indicate that optimal stimulation of class I restricted lymphocytes requires virus-infected cells and, therefore, dominance of CD4 + cells in lines and clones generated may be due to the stimulation conditions.

It was also shown that stimulation of PBMC from individuals in con-valescence from measles expanded primarily CD8 + T cells [46]. This supports the concept that CD8 positive CTLs are important for elimination of measles infection. As depletion of CD8 + T cells in rats promote viral persistence in the brain [27], cytotoxic T cells may also play a role in the control of chronic MV infections such as SSPE.

The role of CD4 positive T cells is still not excluded as protection of rats against a virus challenge by immunization with the MV N protein can be induced even in the absence of neutralizing antibodies and CD8 positive cells [3].

Conclusions

Cell-mediated immunity against MV is important. The recovery from acute measles infection requires a functional MV-specific cellular response as patients with a defective cellular arm of the immunity have difficulties to recover from measles. Specific CMI response seems to be important also for the prevention of a secondary infection although antibodies play a more important role.

Specific CMI to MV has been mainly studied using T cell proliferation assay. The response develops relatively slowly, reaches only a moderate level and wanes soon afterwards. Only a small number of immune individuals react in the assay. All the major structural polypeptides stimulate T cells from responding individuals. Low numbers of memory T cells exist, however, in all immune individuals as MV-specific T cell lines and clones can be estab-lished from subjects without response in the proliferation assay.

Specific cytotoxic T cells are also generated during acute measles but the reaction is weak and detected only in a small number of individuals. Circulating virus-specific cytotoxic cells can not be demonstrated in immune individuals. Cytotoxic T cell lines have been generated and shown to be specific for each of the major structural MV polypeptides. Although class I restricted CD8 positive cytotoxic cells may be the major effector cell in acute measles, class II restricted CD4 positive cytotoxic T cells have also been described.

Acknowledgments

The financial support by grants from the Sigrid Juselius Foundation is acknowledged.

References

1. Bakouche O, Mougin B, Gerlier D (1987) In vitro cellular immune response to measles viral glycoproteins: role of the antigen vector. Immunology 62: 605–611
2. Boner AL, Valletta EA, Bellanti JA (1985) Improvement of atopic dermatitis following natural measles virus infection. Four case reports. Ann Allergy 55: 605–608

3. Brinckmann UG, Bankamp B, Reich A, ter Meulen V, Liebert UG (1991) Efficacy of Individual Measles Virus Structural Proteins in the Protection of Rats from Measles Encephalitis. J Gen Virol 72: 2491–2500
4. Brody J, Overfield T, Hammes L (1964) Depression of tuberculin reaction by viral vaccines. N Engl J Med 271: 1294–1296
5. Burnet FM (1968) Measles as an index of immunological function. Lancet 2: 610–613
6. Casali P, Oldstone MBA (1982) Mechanisms of killing of measles virus-infected cells by human lymphocytes: Interferon associated and unassociated cell-mediated cytotoxicity. Cell Immunol 70: 330–344
7. Dhib-Jalbut S, McFarlin DE, McFarland HF (1989) Measles virus-polypeptide specificity of the cytotoxic T-lymphocyte response in multiple sclerosis. J Neuroimmunol 21: 205–212
8. Galama JMD, Ubels-Postma J, Vos A, Lucas CJ (1980) Measles virus inhibits acquisition of lymphocyte functions but not established effector functions. Cell Immunol 50: 405–415
9. Gallagher MR, Welliver R, Yamanaka T, Eisenberg B, Sun M, Ogra PL (1981) Cell-mediated immune responsiveness to measles. Its occurrence as a result of naturally acquired or vaccine-induced infection and in infants of immune mothers. Am J Dis Child 135: 48–51
10. Garnier F, Fourquet F, Bertolino P, Gerlier D (1991) Enhancement of in vivo and in vitro T cell response against measles virus haemagglutinin after its incorporation into liposomes: effect of the phospholipid composition. Vaccine 9: 340–345
11. Giraudon P, Buckland R, Wild TF (1992) The immune response to measles virus in mice. T-helper response to the nucleoprotein and mapping of the T-helper epitopes. Virus Res 22: 41–54
12. Greenstein JI, McFarland HF (1983) Response of human lymphocytes to measles virus after natural infection. Infect Immun 40: 198–204
13. Greenstein JI, McFarland HF, Mingioli ES, McFarlin DE (1984) The lymphoproliferative response to measles virus in twins with multiple sclerosis. Ann Neurol 15: 79–87
14. Griffin DE, Johnson RT, Tamashiro VG, Moench TR, Jauregui E, Lindo de Soriano I, Vaisberg A (1987) In vitro studies of the role of monocytes in the immunosuppression associated with natural measles virus infections. Clin Immunol Immunopathol 45: 375–383
15. Griffin DE, Ward BJ, Jauregui E, Johnson RT, Vaisberg A (1990) Natural Killer Cell Activity During Measles. Clin Exp Immunol 81: 218–224
16. Ilonen J (1979) Lymphocyte blast transformation response of seropositive and seronegative subjects to herpes simplex, rubella, mumps, and measles virus antigens. Acta Pathol Microbiol Scand C 87: 151–157
17. Ilonen J, Lanning M, Herva E, Salmi A (1980) Lymphocyte blast transformation responses in measles infection. Scand J Immunol 12: 383–391
18. Ilonen J, Mäkelä M, Ziola B, Salmi AA (1990) Cloning of human T cells specific for measles virus hemagglutinin and nucleocapsid. Clin Exp Immunol 81: 212–217
19. Ilonen J, Reunanen M, Herva E, Ziola B, Salmi A (1980) Stimulation of lymphocytes from subacute sclerosing panencephalitis patients by defined measles virus antigens. Cell Immunol 51: 201–214
20. Ilonen J, Salonen R, Marusyk R, Salmi A (1988) Measles virus strain-dependent variation in outcome of infection of human blood mononuclear cells. J Gen Virol 69: 247–252
21. Ilonen J, Ziola B, Reunanen M, Cantell K, Panelius M, Lund G, Salmi A (1987) Immunological findings during an acute relapse of subacute sclerosing panencephalitis in a patient with unusually prolonged disease course. J Clin Lab Immunol 22: 41-47
22. Jacobson S, Richert JR, Biddison WE, Satinsky A, Hartzman RJ, McFarland HF (1984) Measles virus-specific T4+ human cytotoxic T cell clones are restricted by class II HLA antigens. J Immunol 133: 754–757
23. Jacobson S, Rose JW, Flerlage ML, McFarlin DE, McFarland HF (1987/1988) Induction of measles virus-specific human cytotoxic T cells by purified measles virus nucleocapsid and hemagglutinin polypeptides. Viral Immunol 1: 153–162
24. Jacobson S, Sekaly RP, Jacobson CL, McFarland HF, Long EO (1989) HLA class

II-restricted presentation of cytoplasmic measles virus antigens to cytotoxic T cells. J Virol 63: 1756–1762

25. Linnemann CCJ, Dine MS, Roselle GA, Askey PA (1982) Measles immunity after revaccination: results in children vaccinated before 10 months of age. Pediatrics 69: 332–335

26. Lucas CJ, Ubels-Postma J, Galama JM, Rezee A (1978) Studies on the mechanism of measles virus-induced suppression of lymphocyte functions in vitro. Lack of a role for interferon and monocytes. Cell Immunol 37: 448–458

27. Maehlen J, Olsson T, Love A, Klareskog L, Norrby E, Kristensson K (1989) Persistence of measles virus in rat brain neurons is promoted by depletion of CD8+ T cells. J Neuroimmunol 21: 149–155

28. McChesney MB, Altman A, Oldstone MB (1988) Suppression of T lymphocyte function by measles virus is due to cell cycle arrest in G1. J Immunol 140: 1269–1273

29. McFarland HF, Pedone CA, Mingioli ES, McFarlin DE (1980) The response of human lymphocyte subpopulation to measles, mumps and vaccinia viral antigens. J Immunol 125: 221–225

30. Mäkelä MJ, Lund GA, Salmi AA (1989) Antigenicity of the measles virus heamagglutinin studied by synthetic peptides. J Gen Virol 70: 603–614

31. Mäkelä MJ, Smith RH, Lund GA, Salmi AA (1989) T-cell recognition of measles virus hemagglutinin. Scand J Immunol 29: 597–607

32. Nuchtern JG, Biddison WE, Klausner RD (1990) Class II MHC molecules can use the endogenous pathway of antigen presentation. Nature 343: 74–76

33. Partidos CD, Stanley CM, Steward MW (1991) Immune responses in mice following immunization with chimeric synthetic peptides representing B and T cell epitopes of measles virus proteins. J Gen Virol 72: 1293–1299

34. Partidos CD, Steward MW (1990) Prediction and identification of a T cell epitope in the fusion protein of measles virus immunodominant in mice and humans. J Gen Virol 71: 2099–2105

35. Perrin LH, Tishon A, Oldstone M (1977) Immunologic injury in measles virus infection. III. Presence and characterization of human cytotoxic lymphocytes. J Immunol 118: 282–290

36. Richert JR, Rose JW, Reuben BC, Kearns MC, Jacobson S, Mingioli ES, Hartzman RJ, McFarland HF, McFarlin DE (1986) Polypeptide specificities of measles virus-reactive T cell lines and clones derived from a patient with multiple sclerosis. J Immunol 137: 2190–2194

37. Rose JW, Bellini WJ, McFarlin DE, McFarland HF (1984) Human cellular immune response to measles virus polypeptides. J Virol 49: 988–991

38. Salonen R, Ilonen J, Salmi AA (1989) Measles virus inhibits lymphocyte proliferation in vitro by two different mechanisms. Clin Exp Immunol 75: 376–380

39. Sanchez-Lanier M, Guerin P, McLaren LC, Bankhurst AD (1988) Measles virus-induced suppression of lymphocyte proliferation. Cell Immunol 116: 367–381

40. Simpanen E, von Essen R, Isomäki H (1977) Remission of juvenile rheumatoid arthritis (Still's disease) after measles. Lancet 2: 987–988

41. Sissons JGP, Colby SD, Harrison WO, Oldstone MBA (1985) Cytotoxic lymphocytes generated in vivo with acute measles virus infection. Clin Immunol Immunopathol 34: 60–68

42. Suez D, Hayward AR (1985) Phenotyping of proliferating cells in cultures of human lymphocytes. J Immunol Meth 78: 49–57

43. Sullivan JL, Barry DW, Albrecht P, Lucas SJ (1975) Inhibition of lymphocyte stimulation by measles virus. J Immunol 114: 1458–1461

44. Tamashiro VG, Perez HH, Griffin DE (1987) Prospective study of the magnitude and duration of changes in tuberculin reactivity during uncomplicated and complicated measles. Pediatr Infect Dis J 6: 451–454

45. Vainionpää R, Hyypiä T, Åkerman KEO (1991) Early Signal Transduction in Measles Virus-Infected Lymphocytes Is Unaltered, But Second Messengers Activate Virus Replication. J Virol 65: 6743–6748

46. van Binnendijk RS, Poelen MC, de Vries P, Voorma HO, Osterhaus AD, Uytdehaag FG (1989) Measles virus-specific human T cell clones. Characterization of specificity and function of CD4+ helper/cytotoxic and CD8+ cytotoxic T cell clones. J Immunol 142: 2847–2854
47. von Pirquet CP (1908) Das Verhalten der kutanen Tuberculin reaktion während der Masedern. Dtsch Med Wochenschr 34: 1297–1300
48. Vydelingum S, Ilonen J, Salonen R, Marusyk R, Salmi A (1989) Infection of human peripheral blood mononuclear cells with a temperature-sensitive mutant of measles virus. J Virol 63: 689–695
49. Ward BJ, Johnson RT, Vaisberg A, Jauregui E, Griffin DE (1990) Spontaneous proliferation of peripheral mononuclear cells in natural measles virus infection: identification of dividing cells and correlation with mitogen responsiveness. Clin Immunol Immunopathol 55: 315–326
50. Ward BJ, Johnson RT, Vaisberg A, Jauregui E, Griffin DE (1991) Cytokine Production Invitro and the Lymphoproliferative Defect of Natural Measles Virus Infection. Clin Immunol Immunopathol 61: 236–248

Part II
Towards Poliomyelitis Eradication

14

Lessons from poliovirus control: strategies for eradication

J. L. Melnick

Division of Molecular Virology, Baylor College of Medicine, Houston, Texas, USA

Summary

Resolutions of the questions of how and where eradication of poliomyelitis can be achieved will continue to be major objectives in the next few years. The World Health Organization Expanded Programme on Immunization was initiated with the aim of reducing morbidity and mortality rates from seven target diseases, including poliomyelitis, by providing immunization against them for every child in the world. The program depends heavily upon technical cooperation with and among developing countries, particularly those in tropical regions. Results are already being seen: increasing numbers of countries now participate in the EPI program or are otherwise enhancing their polio vaccination efforts. Better and more complete surveillance programs have been implemented to detect and assess cases that occur, as global efforts increase to control polio.

Eradication of polio is part of the EPI goal of universal immunization of children. There have been significant increases in recent years with regard to OPV – from the situation in 1986, when 45% of all children in the world received the required 3 doses of OPV in the first year of life, to that in April, 1992, when this percentage had increased to over 84%. Based on an expected worldwide annual polio incidence of 5 cases per 1000 infants, the global OPV program is currently preventing an estimated 528 000 cases of paralytic disease per year. The disease has almost been eliminated from the Western Hemisphere since universal vaccination has led to a curtailment of the circulation of wild poliovirus.

In some countries previously recorded as places of high prevalence, the use of a combined OPV/IPV schedule early in life has led to the *prompt* elimination of polio in the area. The combined schedule takes advantage of the desirable properties of both vaccines and overcomes the disadvantages of each.

Introduction

There are some general principles that apply to most virus vaccines used in the prevention of human disease [1]. Vaccination does not always result in complete immunity against infection following a subsequent exposure to wild virus. But while vaccine-induced resistance does not totally block infection, it can serve to limit the multiplication of wild virus within the individual and prevent spread to the target organ, the site of severe pathologic damage. Thus wild poliovirus may sometimes replicate to a limited degree in the gut of a vaccinated person, but the virus is prevented from invading the brain and spinal cord.

In many instances when vaccination coverage may not reach 100% of the susceptibles in a population, the herd immunity, even though based on this less-than-total immunization, may serve to break the chain of transmission to a degree that wild virus can no longer sustain endemic circulation in the population. However, herd immunity varies for different viral infections. Therefore, to achieve effective and dependable control, herd immunity should be maintained at high levels.

Killed-virus vaccines prepared from whole virions generally elicit circulating antibody against the coat proteins of the virus. For some diseases, killed-virus vaccines or viral subunit vaccines are currently the only ones licensed. The duration of immunity varies for different viral vaccines; it may be short-lived and required periodic boosting.

Attenuated live-virus vaccines have the advantage of behaving like the natural infection as regards immunity. They multiply in the host and not only stimulate production of long-lasting humoral antibody but also induce cellular immunity and resistance at the portal of entry. However, some disadvantages – particularly that of genetic mutability – are associated with live attenuated vaccines.

Unrecognized adventitious agents may cause latent infections of cell cultures in which virus is cultivated and thus may enter vaccine stocks. Adventitious viruses found in vaccines have included avian leukosis virus, simian papovavirus SV40, and simian cytomegalovirus. The problem of adventitious contaminants has been circumvented through the use of pretested normal cells serially propagated in culture as substrates for cultivation of vaccine viruses. Vaccines prepared in such cultures have been in·use for years and have been safely administered to many millions of persons.

The storage constraints and limited shelf-life of live attenuated vaccines present problems, but these can be overcome by extraordinary efforts to maintain the "cold chain" of refrigeration even under difficult field conditions, and also by the use of viral stabilizers (e.g., molar $MgCl_2$ for polio-vaccine).

Until recent years, virus strains suitable for live-virus vaccines were developed chiefly by selecting naturally attenuated strains or by cultivating the virus serially in various hosts and cell cultures with the aim of deriving an attenuated strain. The selection of such strains is now being approached by laboratory manipulations aimed at specific genetic alterations in the virus. Newer vaccines and vaccine candidates (derived from deletion mutants, from reassortants, or from genetic recombinants) nonetheless must go through the same (1) demonstration of uniformity of manufacture, (2) demonstration of a safe and potent product by laboratory tests, and (3) demonstration of efficacy and safety in the field – the same requirements that have to be met in the development of conventional viral vaccines.

One fact cannot be overemphasized: An effective vaccine does not protect against disease until it is administered at the proper time and in the proper dosage and potency to the proper target population. With the introduction of a new vaccine, there often follows not only a marked decrease in incidence of the target disease, but also a significant change in its epidemiology, brought about by the new age distribution of susceptibles in the population. This phenomenon may result in new challenges for the control of the targeted disease [26].

Control of poliomyelitis

Great strides have been made toward controlling poliomyelitis since the introduction of the two poliovaccines – inactivated poliovirus vaccine (IPV) [18,23,41], which was licensed in the United States in 1954, and live attenuated oral poliovirus vaccine (OPV), licensed in 1961 [3,18,23,25,40]. Today a large majority of physicians and other health care workers in industrialized countries never see a patient with paralytic poliomyelitis. Unfortunately, this is far from the situation in many developing countries, particularly in tropical and subtropical climates, where thousands of children still become paralyzed victims, year in and year out.

It is rare for a serious disease to be controlled so quickly and dramatically as was poliomyelitis, especially in the developed countries. In 1955, a total of more than 76 000 cases of poliomyelitis were reported from 28 industrialized countries. By 1967, the number recorded in these same countries fell to 1013 cases – a reduction of 99%. Numbers of poliomyelitis cases have continued to fall; in most industrialized countries, poliomyelitis is now a rare disease.

In the United States, before the inactivated virus vaccine became available, there were 10 000 to 21 000 paralytic cases per year, and the annual rates were between 5 and 10 cases per 100 000 population. Nevertheless, this meant that significant numbers of cases were still occurring; in 1960 there were more than 2500 paralytic patients. And some of these cases were in fully vaccinated individuals: in a study of several thousand paralytic cases, 17% were in children who had received three injections of the inactivated vaccine. Some of the disappointing results were due to potency problems which have since been corrected, particularly in the few small countries that have used inactivated virus vaccines solely through most of the vaccine era.

After live attenuated vaccine was introduced and came into widespread use in the United States, the numbers of cases were reduced precipitously. The annual number of cases is now less than 10 per year. This translates into case rates as low as 0.001 per 100 000 population.

For many years, the United States has relied almost completely on live poliovirus vaccine. There is no longer any endogenous reservoir of wild polioviruses within the country, and a true break in the chain of infection has been achieved. Wild strains may continue to be introduced, but such imported cases are sporadic and almost never result in secondary cases. The use of live poliovirus vaccine has achieved this result by establishing widespread intestinal resistance to the wild virus, thus reducing the pool of susceptible individuals to a level below that required for perpetuation of the virus in nature [50]. Many other countries with extensive and continuing live vaccine programs are also no longer reporting cases.

In Sweden, Finland, and the Netherlands, where the inactivated polio-vaccines have been used almost exclusively, good results also have been obtained. These are countries with relatively small populations (a total of 28 million persons), culturally homogeneous and socially advanced. They have excellent public health services, and inactivated virus vaccine has been administered in intensive and regularly maintained immunization programs, achieving vaccine coverage and boosting among children that has approached 100%.

But there can be important deficiencies in protection, even in well-vaccinated countries. This vulnerability was demonstrated in 1978 and 1979 by outbreaks of poliomyelitis among members of closely-knit, interconnected religious groups who refused vaccine on religious grounds. Imported into the Netherlands from the Middle East, a virulent type 1 strain spread to related religious groups in Canada and the United States. In none of the countries involved did paralytic cases occur beyond the unvaccinated members of these interconnected religious communities. However, subclinical infections with the epidemic virus did occur in significant numbers of children who had been vaccinated with IPV [42]. In nursery and primary schools in some of the affected Netherlands communities, 71% of the unvaccinated children excreted the poliovirus, and 24% of those vaccinated with the enhanced IPV were also observed to be excreting the epidemic virus.

In Finland, where the use of standard IPV had produced 20 years of freedom from poliomyelitis, 10 cases occurred between August 1984 and January 1985. On the basis of virus isolations from healthy individuals and from sewage, it was estimated that at least 100 000 persons in the general population were infected. The 1984 epidemic strain was a wild type 3 variant that differed in both immunological and molecular properties from the type 3 vaccine strains [18]. Analysis of sera collected before the outbreak indicated that neutralizing antibodies against new type 3 isolates were much less prevalent in the population than antibodies to the type 3 strain (Saukett) used in the killed vaccine.

The precipitating factor in the outbreak was judged to be the appearance of a wild strain of poliovirus type 3 that was sufficiently aberrant to break through the type 3 immunity of the population. However, immunization with the new enhanced IPV or with standard OPV induced high titers of serum antibody against all known type 3 strains. Thus, although intratypic differences among strains from various parts of the world have indeed been demonstrated for many years, the major neutralizing antigen of each poliovirus serotype has proved to be remarkably stable.

The study of the 1988 outbreak in Israel also yielded pertinent information [43]. The cases were often in contacts of IPV-vaccinated children who themselves were protected but who participated in spreading the wild virus through the community. Most cases occurred in young adults who had been given OPV during the first year of life but who had not received any booster doses. This suggested an age-related deficit in immunity against the 1988 wild virus. Neutralizing antibody assays on sera collected from healthy persons under 30 years of age in Israel prior to the outbreak indicated high antibody levels against the Sabin strains in OPV, but very low levels against the wild virus. A booster dose of OPV given in 1988 brought the antibodies to high levels against both the Sabin and the wild strains [13]. In their pre-epidemic sera, nonvaccinated adults over 30 years of age – who had lived through the prevaccine period and who had been exposed to wild viruses as children – were found to have high levels of antibody both to the Sabin strains and to the wild virus [13]. After receiving a single dose of OPV in 1988 they also responded with an increase in antibody titer. These findings indicate that a gap in immunity against wild polio may occur in persons who are vaccinated with OPV in the first months of life and who are not given booster vaccine or who are not exposed to wild virus in the early years after vaccination when antibody levels are highest. This gap in immunity against wild poliovirus strains can be overcome by a booster dose of OPV later in life.

While cases have almost vanished from the industrialized world, polio continues to be an urgent problem, particularly in developing nations in tropical and subtropical zones. In India alone, more than 100 000 cases occurred annually throughout the 1980s. Of these cases, 98% were in children under 5 years of age; 42% of the patients were infants in the first year of life. As has been shown by surveys of lameness indicative of past paralytic poliomyelitis, in many developing countries the reported cases may represent no more than 10% of the actual number of cases. However, with a stimulus provided by WHO, there has been an increase in vaccine coverage; as of April, 1992, 84% of the world's children had received 3 doses of OPV. This has resulted in a decrease in cases, globally, but as of this writing, about 130 000 cases of paralytic poliomyelitis still occur annually [49].

The poliovirus vaccines

Two good vaccines against poliomyelitis are available. Each has its advantages and disadvantages [25,40,41].

Inactivated poliovirus vaccine (IPV), if properly prepared and administered, can confer humoral immunity if sufficient doses are given – particularly if the new vaccine of enhanced potency (eIPV) is used. IPV can be incorporated into a regular pediatric immunization schedule (along with other injectable vaccines such as diphtheria–pertussis–tetanus). In certain tropical areas where live vaccine has failed to "take" in some young infants, IPV has proved useful. Because living virus is not present, the use of IPV excludes the possibility that the progeny of the vaccine virus can revert toward virulence. Also for this reason it can be given safely to immunodeficinent or immunosuppressed individuals and to the household contacts of these immuno-compromised persons.

Among the disadvantages of IPV as originally prepared was its low potency. The need for repeated booster injections added to higher costs, and in developing countries also presented logistic problems of injecting several doses of vaccine at intervals into large numbers of infants and pre-school children. The immunity conferred by IPV impedes pharyngeal and fecal shedding of virus to some extent, but it does not provide a high degree of intestinal resistance. When exposed to wild poliovirus, children even when vaccinated with the new enhanced IPV (eIPV) become infected and excrete the wild virus. Thus they become a source of infection to others. Even in countries with highly developed health care systems and populations very well-vaccinated with IPV or eIPV, imported wild virus can circulate not only in unvaccinated sectors of the population but also through those who have been vaccinated.

Live oral trivalent poliovirus vaccine (OPV) has been widely used because of its ease of administration, its ability to induce not only serum anti-bodies but also intestinal resistance, and its rapid induction of an enduring immunity, similar to that induced by the natural infection. OPV rapidly infects and colonizes the alimentary tract of susceptibles, thus blocking spread of wild virus. Lower cost is an important factor for many countries; not only in the vaccine itself less expensive, but also its administration does not require use of highly trained personnel, thus further reducing program costs, and continuing boosters have not been required.

All living creatures undergo some degree of mutation, and polioviruses are no exception. The mutations that occur during replication of OPV have produced, in very rare instances, virus progeny with neurovirulence sufficiently increased to cause paralysis in vaccine recipients or their susceptible contacts. The proven risks of paralytic polio associated with OPV are exceedingly small. Three sequential 5-year studies of polio cases have been conducted collaboratively among 12 to 15 countries, under the auspices of WHO. In these and other studies, live poliovirus vaccine has been judged repeatedly to be an extraordinarily safe vaccine, with less than one reported case for every million babies vaccinated. These estimates were based on the maximum risk (the risk calculated as though every "vaccine-associated" case – i.e., a case occurring within 60 days of administration of vaccine – were indeed vaccine-caused). In the United States, the overall frequency of vaccine-associated poliomyelitis has been one case per 2.6 million vaccine doses distributed. However, the associated frequency of paralysis was one case per 520 000 first doses, versus one case per 12.3 million subsequent doses. These rates include recipients with immune deficiencies [32].

In some tropical countries live vaccines have not induced antibody production in a satisfactorily high percentage of vaccinees. This lower rate of vaccine "take" has been ascribed to various factors such as interference from other enteroviruses already present in the intestinal tract, the presence of antibody in breast milk, the presence of cellular resistance in the intestinal tract due to previous exposure to wild polioviruses (or perhaps to related viruses), and the presence of an inhibitor in the saliva (alimentary tract) of infants that blocks multiplication of the vaccine virus. Some feel that this low response, especially for type 3, can be overcome by proper and repeated use of live vaccine. In addition, a significant improvement in the rate of seroconversion to type 3 has been obtained in Brazil with a trivalent OPV containing a twofold increase in concentration of the type 3 vaccine strain [34].

Immunization schedules

Immunization at the time of birth

The problems of controlling paralytic poliomyelitis by use of vaccine are often exacerbated by the fact that in some areas the newborn infant has only a short time in which to acquire protective immunity because exposure to wild viruses comes so early, in their first months of life [21]. Various plans have been proposed [22,27,38,40] and some are now being pursued in different developing and tropical countries.

The Global Advisory Group of WHO's Expanded Programme on Immunization (EPI) has concluded that immunization of the newborn with OPV vaccine is a safe and effective means of improving protection and that OPV may be administered simultaneously with BCG vaccine. Although the serological response to OPV in the first week is less than that observed after immunization of older infants, 70% of more of neonates benefit by developing local immunity in the intestinal tract. In addition, 30–50% of the infants develop serum antibodies to one or more poliovirus types. Many of the remaining infants have been immunologically primed and they respond promptly to additional doses later in life.

The EPI Advisory Group emphasizes that for those infants in many countries whose only encounter with preventive services is at the time of birth, this single dose of vaccine will offer some protection, and also they will be less likely to be a source of transmission of wild polioviruses during infancy and childhood. For the infants who receive only one or two additional doses of poliovirus vaccine, the initial dose at birth will help ensure higher levels of immunity. The EPI schedule designed to provide protection at the earliest possible age is: at birth, and then at 6, 10, and 14 weeks of age. If the vaccine is not given at these ages, it should be given as soon as possible afterwards. Intervals between doses greater than those listed do not require restarting the series.

Incorporation of poliovirus vaccine into routine immunization schedules

OPV has been included in the regular childhood immunization programs without necessarily being preceded by mass vaccination campaigns. This has led to a gradual increase in vaccine coverage and an associated gradual decrease in poliomyelitis incidence. In a number of countries using this approach, reported polio cases have significantly declined since the early 1970s [49,50].

An example of recent progress can be seen in special programs conducted in three areas in tropical Africa that in the mid- to late 70s were experiencing incidence rates (as estimated from lameness surveys [33]) ranging from 25 to 62 per 100 000 population. The programs emphasized intensive and carefully targeted schedules of routine vaccination. Programs were instituted in Yaounde, Cameroon, in The Gambia, and in Abidjan, Ivory Coast [11]. They included administration of three doses of OPV during the first year of life, one month apart starting at 2 to 3 months of age. The vaccine coverage achieved was evaluated, and surveillance of poliomyelitis was conducted both before and after the vaccination program [14, 15]. Within a few years the proportion of the children who received 3 doses of vaccine – which had been virtually zero except for low coverage in Abidjan – was increased up to 50–70%. Even with this less than-optimal vaccine coverage, the incidence of poliomyelitis decreased significantly. For example, in Yaounde by 1980, 50% of the children 12–13 months old had received 3 doses of vaccine, and from a pre-immunization average of 62 cases per 100 000 in 1974 and in 1995, the incidence of polio decreased 88%, to 7.5 per 100 000 in 1981. In a later report from Yaounde [16], among children 12–23 months of age only 35% had received 3 doses of OPV, but the incidence of paralytic polio decreased by 85%. Maintaining a routine immunization schedule that reaches virtually all of the target population requires the maintenance of a supply of viable vaccine constantly at the point of contact. This can be difficult in warm climates with limited cold-chain facilities, particularly if the vaccine has not been treated with a stabilizer equal or superior to molar $MgCl_2$ [10,28,30,35].

Mass vaccination campaigns

Sabin [40] advocate that paralytic poliomyelitis in tropical countries might best be eliminated by repeated mass administration of OPV. Annual campaigns should include all age groups in which cases are occurring. All children from birth to 3, 4, or 5 years of age (depending on the epidemiological situation in a given country) should be given two doses of OPV twice each year regardless of how many doses of live vaccine they may have received previously. This mass administration would reach those missed in previous campaigns and would avoid the problems of record-keeping which often has been a barrier to full coverage in administering the vaccine.

Mass campaigns with OPV have been conducted successfully in a number of countries, not only bypassing the need for constantly maintaining refrigeration – as well as other logistic problems – but also providing a high level of coverage. The introduction of massive quantities of live vaccine virus colonizes the alimentary tracts of susceptible hosts, thus blocking wild virus circulation. In Cuba since 1962, all children from birth to three or five years (depending on concurrent serosurvey findings) regardless of individual immunization history have been given OPV on two Sundays, two months apart, each year [40]. Serosurveys show almost universal immunity by the age of three, and since the end of the first campaign year, more than a quarter of a century ago, only six cases have been reported. Czechoslovakia, using the same strategy, has had similar success, with the total disappearance of paralytic poliomyelitis [40]. High levels of public cooperation, discipline, and dedication seem to be vital factors in these successes – levels that may not necessarily be present in all situations.

This strategy of periodic mass campaigns, sometimes alone, sometimes combined with a regularly maintainted schedule for immunizing infants, has been highly successful in some areas. Brazil in 1980 instituted such a program, entailing a huge national effort in which, regardless of any vaccine history, each child in the selected age groups is given a dose of live vaccine twice each year, with a two-month interval between doses. Some 90,000 vaccination stations (10 times the number of regular health stations) were established and 400 000 volunteers were mobilized. From several thousand cases annually in the late 1970s, paralytic cases of poliomyelitis in Brazil have been reduced dramatically, to an annual average of 80 during 1981–1984 [37]. However, there was an increase to 461 cases in 1985 and to 612 in 1986. This seems to have been a temporary setback due to low potency of the type 3 component of the vaccine. This problem is being rectified by increasing the ratio of type 3 to the other types [34].

Such a program poses important questions: Would developing·nations be able and willing to make such an annual commitment of two days for each of two doses, with a two-month interval between them? Would these nations have the resources to conduct such an annual repetitive mass vaccination program? There also had been some misgivings as to whether the absence of record-keeping could be detrimental to the program and whether concentration on OPV could detract from other needed health services. In practice, in Brazil and elsewhere, the annual polio program has led to an increased awareness and activity for other immunizations.

Mexico likewise has had a long struggle to control polio through routine vaccinations, and still had an annual average of almost 70 cases during 1976–1980. In 1981, mass national immunization campaigns were instituted. Monovalent type 1 vaccine – the type that has long predominated in their paralytic cases – was used for the first dose, followed by trivalent vaccine two months later. With the higher vaccination coverage, polio has been drastically reduced, to 186 cases in 1981, 80 cases in 1987, 25 in 1989, and almost to extinction by 1990 [8].

The effectiveness of OPV has been particularly striking in the Western Hemisphere as a result of the Expanded Programme on Immunization and the continuing surveillance for specific virus and antibody. Not only has the number of paralytic cases caused by wild polioviruses decreased, but also there has been a precipitous decrease in the circulation of wild polioviruses. Only 11 of the 2456 cases of acute flaccid paralysis (AFP) reported in 1990 have been confirmed as polio. In addition, 57 of the AFP cases were regarded as compatible, 477 were still under investigation, and 1,911 were categorized as nonpolio [8]. In the few areas in the Americas where cases due to wild polioviruses occur, the EPI activities are focussing on mopping-up operations, consisting of intensive use of OPV in these localities. In 1991, only 2 cases of paralytic polio were identified in the Americas (both of them in Colombia), and none have been reported in 1992. Circulation of wild poliovirus no longer occurs in the Western Hemisphere, and it is anticipated that the entire Region will soon be declared free of all indigenously transmitted paralytic polio-myelitis [8].

In India, despite many efforts at controlling poliomyelitis, thousands of cases still occur, the highest incidence being in infants 6 to 18 months old [6]. It is estimated that in the 1980s more than 100 000 cases were occurring annually. To increase the level of immunity, a "pulsed" mass vaccination program has been proposed [19], i.e., OPV is given to all the target children in each community at the same time, rather than following individual routine immunization schedules. This proposal recommends three doses of OPV before the first birthday, and three more between the first and second birthdays; a village would be visited three times in the course of the year at intervals of 4 to 6 weeks, and then no more vaccine would be given over a nine-month period.

Insurance against poliomyelitis through vaccination

An innovative approach to obtain better delivery of vaccine – not only of OPV but also of other pediatric immunizations – has proven productive in one area of China (Gaoyi County, Shijiazhuang Prefecture, Hebei Province) – a county with a total population of 140000 in 107 villages). To increase family cooperation in immunization programs and to encourage health staff members to enhance their preventive health efforts, the new plan, introduced in 1984, was expanded by early 1986 to include more than 12,000 children (62% of children under 7 years of age in the county) [48].

The average family income in Gaoyi is 438 yuan (about $220). Parents pay 10 yuan to the health center in the first year of their child's life, with a contract that "guarantees" that their child will not contract measles, pertussia, poliomyelitis, tetanus, or diphtheria if he or she receives the full recommended course of immunization. If the insured child suffers from measles or whooping cough, the parents recive an indemnity of 50 yuan; if poliomyelitis, diphtheria or tetanus, 300 yuan.

The insurance premiums received (nearly 100000 yuan as of 1985) are distributed among the village doctor, county health centers, and the county epidemic station, which uses them for immunization equipment and education, reserving funds to compensate parents when an enlisted child contracts any of the designated diseases. In turn, the village doctor and the county agencies share the costs of compensation. Up to the end of January, 1986, a total of only 750 yuan had been paid in compensation, covering 15 cases of measles or whooping cough. There were no cases of poliomyelitis.

The advantages of this system have been discribed as follows [48]:

"1. It stimulates staff of health centres and village doctors to implement immunization activities.
2. Parents are enthusiastic to have their child immunized because they feel that they have paid a lot of money.
3. The primary health care setting has been strengthened because the staff can earn one-quarter to one-half more than their regular income.
4. Underreporting of cases has been reduced because parents are motivated to report a suspected case in order to receive compensation.
5. Wastage of vaccine has been minimized because the activities have been carried out strictly according to the schedule due to the enthusiasm of staff and parents."

Combined schedules using both live and killed poliovirus vaccines

In some localities, it has been found advantageous to establish immunization schedules using both killed and live poliovirus vaccines. One type of situation in which this strategy has been successfully carried out has been the relatively recent addition of killed poliovaccine to supplement live poliovaccine immunization schedules in some high-risk localities where there is very early and repeated exposure of infants to challenge by importations of wild virulent viruses [12,21,24]. In such high-risk areas, IPV alone has been inadequate, as in Israel in 1988 [43]. The lessons of the 1988 outbreak in Israel are (i) that IPV alone does not interrupt the circulation of wild virus, which singles out susceptible contacts as targets; (ii) that OPV alone, administered in infancy, is not completely effective for life. A combined schedule [22,45] offers substantial benefit, with the optimal times of vaccination yet to be determined for different regions. For ease of administration, giving parenteral IPV and oral OPV simultaneously offers advantages. This procedure has been used regularly starting in 1978, in the West Bank and Gaza where some cases had been occurring, particularly in infants, despite extensive campaigns of immunization with OPV. The combined schedule included administration of OPV (type 1 monovalent) during the first month of life; then at $2\frac{1}{2}$ months, and again at 4 months, trivalent OPV is given together with a quadruple vaccine consisting of DTP plus IPV; trivalent OPV is given at $5\frac{1}{2}$ months, and again at 12 months. The rationale for the combined schedule is that, under conditions of regular and heavy importation of virus resulting in frequent challenge from virulent wild polioviruses early in infancy, features of both types of vaccine are needed. OPV acts by inducing protective immunity both in the form of circulating humoral antibodies and in the form of intestinal immunity. Furthermore, the immunity that ensues is long-lasting, like that which follows the natural infection. IPV, on the other hand, provides an immediate immunogenic stimulus that is not subject to the interfering or inhibiting factors described above, that may prevent live vaccine "takes" in some young infants. By administering both vaccines in a

Table 1. A combined vaccination schedule proposed for EPI

Age	Vaccine Doses*
At birth	BCG/OPV
6 weeks	DPT1/OPV1
10 weeks	DPT2·IPV1/OPV2
14 weeks	DPT3·IPV2/OPV3
6–12 months	Measles

* The first, second, and third doses are indicated by the numbers. A nonscheduled dose is also recommended at birth when the infant is readily available. After the circulation of wild poliovirus in the community is sharply reduced by the above schedule, progress toward eradication can be achieved by use of OPV alone

combined schedule, immediate protection can be provided in the critical first weeks or months of life, and long-lasting protection – both humoral and intestinal – also is provided. A combined immunization schedule compatible with that of EPI is shown in Table 1.

Studies of the combined vaccine program described above indicate: (i) Protection provided by OPV alone was about 90 percent effective, that is, the case rate in those who received OPV alone was one-tenth the rate in those who were not vaccinated or who were incompletely vaccinated. Similar protection rates were subsequently recorded for OPV in Oman [44] and in The Gambia [7]. (ii) The children in the Middle East who were fully vaccinated with OPV and had received IPV in addition were even more effectively protected; virtually 100 percent protection was seen in those who received both vaccines. Serum surveys in children aged 9 to 36 months revealed high antibody levels, substantiating their protection [22,45]. Another type of combined program has been the addition of live poliovirus vaccine in 1968 to supplement inactivated poliovirus vaccine (IPV) in Denmark where, despite extensive coverage with IPV alone, epidemics still occurred, and antibody prevalence was less than desired [46]. The combined IPV/OPV program has resulted in the elimination of poliomyelitis from Denmark, as well as from Israel, Gaza, and the West Bank. In earlier years the incidence of polio in the latter region had been one of the highest in the world. Today paralytic polio continues to be reported in the neighboring countries, but the last reported case in the Israel, Gaza, West Bank region occurred in 1988.

OPV is now in use in most countries, and wherever it has been used, polio cases have decreased, but persist to different degrees in different areas. From the results to date, public health workers are now able to take advantage of the assets of both IPV and OPV to quickly bring the disease under total control. The schedule shown in Table 1 is proposed as one which would be compatible with current EPI recommendations. This schedule should (i) introduce IPV into the current OPV immunization program so that in the first year of life both *IPV* and *OPV* are given to all infants, and (ii) in subsequent years, continue with the current EPI program which uses OPV alone for polio immunization.

For countries adopting mass campaigns, there would also be benefits from incorporating both IPV and OPV into the program for the first year. For subsequent years, the program with OPV alone would be sufficient to prevent wild poliovirus from colonizing the vaccinated population.

Continuing basic research on poliovirus

Related research on poliovirus is leading to increased knowledge of the structure of the agent [17] and its molecular biology [29,41]. It is noteworthy that poliovirus continues to serve as an endless frontier for scientists engaged in basic research.

One important outcome is the detection of an increase in neurovirulence of the vaccine virus propagated either in cell culture or in the human gut by monitoring a change in the nucleotide at position 472 from uridine (U), found in the genome of the type 3 vaccine strain, to cytosine (C), found in type 3 wild-type strains [2]. This finding may lead to the development of methods whereby the burdensome monkey neuro-virulence test required in OPV manufacture would be replaced by a simple biochemical test [4].

A key determinant of poliovirus infection is the cell receptor, upon which the restricted tropism of the virus depends. Molecular clones of the poliovirus receptor have been isolated, and the encoded protein identified as a new member of the immunoglobulin family [36]. Modification of the receptor may lead to control of disease by a new avenue.

Transgenic mice have been developed in which the human gene encoding cellular receptors for poliovirus have been introduced into the mouse genome [20, 36]. The new mice have proven to be susceptible to all three poliovirus types, and are being investigated as models for testing OPV lots for neurovirulence, which currently require the monkey test.

The diagnosis of infections by poliovirus and other enteroviruses is beginning to shift from the cell culture laboratory to the biochemical laboratory. Nucleic acid hybridization with specific probes offers quick and reproducible methods for detecting these viruses [5,39]. Molecular biology is providing new insights into pathogenesis [9,31].

Striking advances have been achieved recently in understanding the nature of poliovirus, leading to the preparation of modified live vaccine candidates of potentially greater genetic stability. However, it will be difficult to test such new vaccine candidates in healthy children under natural conditions, as it will be necessary to prove that the vaccines produce fewer than one vaccine-associated case per million susceptible recipients. The global application of the currently licensed vaccines is fast achieving an interruption of the circulation of wild poliovirus, closing the window during which any newly developed vaccine strains can be properly field-tested.

References

1. Brown F (ed) (1990) Modern approaches to vaccines. In: Seminars in Virology, 1:1–74
2. Burke KL, Almond JW, Evans DJ (1991) Antigen chimeras of poliovirus. Prog Med Virol 38:56–68
3. Centers for Disease Control (1990) Progress toward eradicating poliomyelitis from the Americas. MMWR 39:557–561
4. Chumakov KM, Powers LB, Noonan KE, Robinson IB, Levenbook IS (1991) Correlation between amount of virus with altered nucleotide sequence and the monkey test for acceptability of oral poliovirus vaccine. Proc Natl Acad Sci USA 88:199–203
5. da Silva EE, Pallansch MA, Holloway BP, Oliveira MJC, Schatzmayr HG, Kew OM (1991) Oligonucleotide probes for the specific detection of the wild poliovirus types 1 and 3 endemic to Brazil. Pediatr Inf Dis J 19:222–229
6. Dave KH (1986) Report of the Enterovirus Research Centre, Bombay
7. Deming MS, Jaiteh KO, Otten MW, Jr, Flagg EW, Jallow M, Cham M, Brogan D, N'jie H (1992) Epidemic poliomyelitis in The Gambia following the control of poliomyelitis as an endemic disease. II. Clinical efficacy of trivalent oral polio vaccine. Am J Epidemiol 135:393–408
8. de Quadros C, Andrus JL, Olive J-M, da Silveira M, Eikhof RM, Carrasco P, Fitzsimmons JW, Pinheiro FP (1991) Eradication of poliomyelitis: progress in the Americas. Pediatr Inf Dis J 19:222–229
9. Eggers HJ (1990) Notes on the pathogenesis of enterovirus infections. Med Microbiol Immunol 179:297–306
10. Finter NB, Ferris R, Kelly A, Prydie J (1978) Effects of adverse storage on live virus vaccines. Vaccinations in the Developing Countries. In: Developments in Biological Standardization, vol 41. Karger, Basel, pp 61–66
11. Foster SO, Kesseng-Maben G, N'jie H, Coffi E (1984) Control of poliomyelitis in Africa. Rev Infect Dis 6 [Suppl 2]:433–437
12. Goldblum N, Swartz T, Gerichter CB, Handsher R, Lasch EE, Melnick JL (1984) The natural history of poliomyelitis in Israel, 1949–1982. Prog Med Virol 29:115–123
13. Green MS, Handsher R, Cohen D, Melnick JL, Slepon RS, Mendelsohn E, Danon YL (1993) Age differences in immunity against wild and vaccine strains of poliovirus prior to the 1988 outbreak in Israel: Evidence supporting the need for a booster immunization in adolescents. Vaccine, in press
14. Henderson RH, Sundaresan T (1982) Cluster sampling to assess immunization coverage: a review of experience with a simplified sampling methodology. Bull WHO 80:253–260
15. Heymann DL, Floyd VD Luchnevski M, Kesseng-Maben G, Mvongo F (1983) Estimation of incidence of poliomyelitis by three survey methods in different regions of the United Republic of Cameroon. Bull WHO 61:501–507
16. Heymann DL, Murphy K, Brigaud M, Aymard M, Tembon A, Maben GK (1987) Oral poliovirus vaccine in tropical Africa. Bull WHO 65:495–501
17. Hogle MJ, Chow M, Filman JD (1985) Three-dimensional structure of poliovirus at 2.9 A resolution. Science 229:1358–1365
18. Hovi T (1991) Remaining problems before eradication of poliomyelitis can be accomplished. Prog Med Virol 38:69–95
19. John TJ (1984) Poliomyelitis in India: Prospects and problems of control. Rev Infect Dis 6 [Suppl 2]:438–441
20. Koike S, Taya C, Kurta T, Abe S, Ise I, Yonekawa H, Nombto A (1991) Transgenic mice susceptible to poliovirus. Proc Natl Acad Sci USA 88:951–955
21. Lasch EE, Abed Y, Abdulla K, El Tibbi AG, Marcus O., El Massri M, Handscher R, Gerichter CB, Melnick JL (1984) Successful results of a program combining live and inactivated poliovirus vaccines to control poliomyelitis in Gaza. Rev Infect Dis 6 [Suppl 2]:467–470

22. Lasch EE, Abed Y, Marcus O, Gerichter CB, Melnick JL (1986) Combined live and inactivated poliovirus vaccine to control poliomyelitis in a developing country – five years after. Development in Biological Standardization 63: 137–143
23. Lemon SM, Robertson SE (1991) Global eradication of poliomyelitis: Recent progress, future prospects, and new research priorities. Prog Med Virol 38: 42–55
24. Melnick JL (1981) Combined use of live and killed vaccines to control poliomyelitis in tropical areas. In: Hennessen W, van Wezel AL (eds) Reassessment of Inactivated Poliomyelitis Vaccine. In: Developments in Biological Standardization. Karger, Basel, 47: 265–273
25. Melnick JL (1988) Live attenuated poliovaccines. In: Plotkin SA, Mortimer EA (eds) Vaccines. Saunders, Philadelphia, pp 115–157
26. Melnick JL (1990) Conventional viral vaccines and their influence on the epidemiology of disease. In: Dimmock NJ, Griffiths PDF, Madeley CR (eds) Control of Virus Diseases. Cambridge University Press, pp 3–19
27. Melnick JL (1992) Poliomyelitis: Eradication in sight. Epidemiol Infect 108: 1–18
28. Melnick JL, Ashkenazi Al Midulla VC, Wallis C, Bernstein A (1963) Immunogenic potency of $MgCl_2$-stabilized oral poliovaccine. J Am Med Ass 185: 406–408
29. Minor PD, Ferguson M, Evans DMA, Almond JW, Icenogle JP (1986). Antigenic structure of polioviruses of serotypes 1, 2, and 3. J Gen Virol 67: 1283–1291
30. Mirchamsy H, Shafyi A, Mahinpour M, Nazari P (1978) Stabilizing effect of magnesium chloride and sucrose on Sabin live polio vaccine. In: Vaccinations in the Developing Countries: Developments in Biological Standardization, vol 41. Karger, Basel, pp 255–257
31. Morrison LA, Fields BN (1991) Parallel mechanisms in neuropathogenesis of enteric virus infections. J Virol 65: 2767–2772
32. Nkowane BM, Wassilak SGF, Orenstein WA, Bart KJ, Schonberger LB, Hinman AR (1987) Vaccine-associated paralytic poliomyelitis. United States: 1973 through 1984. J Am Med Ass 257: 1335–1340
33. Ofosu-Amaah S, Kratzer JH, Nicholas DD (1977) Is poliomyelitis a serious problem in developing countries? Lameness in Ghanaian schools. Br Med J i: 1012–1014
34. Patriarca PA, Palmeira G, Filho JL, Cordeiro MT, Laender F, Couto Oliveira MJ, deSouza Dantes MC, Risi JB Jr, Orenstein WA (1988) Randomised trial of alternative formulations of oral poliovaccine in Brazil. Lancet i: 429–433
35. Peetermans PA, Colinet G, Stephenne J (1976) Activity of attenuated poliomyelitis and measles vaccines exposed at different temperatures. In: Proc. Symposium on Stability and Effectiveness of measles, Poliomyelitis, and Pertussis Vaccines. Zagreb, Yugoslav Academy of Sciences and Arts, pp 61–66
36. Ren R, Costantini F, Gorgacz EJ, Lee JJ, Racaniello VR (1990) transgenic mice expressing a human poliovirus receptor: A new model for poliomyelitis. Cell 63: 353–362
37. Risi JB Jr. (1984) The control of poliomyelitis in Brazil. Rev Inf Dis 6 [Suppl 2]: 400–403
38. Robinson DA (1982) Polio vaccination – a review of strategies. Trans Roy Soc Trop Med & Hyg 76: 575–581
39. Rotbart HA (1991) New methods of rapid enteroviral diagnosis. Prog Med Virol 38: 96–108
40. Sabin AB (1985) Oral poliovirus vaccine: History of its development and use and current challenge to eliminate poliomyelitis from the world. J Infect Dis 151: 575–581
41. Salk J, Drucker J (1988) Noninfectious poliovirus vaccine. In: Plotkin SA, Mortimer EA (eds) Vaccines. Saunders, Philadelphia, pp 158–181
42. Schaap GJP, Bijkerk H, Coutinho RA, Kapsenberg JG, van Wezel AL (1984) The spread of wild poliovirus in the well-vaccinated Netherlands in connection with the 1978 epidemic. Prog Med Virol 29: 124–149
43. Slater PE, Orenstein WA, Morag A, Avni A, Handsher R, Green S, Costin C, Yarrow A, Rishpon S, Havkin O, Ben-Zvi T, Kew OM, Rey M, Epstein I, Swartz TA, Melnick JL (1990) Poliomyelitis outbreak in Israel in 1988: A report with two commentaries. Lancet 335: 1192–1198

44. Sutter RW, Patriarca PS, Brogan S, et al (1991) An outbreak of poliomyelitis in Oman: evidence for widespread transmission among fully vaccinated children. Lancet 338: 1053–1055
45. Tulchinsky T, Abed Y, Shaheen S, Toubassi N, Sever Y, Schoenbaum M, Handsher R (1989) A ten-year experience in control of poliomyelitis through a combination of live and killed vaccines in two developing areas. Am J Public Health 79: 1648–1652
46. von Magnus H, Petersen I (1984) Vaccination with inactivated poliovirus vaccine and oral poliovirus vaccine in Denmark. Rev Infect Dis 6 [Suppl 2]: S471–S474
47. Wimmer E, Emini EA, Diamond DC (1986) Mapping neutralization domains of viruses. In: Notkins AL, Oldstone MBA (eds) Concepts in Clinical Pathogenesis, vol 2. Springer, Berlin Heidelberg New York Tokyo, pp 159–173
48. World Health Organization Expanded Programme on Immunization (1987) Contract system tested. Weekly Epidemiol Rec 62: 142–143
49. World Health Organization, Expanded Programme on Immunization (1987–1992) Reports and notices on poliomyelitis cases, vaccine trials, etc. Wkly Epidemiol Rec 1988–1992
50. World Health Organization Expanded Programme on Immunization (1992) Information System, April, 1992

15

Eradication of poliomyelitis: countdown or slowdown

B. D. Schoub

National Institute for Virology and Department of Virology, University of the Witwatersrand, Johannesburg, South Africa

Summary

Vast obstacles remain in the path to the global eradication of poliomyelitis, particularly on the African continent. These include huge deficiencies in primary health care aggravated by the competition with other daunting health and social priorities, political difficulties and ignorance. Primary health care remains the cornerstone of the drive to eradicate polio and needs to be strengthened by greater community involvement, increased mobility and therefore reach, and increasing access to newborns. However, polio eradication cannot be achieved without mass immunization campaigns additional to primary health care programmes. The immense cost savings of mass immunization and disease eradication need to be strongly emphasized.

Introduction

Less than a decade remains before reaching the World Health Organization (WHO) target date for the eradication of poliomyelitis by the end of this century. Major obstacles remain before this goal is achieved and endemic polio with periodic epidemics remains a feature of many Third World countries in Asia and especially Africa. The comprehensive costs of the eradication campaign, estimated at $ 155 million from 1989 to 2000, additional to routine EPI operations, raises immediate questions of medical philosophy. With competing health priorities of gigantic proportions in the Third World, especially the latter-day medical disaster of AIDS, the justification for a poliomyelitis eradication programme needs to be clearly defined.

In this paper I would like to deal with five issues regarding the campaign to eradicate poliomyelitis from the globe:

1. Impetus for the development of a global poliomyelitis eradication campaign.
2. Poliovirus vaccines and the control of poliomyelitis by immunization.
3. Status of poliomyelitis: 1992.

4. Obstacles to the final push to eradication in 2000.
5. Approaches to the final push to eradication.

1. Impetus for the development of a global poliomyelitis eradication campaign

It is very unlikely that many people would take issue with the oft-quoted aphorism that smallpox represents the greatest triumph of 20th century public health and preventive medicine. A discussion of the elements of the smallpox eradication campaign is beyond the scope of this paper, although the details of that programme are essential knowledge to anyone involved in the poliomyelitis eradication campaign. In my opinion there were, in essence, three major benefits which resulted from the successful completion of that programme:

(i) *Cost benefit*
   The obvious one is the dramatic demonstration of the cost benefit of immunization.
(ii) *Eradicability of infectious diseases*
   The programme demonstrated that vaccine-preventable infectious diseases could be totally eradicated by immunization.
(iii) *Further programmes*
   The smallpox eradication campaign gave impetus to three further major public health initiatives:

   (a) The Expanded Programme on Immunization (EPI) and the Programme for Universal Childhood Immunization by the year 1990 (UC90)
   (b) The Alma Ata declaration of health for all by the year 2000 and the GOBI-FFF initiatives for primary health care
   (c) The Global Poliomyelitis Eradication Campaign

Before examining the polio eradication campaign I would like to very briefly review some of the pioneering work in polio control carried out in my part of the world.

2. Poliovirus vaccines and the control of poliomyelitis by immunization

The romance and nostalgia of the early pioneering days of the development of poliovirus vaccines as well as the implementation of universal immunization campaigns, still stir the emotions and provide inspiration to present-day virologists. My own Institute, the National Institute for Virology, is the successor to the Poliomyelitis Research Foundation which was established to research and to develop poliomyelitis vaccines in the early 1950s. Its first Director, Professor James Gear, was one of the trail-blazing giants in the world of poliomyelitis at that time. Together with his brother, Dr Harry Gear, they played a leading role in the founding of the WHO in the years immediately after World War 2. (Dr Harry Gear incidentally served as Chairman of the Executive Board of the WHO for its sixth and seventh sessions, following which he served as Assistant Director-General for seven years.) James Gear played a critical, perhaps somewhat under-recognized role, in the development of poliomyelitis vaccines. An inactivated vaccine was developed at the laboratories of the Poliomyelitis Research Foundation at Rietfontein in South Africa, together with, and at the same time, as that of Salk in the United States. Three months after the announcement of successful trials of vaccine in the United States inactivated vaccine was released in South Africa, and from the 19th of September 1955 thousands of children were immunized, making South Africa one of the first countries in the world to institute routine polio immunization. In 1960, soon after locally developed live attenuated vaccine, derived from strains obtained from Sabin, were released in South Africa, a quarter of a million doses of vaccine were used in Mauritius making South Africa, perhaps, the first country in the world to apply its live attenuated vaccine to an entire population.

Following worldwide routine immunization against polio the incidence of the disease plummeted and dramatic graphs depicting the striking fall-off of the disease became a feature of country after country. In the developed world in less than two decades, the once feared plague was reduced to fading pictures of a disease which the majority of medical students will probably never again encounter. In much of the developing world, however, inadequate immunization, inadequate storage and delivery of vaccine and interference have ensured that it still remains a significant cause of disability and a consumer of precious health-care resources.

3. Status of poliomyelitis: 1992

Over the last decade a number of milestones towards the eradication of poliomyelitis have been, or are soon to be, achieved:

(a) A number of Western countries have announced that wild-type poliomyelitis has been eliminated from within their borders. Indeed, publications from the Centers for Disease Control dating as far back as 1984 maintain that no wild-type poliovirus had been isolated in the USA since 1979 [1] and similar reports have also emanated from a number of European countries.

(b) The goal of the Pan American Health Organization (PAHO) to eliminate wild-type poliomyelitis from the Western Hemisphere by the year 1990 now appears to be imminent [2].

(c) The EPI goal of universal childhood immunization by the year 1990, which aimed to achieve a level of 80% immunization against the six EPI diseases, was reached in early October 1991 [3].

The attaining of these goals which, when they were set, appeared to be extravagantly ambitious and perhaps almost unreachable, represents major, highly impressive achievements by public health bodies and the WHO. Nevertheless, the fragility and fluidity of these milestones must be recognized. As long as wild-type virus circulates in human populations anywhere in the world, countries will remain vulnerable to individual infections and to outbreaks. Even developed countries with high levels of vaccine coverage have experienced outbreaks in ostensibly well immunized populations.

Global immunization coverage figures, as reassuring as they may sound initially, may be quite deceptive in reference to many of the poorer developing countries which fall substantially below the global averages. In addition, immunization coverage figures may well be calculated and based on centralized distribution information rather than on the basis of actual administration at the level of the child receiving its vaccine.

Other factors such as problems in transport and storage due to cold-chain inadequacies or failure of vaccine to "take" may also contribute to the deception of immunization figures.

Furthermore, immunization coverage or even seroprevalence data, which are the more objectively precise measures of community immunity, determine only one of the two factors responsible for the vulnerability of populations to polio outbreaks – i.e. they determine the level of community immunity. However, the viral load present or introduced into a community from its reservoir in the human gastrointestinal tract is not readily quantitated. Clearly, control programmes need to take cognisance of both facets in order to avert polio outbreaks.

Thus, immunization coverage or the publicized attainment of national or regional elimination of wild-type virus could, not infrequently, be overly optimistic. In 1992 we may well not have reached as far a level towards eradication as we may wish to think.

4. Obstacles to the final push to eradication in 2000

The poliomyelitis eradication campaign, having achieved elimination of the virus in the Americas as well as the developed world and a number of Asian countries, still needs to overcome its most formidable obstacles in the final push to eradication, particularly on the African continent. Clearly, vaccine failure in developing and especially tropical and subtropical countries, is a major issue to be addressed. Immunogenicity of poliovirus vaccine, in my opinion, and vaccination failure in developing countries are of, at least, equal importance and I would like to look at four of these in particular:

i. The drive by health authorities to establish a hierarchy of health priorities

The severe restraints on budgets for health-care in Third World countries and in particular on the African continent with the disastrous consequences of the AIDS epidemic, has had a marked effect in promoting the diversion of health-care resources away from other major causes of death and disability such as tuberculosis, nutritional programmes and others. Against these gigantic health problems very powerful and cogent motivation will be needed for a programme to eradicate a disease which is a relatively minor cause of mortality and disability.

ii. Cost/accessibility of primary health-care

The inaccessibility and unaffordability of health-care remains the lot of a large percentage of rural inhabitants of developing Africa. In many African villages a visit to a primary health-care clinic is a major drain on meagre family resources. Usually the male breadwinner is a migrant labourer living for most of the year in the cities. An African mother would need to obtain the services of a child-minder for the other siblings in order to take a child to a clinic for immunization. This may thus often involve considerable costs additional to those for transportation. The journey may well take a number of hours and not infrequently breakdowns may occur on the way, with mother and child perhaps reaching their distination after the clinic has closed for the day.

In most of rural Africa, primary health-care clinics and facilities are based on fixed stations – stand-alone clinics, church halls, schools, shops etc. Frequently, however, settlements are far even from these basic facilities with roads, when they do exist, difficult to traverse or even impassable in the rainy season.

iii. Political problems

The provision of primary health-care facilities by central authorities may give rise to suspicion and avoidance of utilizing them by communities unfavourably disposed to governmental authorities. Although this is usually directed particularly against family spacing and AIDS control programmes, politically inspired hostility to polio immunization programmes has also been encountered with rumours that oral polio vaccine was being used for sterilization purposes.

iv. Ignorance

Ignorance remains a major obstacle to the implementation of polio immunization programmes. It may be manifest at the level of clinic sisters with respect to storage conditions of vaccine or false contraindications to administration of vaccine, or erroneous deviations from routine schedules. In the community, suspicion directed against Western medicine still remains, although it is fortunately being rapidly overcome. Traditional healers who form such an intrinsic part of African society are now being recognized and involved as partners rather than as competitors or opponents, with increasingly gratifying results.

5. Approaches to the final push to eradication

The seven-point plan of action formulated by WHO in 1988 when the eradication programme was announced [4], was based, to a large extent, on the knowledge gained from the smallpox eradication experience. There is little doubt that this blueprint must still remain the cornerstone of the final eradication campaign. There are, however, a few supplementary remarks which I would like to make from the vantage of our experience on the southern tip of the African continent:

i. The importance of primary health care

The major key to the final success of the polio eradication campaign lies, in my opinion, in the penetration and efficacy of delivery of primary health-care services to the entire population. Since the 1978 Alma Ata conference, primary health care has become the buzzword of health departments and health agencies involved in Third World medicine. Unfortunately, as with all hackneyed terms, it may well also be in danger of losing its true meaning. Serious endeavours must be made, not merely to pay lip service to the term "primary health care", but to strengthen and formalize the structures for primary health care systems for each country. The general strengthening of primary health care services must not compete with or displace or be displaced by any specific primary health care initiatives. – AIDS prevention initiatives, tuberculosis control programmes or the poliomyelitis eradication programme. It is imperative to secure the essential community, social and political will, that all of these initiatives be accepted and appreciated as forming interdependent components of preventative medicine.

ii. Community involvements in primary health care

If primary health care is to acquire acceptability in the community and be a worthwhile facility for which mothers are prepared to make a not inconsiderable sacrifice in order to reach them, it is vital that the community be intimately involved in the establishment and administration of primary health care facilities. The warmth of feeling of possession of "our" facility will go an enormous way to encouraging community participation and patronising of primary health care clinics.

A suggested system for a primary health care programme in a developing country based on a federal structure may consist of the following:

A *regional primary health care committee* which could consist of sociological, social anthropological and demographic skills. The function of this body would be to delineate *peripheral, community-based primary health care committees* consisting of trained staff and, most importantly, participants from the community to be served. The regional primary health care committees would, themselves, form a major component of the Ministry of Health of that country.

iii. Mobility

Critical to the success of primary health care delivery is mobility. Stationary primary health care facilities based on stand-alone buildings or in borrowed structures such as churches, schools, shops and others, do form important sites for the dispensing of primary-health care to rural populations. However, for large sectors of the rural African community, and in some countries a major portion of the population, static sites are inaccessible or inadequately accessible for satisfactory delivery of basic health services. A number of mobile primary health care systems have been developed, varying from the nurse on a bicycle with a cold pack of vaccine to comprehensively equipped mobile clinics able to perform minor surgical procedures, radiology and laboratory diagnostic tests. These latter units can function as the nucleus of the mobile primary health care networks controlled by a regional primary health care committee which itself could be based on a fixed primary health care clinic.

iv. Access to newborns

Not only is access to newborns a major component of the delivery of a number of primary health care services but it is a particularly important stage of life when polio vaccine should be administered.

v. Mass immunization

One of the most important issues that still needs to be urgently addressed before eradication will become a global feasibility is the successful marketing of mass immunization for polio to the developing world, over and above routine immunization programmes. It will indeed be a daunting task to persuade health administrations in countries where resources for health are meagre and health needs are gigantic, to now start yet a new health initiative for a disease which is, at most, a relatively minor cause of mortality and morbidity. The concept of mass immunization is, however, one which is really directed to the Third World in particular. Poliomyelitis has been eliminated from the developed world by routine EPI immunization prog-rammes alone. However, in the developing world, the most cogent argument for mass immunization is simply that without it poliomyelitis cannot be eradicated. At best it can only be brought under temporary control. Observations of recent polio eqidemics in Oman [5] and the Gambia [6] and also in South Africa in the 1987/1988 KwaZulu/Natal epidemic [7], have graphically demonstrated how substantial outbreaks can still occur even in the face of immunization coverage of over 80% and high levels of immunity in the population. Clearly, given the high levels of faecal-oral transmission of enteric pathogens in developing countries and the mass of wild-type poliovirus which could build up in susceptible pockets, this could be sufficient to rapidly "ferret out" the remnant susceptibles in the population and cause substantive outbreaks. It would therefore appear that no matter how successful and extensive routine immunization may be, it may still be inadequate to provide sufficient herd immunity to prevent outbreaks occurring if wild-type virus is introduced. At best temporary control could be achievable with high immunization coverage but there would continue to be a vulnerability to outbreaks from imported wild-type virus.

It is really only through mass immunization, which can be carried out during the season of low transmission of enteric organisms, that interruption of transmission can be effected. Nonetheless, mass immunization does add significantly to routine costs and also requires considerable motivation and energy to initiate and to maintain regularly on an annual basis. Much preparation would need to be done as conditions in Latin America where mass immunization has been so successful, are markedly different in many respects to those in Africa.

Against the initial investment of mass immunization must be taken into account the future savings of dispensing with routine polio immunization in perpetuity, a cost which is currently estimated in the USA to be of the order of $114 million per year [8]. In addition to this must be counted the costs of treatment and rehabilitation of the 120 000 or so estimated cases of polio still occurring worldwide per year, as well as the rare, but often visible and potentially demoralizing, cases of vaccine-associated paralysis.

v. Integration of poliomyelitis mass immunization into the primary health care structure

Clearly the planning and the budgeting for mass immunization initiatives needs to be additional to routine immunization programmes. However, the presence of an establish primary health care network with provision for highly mobile peripheral elements could ideally be utilized for mass immunization programmes. In the Latin American programmes dedicated and independent eradication teams operated highly successfully. However, on the African continent where health resources are even more restricted, mass immunization programmes for poliomyelitis, while they may need to be supplementary and independent of routine immunization, would still have to be carried out by personnel and infrastructures which are used for primary health care programmes. To this end, a formal structure with a peripheral mobile element could, on a regular annual basis, be dedicated to the task of mass immunization.

Conclusion

The countdown to the eradication of poliomyelitis dare not waver, as more than the elimination of this disease is at stake. Eradication will be, like smallpox, a vastly cost-effective achievement, something which is desperately needed in the AIDS era. The eradication of a second disease after smallpox will have a tremendous medical psychological impact, demonstrating that smallpox eradication was not merely the result of a fortuitous conjunction of "one-off" circumstances, and that in future even further infectious diseases could be eradicable. Politically it could demonstrate that the First World had the means, the capability and the will to assist those less fortunate members of our planet. Lastly, poliomyelitis eradication would be a dramatic and striking demonstration of the power of primary-health care and preventative medicine.

References

1. Kim-Farley RJ, Bart KJ, Schonberger LB, Orenstein WA, Nkowane BM, Hinman AR, Kew OM, Hatch MH, Kaplan JE (1984) Poliomyelitis in the USA: Virtual elimination of disease caused by wild virus. Lancet ii: 1315–1317
2. de Quadros CA, Andrus JK, Olivé J-M, da Silveira CM, Eikhof RM, Carrasco P, Fitzsimmons JW, Pinheiro M, and FP (1991) Eradication of poliomyelitis: progress in the Americas. Pediatr Infect Dis J 10: 222–229
3. World Health Organization (1988) Global eradication of poliomyelitis by the year 2000. Wkly Epidemiol Rec 63: 161–168
4. World Health Organization (1988) Global poliomyelitis eradication by the year 2000: Plan of action. Expanded Programme on Immunisation (EPI/POLIO/88–1) 1–21
5. Sutter RW, Patriarca PA, Brogan S, Malankar PG, Pallansch MA, Kew OM, Bass AG, Cochi SL, Alexander JP, Hall DB, Suleiman AJM, Al-Ghassany AAK, El-Bualy MS (1991). Lancet 338: 715–720
6. World Health Organization (1989) Expanded Programme on Immunization (EPI). Poliomyelitis in 1986, 1987 and 1988, Part I. Wkly Epidemiol Rec 64: 273–279
7. Schoub BD, Johnson S, McAnerney JM, van Middelkoop A, Küstner HGV, Windsor I, Vinsen C, McDonald K (1992) Poliomyelitis outbreak in Natal/KwaZulu, South Africa, 1987–1988. II. Immunity aspects. Trans R Soc Trop Med Hyg 86: 83–85
8. Wright PF, Kim-Farley RJ, de Quadros CA, Robertson SE, Scott R McN, Ward NA, Henderson RH (1991) Strategies for the global eradication of poliomyelitis by the year 2000. N Eng J Med 325: 1774–1779

16

The potential for poliovirus eradication

N. A. Ward

Expanded Programme on Immunization, World Health Organization, Geneva, Switzerland

Summary

"Poliomyelitis has the potential to be globally eliminated and the causative wild polioviruses to be totally eradicated. Excellent vaccines with high levels of safety are available. Sound policies with the capacity to eradicate the virus have been developed and, throughout the Americas and in other areas, demonstrated to be highly effective. These policies require that routine immunization be supplemented by the mass administration of oral polio vaccine to a wide age range within a limited period of time throughout high risk areas. Strong political will for polio eradication has been expressed through various international fora, including by Heads of State in the 1990 World Summit for Children. When implemented through the correct policies, activities aimed at polio eradication have the potential to further develop primary health care in the fields of immunization, surveillance, epidemiology and laboratory services. The major limitations, which are threatening the success of polio eradication are a shortage of resources, especially available vaccine, insufficient commitment to the target, competing interests and civil or military disturbance."

Introduction

The potential for the eradication of poliovirus from the world, as for any eradication programme, depends on the answers to six questions:

  (i) is it theoretically possible?
  (ii) is there any experience to show that eradication can actually be achieved?
  (iii) have basically simple techniques, proven to be effective in achieving virus eradication, been developed, which can be translated into effective operational strategies?
  (iv) is there a political will to target for eradication?
  (v) is there sufficient commitment from planners and health professionals to allow achievement of eradication?
  (vi) is progress on course for achievement of the target?

Each of these questions needs to be considered and a positive answer obtained for each question, before it can be stated with any confidence that global poliovirus eradication is a practical proposition.

Even with positive answers to these six questions, a seventh needs then to be asked, before the practicality of eradication is translated into reality:

What might stop achievement of the target?

1. Is global poliovirus eradication theoretically possible?

For any disease eradication initiative to succeed, the causative organism and the clinical disease it causes must possess certain characteristics:

i. The causative organism

- in natural conditions, only affects human beings
- is only transmitted from human to human, with no animal or insect vector
- under normal circumstances, only has short term survival outside the human body
- is immunologically stable

These characteristics are intrinsic properties of the organism. For practical purposes, wild polioviruses meet all these criteria.

ii. The clinical disease

- is a serious clinical problem with significant morbidity and mortality
- is a real or potential cause of economic problems at the country, community or family level
- is known and feared by the public
- can be prevented by effective vaccines or therapy, which, if correctly applied, have the potential to stop continuing transmission of the organism.
- is typically characteristic and, as a basis for surveillance, can be readily recognized by trained staff.

The first three of these characteristics can potentially change as the incidence of the disease declines and a new generation forgets its impact. Equally, the development of new and improved vaccines may move a previously non-eradicable disease to one with a realistic chance of eradication.

Few organisms and clinical diseases fulfil all these criteria. Smallpox did and, one hundred and seventy years after the discovery of an effective vaccine, was finally eradicated. Poliomyelitis and wild poliovirus meet the criteria for eradication, but thirty years after the development of effective vaccines, remain widespread across the world.

2. Is there any experience to show that poliovirus eradication can actually be achieved?

There are three characteristic geographical examples to show the impact of the correct administration of poliovirus vaccines on the survival of the wild poliovirus, i.e. in the countries of the industrialized world, in a number of tropical developing countries and in the Americas.

(i) In the *industrialized countries* before the mid-1950s, poliomyelitis, widely known as infantile paralysis, was a much feared disease. The development of effective vaccines was followed by their wide distribution to the public through well developed health services. The public, generally well educated and fearful of the disease, ensured that virtually all children received the vaccine within a short time of its initial availability.

   The impact was dramatic, with an immediate decline in the number of cases. Frequently, low levels of virus transmission persisted, eventually resulting in a small epidemic. When this was vigorously controlled, a decline to zero cases followed. Figure 1, showing polio incidence in the United Kingdom is typical of such countries.

(ii) Similar experience was also achieved in a number of *less industrialized countries located in the tropics*. The lesson was clear. When oral poliovirus vaccine (OPV) was administered over a short period of time to all susceptible children, as it was in Malaysia in response to an epidemic, the disease rapidly disappeared, frequently after a subsequent period of low transmission, characterized by the energetic control of any persistent foci (Fig. 2).

(iii) The most striking experience has been in the *continents of the Americas*, where over a period of seven years, as a result of an approach, strongly

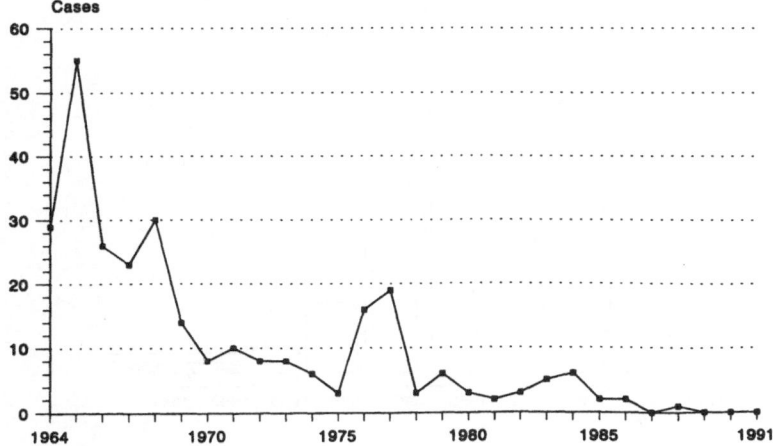

Fig. 1. Reported incidence of poliomyelitis, England and Wales, 1964–1991

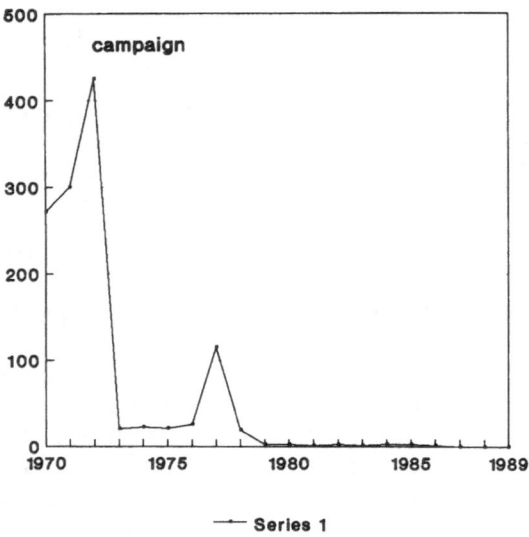

Fig. 2. Reported cases of poliomyelitis, Malaysia, 1971–1989

targeted at polio eradication, the disease has been reduced from a state of widespread endemicity, to one where interruption of wild virus transmission and eradication of the virus appears imminent (Fig. 3).

From practical experience, there is strong and conclusive evidence that poliovirus eradication over large geographical areas is possible, within a short period of time. This is true in industrialized and in developing countries, in the tropics and in colder climates.

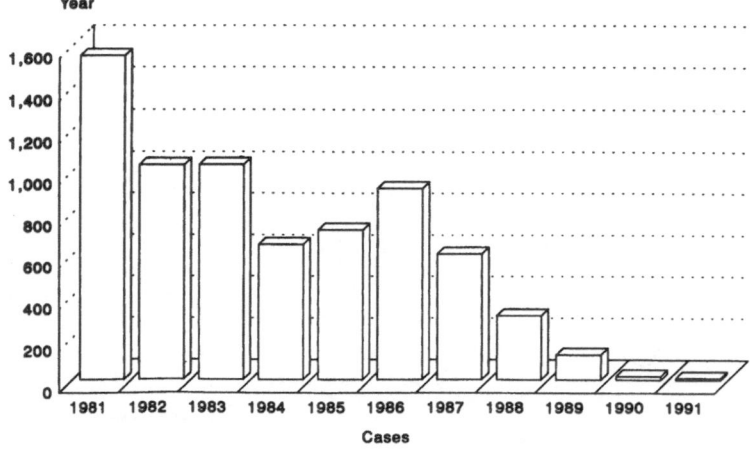

Fig. 3. Reported incidence of poliomyelitis, Americas, 1981–1991

3. Have basic simple techniques, proven to be effective in achieving virus eradication been developed, which can be translated into effective operational strategies?

Over the past five years, the basic principles on which wild poliovirus eradication can be achieved have been defined. Effective strategies, based on experience globally, but especially on progress in the Americas in the 1980s, were recommended by the Third Consultation on the Eradication of Poliomyelitis, meeting in Geneva in September 1990.

The principles of the strategy are very simple. High immunization coverage, using OPV, achieves a marked reduction in wild virus transmission; persisting virus is detected through surveillance of flaccid paralysis and all possible contact susceptibles receive OPV to colonize their gut during the risk period and, hopefully, to produce immunity.

In practice, these principles translate into a series of achievable activities:

- *high immunization coverage* is achieved, using OPV, stored optimally in an effective cold chain. The most marked impact is achieved by the use of supplementary immunization days, either nationally or subnationally, ideally conducted during the cold dry season, when wild virus transmission is at its lowest and competing enteroviruses are also at a low prevalence. Any areas of low coverage must be identified and corrective measures, e.g. sub-national immunization days conducted. The immunization days are repeated at least once after a minimum four week interval.
- *effective surveillance* is developed to identify all cases that might be poliomyelitis due to poliovirus. When suspect cases are detected, the diagnosis is confirmed, but, in the meantime, scrupulously thorough immunization of all possible susceptibles, house to house, over a wide area is conducted and repeated after four weeks. This strategy is designated as "mopping-up".
- These two key policies are supported by additional strategies, e.g. development of a poliovirus laboratory capacity, expert team diagnosis of suspect cases, an immediate outbreak response, even before mopping-up, effective reporting and monitoring of progress, sampling among high risk populations, rehabilitation and research.

The strategies developed, when correctly and thoroughly applied, have proved effective. Each can be reduced to simple, manageable activities, requiring much careful planning and attention to detail, but, given adequate laboratory support, well within the capacity of all countries.

4. Is there a political will to target for poliomyelitis eradication?

The World Health Assembly, in 1988, committed the World Health Organization to the target of poliomyelitis eradication by the year 2000. All Regional Committees have endorsed the target, some Regions and many Member States having established their own earlier target dates.

No countries have declined to be a part of the global initiative.

In September 1990, over 70 national leaders, meeting in the World Summit for Children, produced a "Declaration on the Survival, Protection and Development of Children". This declaration committed the leaders to a Plan of Action, including the global eradication of polio by the year 2000 and to making available the resources required to meet their commitments.

Nominally, at least, there appears to be strong political will for eradication. That will and the statements declaring it, now need to be made manifest as practical support and strong levels of commitment.

5. Is there sufficient commitment from planners and health professionals to allow achievement of eradication?

Nominal commitment to global poliomyelitis eradication, without the essential provision of resources and finance is not sufficient. Poliomyelitis eradication is highly cost-effective.

The estimated additional cost over the next ten years of eradicating wild poliovirus is US $1.5 billion. At present, the United States alone annually spends US $114 million on poliovirus vaccine. Once immunization against poliomyelitis can be stopped, when the world is declared free of circulating wild virus, the total outlay on eradication will be re-couped within a very few years and the benefits will continue indefinitely.

Even with its high cost-effectiveness, some initial additional outlay is required, if poliovirus is to be eradicated. Without that outlay, the chances of early success are not high. The extra costs are required primarily for purchase of vaccines, for establishing and maintaining the laboratory network, providing support for health staff to work effectively in the fields of surveillance and outbreak responses and in support of the costs involved in "mopping-up" activities.

Most countries, through their experience in EPI and the dramatic improvements they have achieved in global coverage levels, have excellent, experienced health staff, well able to implement the technologies required for polio eradication. A certain amount of re-training, especially in surveillance, epidemiology, diagnostic skills and laboratory virology, is required, but workforce skills already exist. Limited numbers of international epidemiologists will be used to provide advice, training, monitoring and eventually, certification of eradication.

In general, strong support for the eradication initiative has come from health professionals. This is most noticeably the case in those countries where the initiative is being energetically pursued, where specific activities are being developed to a high level of competence and where, consequently, encouraging results are being recorded.

Some concern has been expressed that energetically pursuing polio eradication might delay the development of Primary Health Care. No convincing justification for this concern has yet been documented. In contrast, in the Americas, Europe, the Western Pacific and several sub-Regional areas. promoting polio eradication has strengthened delivery of immunization services. The advantages contributed to health service development by an effective polio eradication programme fall under three headings:

i. "Leading edge" ways in which poliomyelitis eradication can be used as a model system to develop activities which will strengthen EPI and PHC:

– development of plans of action
– development of surveillance
– identification of high risk areas
– donor coordination
– vaccine supply issues
– vaccine quality issues

ii. Specific programmatic benefits of poliomyelitis eradication to improve EPI and PHC

– use of "mopping-up" activities to strengthen EPI and PHC by
  – increasing coverage by catch-up immunization
  – promotion of other interventions
  – disseminating information
– strengthening laboratories to support immunization services and disease diagnosis
– emphasis on rehabilitation services, particularly community based rehabilitation

iii. Social mobilization activities

– heightened awareness of EPI diseases, their prevalence and their potential for prevention
– involvement of communities in reporting diseases and in special activities to increase coverage and improve health
– strong political commitment necessary to achieve poliomyelitis eradication will increase the attention given to EPI and PHC

A capacity to conduct communicable disease control as an essential part of Primary Health Care may prove to be the most critical contribution of poliomyelitis eradication to public health in the next century.

6. Is progress on course for achievement of the target?

In the 1980s, immunization coverage with full courses of poliovaccines in the first year of life, has reached high levels (Fig. 4) and, although the African

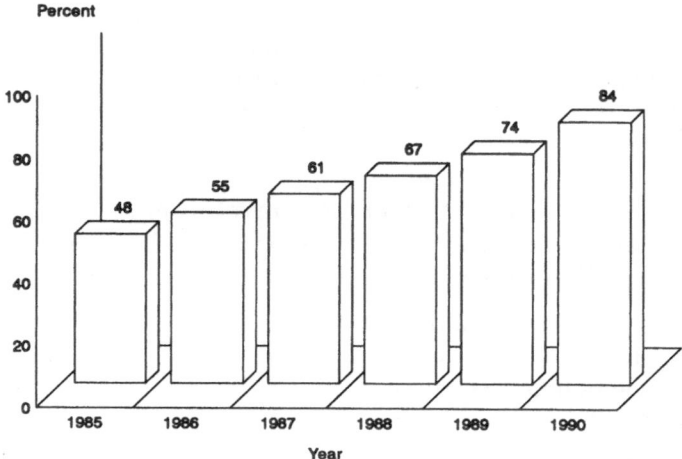

Fig. 4. Global immunization coverage of OPV3 in first year of life by year

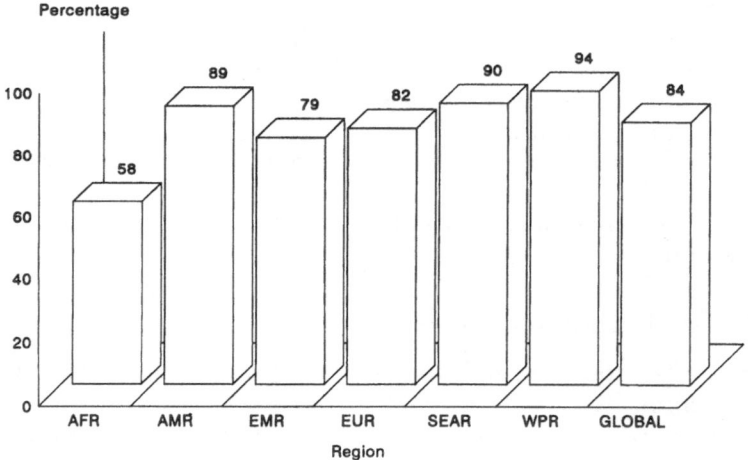

Fig. 5. OPV3 coverage for children under one by WHO region in 1990

Region, on average, lags behind the others, these high levels have been reached in all parts of the world (Fig. 5).

In *the Americas*, the administration of sound policies have reduced polio incidence to the level where, the 1991, only seven cases were reported, with only one in the last eight months, in spite of most extensive and detailed surveillance (Fig. 6).

In *Europe*, persisting areas of poliovirus infection have been increasingly submitted to the energetic control policies of the National programmes. Apparent control has been achieved in Bulgaria and Romania. There is a

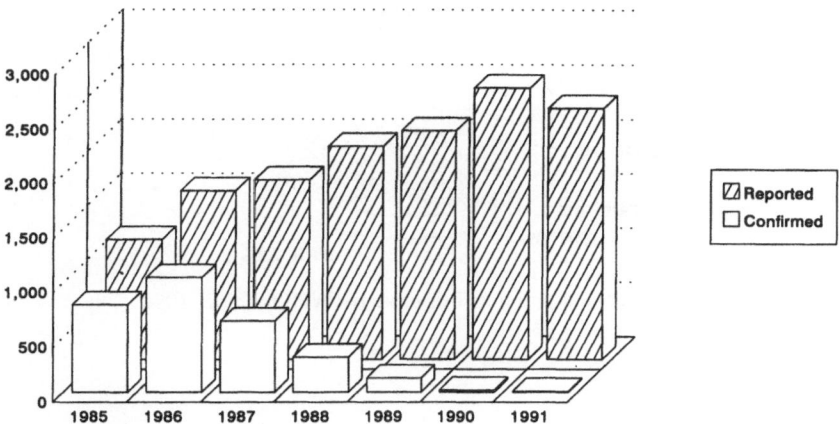

Fig. 6. Reported and confirmed poliomyelitis cases Americas, 1985–1991

much reduced incidence in Turkey, but civil disturbances are proving a concern in some of the Russian Republics and in the former Yugoslavia and threaten an early polio-free Europe.

In the *Western Pacific*, the strong commitment of the six Member States still suffering from endemic wild polio virus transmission has lead to Plans of Action aimed at its early eradication. Reported cases in 1991 are markedly down to half the levels reported in the previous two years, in spite of enhanced surveillance.

Much of the *Eastern Mediterranean* has strong programmes of immunization. Even where many cases are still being reported, as in Egypt, countries have adopted appropriate plans of action, which define strategies that over time will markedly reduce incidence to a point where each case can be treated as an emergency.

In terms of numbers of cases, the most dramatic progress has been seen in the *Region of South-East Asia*, where the highly successful immunization programme in India and other member States has lead to a dramatic fall in polio incidence. If this fall reflects reduced transmission as well as high levels of individual protection, and is backed up by effective surveillance and mopping-up, the gains already recorded will be consolidated into polio free areas. The Region in 1990, still accounted for 52% of reported cases (Fig. 7).

In the *African Region*, progress has been, if anything, more creditable than in other Regions. However, since many African states are starting from a lower base, they have, to date, reached lower immunization coverage levels. Even so, large areas and many countries especially in the Southern and Eastern parts of the continent have recorded high coverage and much reduced polio incidence. There is real hope but much, much more to be done.

As a consequence of all this achievement, the global situation looks promising. Incidence is reduced, certainly for the last 3 years and, based on provisional data for 1991 for 4 years (Fig. 8). However, only in the Americas

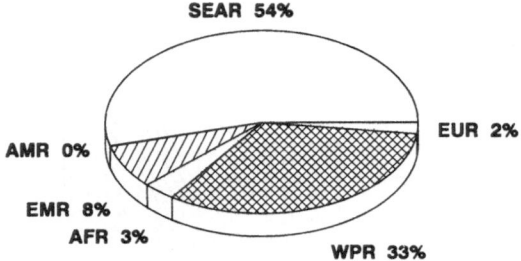

Fig. 7. Proportion of reported poliomyelitis cases by WHO region, 1990

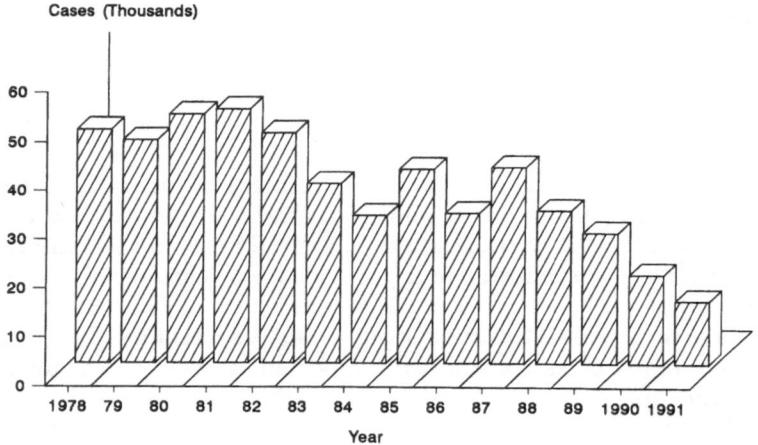

Fig. 8. Reported incidence of acute poliomyelitis, by year

and in the industrialized parts of the rest of the world, is it probable that the gains already made are based on eradication of the wild virus. Only when wild virus transmission is stopped, is it possible to be confident that the virus and the disease will not return. The world is a long way from that situation today.

7. What might prevent achievement of the eradication target?

Four factors may prevent achievement of the Year 2000 target and prevent the world community realising what has been described as "a gift of this century to the children of the next century"

(i) *A lack of adequate resources.* Above everything else an adequate supply of poliovirus vaccine to meet the needs of routine and supplementary immunization. Failure to guarantee this supply would mean that the commitments made in fora like the World Summit for Children were not being honoured. Other costs, such as local expenses to allow highly expert investigation of outbreaks and cases remain largely a commitment of Governments and, in some cases, the international agencies.

(ii) *A lack of continuing commitment.* As progress is made, the number of cases will decline, possibly to zero. Until it is certain that virus eradication has been achieved, any lessening of commitment from countries or agencies will create the dangerous situation where reduced activity in achieving high immunization coverage in all areas, in developing highly effective surveillance and outbreak control will result in a build up of susceptibles. Persisting undetected transmission will then possibly lead to epidemics.

(iii) *Competing interests.* Poliomyelitis eradication will *only* succeed, if its essential components activities are carried out in a meticulous, thorough and professional manner. There are many other, more severe disease problems, each demanding its share of attention and scarce resources and these must also be tackled. However, if poliomyelitis eradication is to succeed, it must be as a supported, high performance activity in all health plans. It cannot be ignored, nor can it be delayed without eventually exacting much increased cost and effort. Competing interests which separate poliomyelitis eradication activities and targets from the rest of EPI and PHC and consequently lower its importance, which advocate vaccines less appropriate for wild virus eradication than that recommended by WHO, do a grave disservice to the potential for achievement of the Year 2000 target.

(iv) *Conflict and civil disturbances.* Are severely limiting factors which prevent or slow the achievement of poliomyelitis eradication. Several examples exist of the willingness of protagonists to stop fighting for an immunization "period of tranquility", as in El Salvador and in Lebanon. These periods are not yet, and by themselves may never be, sufficient to allow the development of public health and preventive medical services in many troubled countries of the world. Our hope must be that taking advantage of time and the ebb and flow of conflict and the possible goodwill of involved leaders will allow sufficient peace at least to allow provision of services for a period.

The dramatic success of immunization services in the 1980s are already showing early signs of impact, in terms of reduced disease incidence. The logical extension of this immunization is to eradicate the disease and to be able, eventually to stop immunization altogether.

Undoubtedly, every present indication is that the potential for the eradication of poliomyelitis exists, is developing and is, to date being sustained.

Conclusion

"The eradication of poliomyelitis and its causative viruses is technically feasible. To achieve success, it is necessary to achieve and maintain strong political and professional will, a reliable supply of essential resources and strong belief from health workers that the target can be reached and is worth it".

References

1. Forty First World Health Assembly. Global Eradication of Poliomyelitis by the Year 2000. Geneva. World Health Organization WHA41.28
2. Robertson SE, Chan C, Kim-Farley R, Ward N (1990) Worldwide status of poliomyelitis in 1986, 1987 and 1988 and plans for its global eradication by the year 2000. World Health Statist. Quarterly 43: 80–90
3. World Declaration on the Survival, Protection and Development of Children and Plan of Action for Implementing the World Declaration on the Survival, Protection and Development of Children in the 1990s. World Summit for Children. September 1990
4. Declaration of New York. The Children's Vaccine Initiative September 1990
5. Eradication of Poliomyelitis. Report on the Third Consultation. World Health Organization September 1990
6. Report of the 13th Global Advisory Group on the Expanded Programme on Immunization. Cairo October 1990
7. Expanded programme on Immunization (1989) Poliomyelitis Review. Weekly Epid. Record 34: 261–264
8. Expanded programme on Immunization (1989) Poliomyelitis in 1986, 1987 and 1988; Weekly Epid Rec 36: 281–285
9. Expanded programme on Immunization (1991) Poliomyelitis in 1987, 1988 and 1989: Weekly Epid Rec 8: 70–71
10. Expanded programme on Immunization (1991) Poliomyelitis in 1988, 1989 and 1990: Weekly Epid Rec 16: 113–117
11. EPI Newsletter; Expanded Program on Immunization in the Americas XIII; 6

17

Poliovirus immunization situation in China

B. Yang

Division of Expanded Programme on Immunization, Department of Epidemic Prevention, Ministry of Public Health, Beijing, China

Summary

Following the introduction of the Expanded Programme on Immunization in China, poliomyelitis outbreaks were successfully controlled and poliomyelitis incidence greatly decreased. By 1988, the reported number of poliomyelitis cases decreased to 667 and China set the target of 1995 for eradication of poliomyelitis. In 1989 and 1990, however, epidemics recurred in many provinces of China. Poliomyelitis mainly occurs in unvaccinated and partially vaccinated children under the age of 4 years old. The geographic distribution of cases indicates that continuous low-level transmission of poliovirus is occurring throughout much of the country.

Introduction

After the Western Pacific Regional Office of World Health Organization and The Forty-first World Health Assembly adopted resolutions on poliomyelitis eradication by the year 1995 and the year 2000 respectively in 1988, China set the goal of poliomyelitis eradication by 1995, and developed a national plan of action accordingly.

The objectives of this paper is to study the epidemiology of poliomyelitis cases and to describe the characteristics of poliomyelitis cases and immunization activities related to poliomyelitis.

Demography and poliomyelitis incidence

Administrative divisions

China is administratively divided into 30 provinces, autonomous regions and municipalities (excluding Taiwan province). These are further divided into 334 prefectures, 2831 rural counties and urban districts, 58 000 townships and finally divided into 776 000 villages.

Demographic data and geography

According to the 1990 population census, the total population of China is 1.13 billion with 80% in rural areas. The provincial population ranges from 2 million in the sparsely populated province to over 100 million in the most densely populated one. From July first of 1989 to June 30th of 1990, the number of birth was 23 500 000, the birth rate was 20.98/1000. The country has a total area of 9.6 million square kilometers, of which 2/3 are mountainous areas, plateaus, or highlands.

Incidence of poliomyelitis

Paralytic poliomyelitis in China was first reported by Zia[1], who noted two cases in 1930. The first outbreak of poliomyelitis was documented in China in Nanton prefecture, Jiangsu province, in 1953. A total of 2607 cases was reported in 1951–1955. The incidence rate was 32.1 cases per 100 000 population in 1955[2].

Poliomyelitis has been a national notifiable disease since 1956 (Table 1). According to this routine reporting system, there was an annual average of 20 000 cases ranged from 43 156 to 9044 in the 1960's and average of 12 000 cases per year ranged from 4645 to 23 271 in the 1970's while the average number of cases per year during the last decade was 4000. Between 1981 and 1988 there was a successive decrease in cases, most cases occurred in a few provinces or autonomous regions, for example in 1986, the number of cases in Guangxi autonomous region accounted for 62% of the total cases for whole country. However the trend of decrease in poliomyelitis cases was reversed with an occurrence of 4628 cases in 1989, the poliomyelitis cases

Table 1. Incidence of poliomyelitis cases, China, 1960–1991

Year	No. of cases	Year	No. of cases
1960	15 799	1976	4645
1961	10 332	1977	7413
1962	9044	1978	10 353
1963	36 383	1979	5520
1964	43 156	1980	7442
1965	28 970	1981	9625
1966	23 859	1982	7741
1967	10 310	1983	3296
1968	11 313	1984	1626
1969	14 370	1985	1537
1970	20 879	1986	1844
1971	18 324	1987	969
1972	23 271	1988	667
1973	17 533	1989	4628
1974	11 070	1990	5065
1975	7815	1991	1926

in five provinces accounted for 63% of the country's total number of poliomyelitis cases, and within those provinces seriously affected, cases tend to be concentrated in a few counties, for example, in 1989 the large outbreak of 779 cases occurred in Jiangsu province with 75 per cent of the cases in one county only. Although 12% of counties in China had confirmed poliomyelitis cases in 1989. In 1990, 5065 poliomyelitis cases were reported and distributed to 673 (24%) counties of 28 (93%) provinces in 1990. The trend of increase in poliomyelitis cases and in number of counties with cases was interrupted, and 1926 poliomyelitis cases were reported from 358 counties in 1991.

Poliovirus vaccine production and immunization schedule

In 1959, China succeeded in preparing the first lot of oral poliovirus vaccine (OPV) with Sabin's strain of poliovirus as seed material and obtained a satisfactory immunological and epidemiological result in 11 major cities [3]. In 1962, China started in preparing OPV in dragee candy. Since 1964, OPV in dragee candy has been gradually and extensively supplied to the whole country. Until to 1982, each child was vaccinated once a year with one dose of OPV for three years. Each dose consisted of two feedings: the first feeding was type 1 poliovirus and the second feeding was type 2 and 3 poliovirus, given one month later. The first dose was given to infants from two to 10 months of age [4]. Vaccination was carried out and completed within a few days during the winter. After 1986, China produced trivalent oral poliovirus vaccine (TOPV) in dragee candy. A unified 4 doses schedule was that TOPV was given to the children at the age of 2 months, 3 months, 4 months and 4 years. The vaccination was carried out by at least every month in urban areas and every two to three months in most of rural areas. The immunization coverage has been increased gradually. According to the results of national EPI review conducted in 1991, the coverage rate of OPV in children under the age of 12 months was reached more than 85% by county.

In this paper are included: (1) information on the number of poliomyelitis cases from 1960 to 1991 (2) information on individual poliomyelitis cases reported in 1991.

Information was available on the number of poliomyelitis cases reported through the national acute communicable diseases reporting system from 1960 to 1991. This system has been in existence since 1956.

Reports on acute communicable diseases are submitted by provincial Epidemic Prevention Stations to the national level every 10 days. Poliomyelitis cases are diagnosed mostly on clinical grounds, but also sometimes by serology, by physicians at township, county, prefecture, and provincial hospitals.

In addition to information through the reportable disease system, information on individual poliomyelitis cases came from a new rapid national reporting system for cases of acute flaccid paralysis began in January 1991. Cases reports of acute flaccid paralysis (suspected poliomyelitis) were received by provincial Epidemic Prevention Stations from county Epidemic Prevention

Fig. 1. Confirmed and suspected reported cases of poliomyelitis, China, Jan.–Oct., 1991

Stations. Acute flaccid paralysis is defined as any case of paralysis with an acute onset for which no other cause (such as injury and hip infection) can be determined. Selected information from these cases is sent to the national level on a case line-listing report from every 10 days. In November 1991, provincial Expanded Programme on Immunization managers were requested to submit a complete list of information on each case of acute flaccid paralysis reported from January to October 1991. Report forms include date of birth, vaccination status, gender, date of onset of paralysis, date of investigation, and method of confirmation.

Poliomyelitis cases

Wild poliovirus infection appears to be endemic throughout China. The geographic distribution of cases over the past years indicates that low-level, continuous transmission is occurring throughout much of the country. The percentage of infected counties has been as high as 24% in 1990, many of which continue to have cases reported year after year.

Poliomyelitis mainly occurs in unvaccinated or partially vaccinated children under the age of 4 years old.

Up to October 1991, 2140 acute flaccid paralysis (AFP) cases were reported through rapid reporting system, among those AFP cases, 323 or 15% have been discarded, 1047 have been confirmed, and 770 are suspected cases that have not had a final determination of confirmed or discarded (Table 2).

For our analysis we included both confirmed cases and suspected cases with no final determination. Although some of the cases that remain suspected

Table 2. Status of reported polio cases in China, Jan.–Oct. 1991

Status	No. of cases	%
Confirmed	1047	49
Suspected	770	36
Discarded	323	15
Total	2140	100

will be eventually discarded, the majority will be probably be confirmed since the vaccination status and age distribution of suspected cases is similar to that of the confirmed cases.

The percent of method of confirmation in the 1047 confirmed cases shows that the percent lost to followup is low (4%) but the percent confirmed by residual paralysis at 60 day exam is high (61%), especially when compared to the percent confirmed by virus isolation from stool culture (8%), the percent confirmed by epidemiological linkage is 23%. 62% of all cases had no stools collected, 29% had one stool collected, and only 8% had two stools collected (Table 3).

The age distribution of cases showed that 33% of the cases were children under the age of 12 months, 39% at the age of 12–23 months, 13% at the age of 24–35 months, 6% at the age of 36–47 months and 91% of cases under the age of 4 years old (Table 4).

Table 3. Percent of methods of confirmation for poliomyelitis cases, China, Jan.–Oct. 1991

Method	No. of cases	%
Virus isolation	84	8
Residule paralysis at 60 day exam.	638	61
Epidemiological linkage	237	23
Death	47	5
Lost to follow up	41	4
Total	1047	100

Table 4. Age distribution of poliomyelitis cases, China, Jan.–Oct. 1991

Age in years	No. of cases	%	Cumulative %
0	557	33	33
1	667	39	72
2	225	13	85
3	91	6	91
4	65	3	94
5 & 5 +	91	5	99

Table 5. Immunization status by number of poliomyelitis cases in China, Jan.–Oct., 1991

Cases per County	0 Dose (%)	1 Dose (%)	2 Doses (%)	3&3 + Doses (%)	Missing (%)
1	38	18	10	27	7
2	41	17	9	27	6
3–10	46	16	9	23	6
> 10	69	13	7	9	2

Information of immunization status of cases showed that 19% reported receiving three or more doses of OPV and 51% reported receiving 0 dose of OPV. The percentage of cases with zero dose is very high in cases which occurred in counties with 10 or more cases. As seems logical, this pattern indicates coverage was probably higher in counties with only 1 or 2 cases and is very low in counties with 10 cases or more. The percentage of cases fully vaccinated with 3 doses has increased from 9% in 1989 to 12% in 1990 and finally to 19% in 1991 (Table 5).

There were 368 counties that had one case as to October in 1991, 138 counties that have reported 2 cases, 41 counties have had 6–10 cases. A total of 665 counties had cases. 55% of 665 counties had only one case, 5% of the 665 counties had 6–10 cases and 3% had 11 or more cases. Cases that occurred in counties with one case per county made up 20% of the total of 1817 cases. Cases that occurred with 2 cases per county made up 15% of the

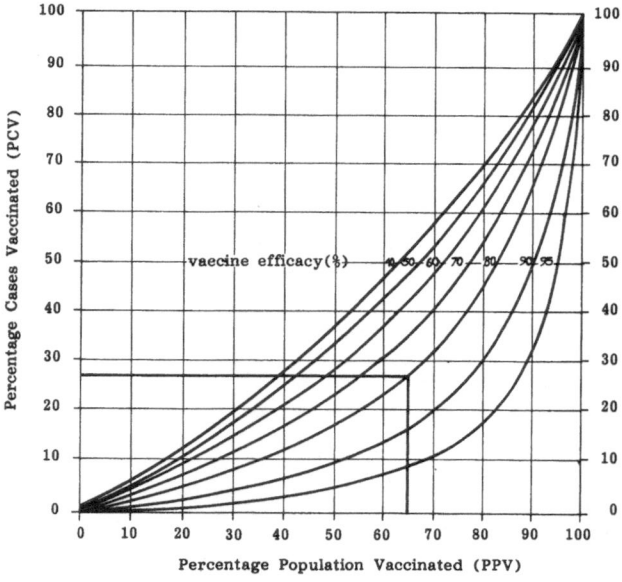

Fig. 2. Use of vaccine efficacy curve to estimate OPV coverage with 3 + doses

Table 6. Distribution of poliomyelitis cases in countries, China, Jan.–Oct., 1991

Cases per county	No. of infected counties	% of infected counties	% of cases
1 Case	368	55	20
2 Cases	138	21	15
3–5 Cases	93	15	19
6–10 Cases	41	5	16
> 11 Cases	25	3	29
Total	665	100	100

total, and cases which occurred in counties with 6 or more cases made up 45% of the total cases. 8% of the counties with 6 or more cases were responsible for 45% of the cases. And, 76% of counties with reported cases had only 1 or 2 cases. (Table 6.)

$$\text{Percentage of cases vaccinated} = \frac{19\% \text{ with 3 doses}}{51\% \text{ with 0 doses} + 19\% \text{ with 3 doses}}$$
$$= 27\%$$

We attempted to estimate to vaccination coverage in areas where the poliovirus was circulating and causing cases. We used vaccine efficacy nomogram and used 19% as the percentage of cases fully vaccinated and assumed a vaccine efficacy of three doses of OPV as 80%. From this estimate, the coverage with OPV3 in the areas in which the poliovirus is circulating the approximately 60 to 70%. As indicated by cases, poliovirus is widespread in China at this time. Therefore, eradication activities cannot be focused in just a few provinces. In addition, coverage with 3 or more doses of OPV is probably 70% on average in townships or counties in which poliovirus is circulating. Although it is encouraging that cases were reduced compared to previous year. It is important to realize that reported cases were even lower in 1987 and 1988 and a large rebound in cases occurred in 1989 and 1990.

In the coming years, we plan to use routine, supplemental, and outbreak responses immunization to decrease the number of cases. We need to solve the problem of undervaccination of children in areas where the poliovirus is circulating both in the routine system but especially in the supplemental sessions.

We are not able at present time to conduct province wide double round supplemental immunization in children 47 months or less in high risk provinces and countywide outbreak response for each suspected case. Although this strategy would be successful and could be easily implemented in our country, we do not have sufficient vaccine to do this at the present time.

Therefore, we have to find alternatives in different provinces which are actually the size of most countries, and further study the epidemiology of the poliovirus in those provinces. And, based on the vaccine available to them, most provinces will have to restrict the age group, the geographic area, and the frequency of supplemental and containment immunizations.

Acknowledgements

The author acknowledge all epidemic prevention station staff who have been involved in poliomyelitis surveillance, Dr. Mac Otten who assisted with the data analysis.

References

1. Zia SH (1930) The occurrence of acute poliomyelitis in north China. National Medical Journal (Peking) 16:135–144
2. Ye ST (1959) Epidemiological analysis of 2607 cases of poliomyelitis (in Chinese). J Epidemiol 4:2–7
3. Ku FC, Chang PJ, Chen HS, Shen YC, Wu MH, Mao CS, Li HT (1963) A large scale trial with live poliovirus vaccine (Sabin's strain) prepared in China. Chin Med J (Engl) 82:131–137
4. Ku Fang-chou (1982) Poliomyelitis in China. J Infect Dis 146(4)

18

Epidemiologic and laboratory classification systems for paralytic poliomyelitis cases

R. W. Sutter, S. L. Cochi, and P. A. Patriarca

Division of Immunization, Centers for Disease Control and Prevention, Atlanta, Georgia, USA

Summary

Classification systems for poliomyelitis cases were first developed in industrialized countries to provide a uniform and standardized method to assign epidemiological categories to individual cases, particularly for vaccine-associated paralytic poliomyelitis cases. The classification of paralytic poliomyelitis cases follows individual case investigations that attempt to collect all available clinical, epidemiologic and laboratory information on each suspected poliomyelitis case. Many countries have convened expert review committees to evaluate, confirm and classify suspected poliomyelitis cases. In the United States, classification systems for paralytic poliomyelitis cases have been in use since 1976. The goals of regional elimination of poliomyelitis from the Americas by the year 1990, and of global eradication of poliomyelitis by the year 2000, provide an additional impetus to extend classification systems to poliomyelitis-endemic areas and to standardize methods for monitoring progress towards these eradication objectives.

Introduction

Widespread use of inactivated (IPV) and oral poliovirus vaccines (OPV) led to marked declines in the incidence of poliomyelitis from many industrialized countries. In the United States, paralytic poliomyelitis cases declined from approximately 16 000 per year in the early 1950s, the immediate pre-vaccine era, to an annual average of less than 10 cases during 1980–1989 [1,2]. Although control of poliomyelitis in developing countries has been considerably more challenging, the World Health Organization (WHO) has documented significant progress during the last decade. WHO has estimated that more than 100 000 cases of poliomyelitis occurred during 1990 [3]; an incidence is approximately 2 to 3-fold lower than had been estimated only a few years earlier [4]. Achievements in the control or elimination of the disease in industrialized countries, and progress toward elimination in many developing countries generated interest in the global eradication of poliomyelitis

[5,6], and prompted the World Health Assembly to adopt the goal in May, 1988, of global eradication of poliomyelitis by the year 2000 [7]. The WHO Region of the Americas, through the Pan American Health Organization (PAHO), had established a target for elimination of indigenous poliomyelitis by the end of 1990, while the WHO European and Western Pacific Regions have set targets for elimination by 1995. Subsequently, substantial progress toward this objective has been documented by the Expanded Programme on Immunization of WHO, as well as improvements in immunization coverage levels with three doses of poliovirus vaccine to more than 80% of children globally in 1990 [8]. Global eradication of poliomyelitis appears feasible by achievement and maintenance of high immunization levels, effective surveillance to detect all new cases, and a rapid vigorous response to the occurrence of new cases [9].

In this chapter, we outline the components for organized surveillance and laboratory activities that should be in place before the full benefits of systems for. classifying reported paralytic poliomyelitis cases can be obtained, and describe two examples of classification systems. One of the two systems will serve as a surrogate for the type of classification system appropriate for industrialized countries [10], and the other system for classifying paralytic poliomyelitis cases in developing countries [11]. While the former system requires many different categories to accurately classify both wild and vaccine-associated paralytic poliomyelitis, the latter system has been, and will continue to be, useful to monitor the control, elimination and eventual eradication of poliomyelitis in the developing world.

Prerequisites for using classification systems

Certain basic criteria should be met to obtain the full benefits of classifying paralytic poliomyelitis cases. These include: 1) adoption and implementation of uniform case definitions for paralytic poliomyelitis; 2) the existence of effective surveillance for poliomyelitis; 3) the availability of laboratory support for the isolation and characterization of polioviruses from stool specimens (including intratypic differentiation); and 4) creation of an expert review committee to evaluate suspected cases of poliomyelitis. Once these steps have been taken in the evolution of a poliomyelitis control program, a classification system for paralytic poliomyelitis cases becomes a relevant tool to achieve and monitor further reductions in poliomyelitis cases.

Role of case definitions

Clinical case definitions

Surveillance for many infectious and non-infectious entities relies on appropriate case definitions [12]. Increasingly, different case definitions have been

Table 1. Screening and confirmatory case definitions for poliomyelitis

Screening case definitions

United States

Physician-diagnosed suspected poliomyelitis.

World Health Organization

A case of poliomyelitis is defined as any patient with acute flaccid paralysis (including any child less than 15 years of age diagnosed to have Guillain-Barré syndrome) for which no other cause can be identified [11]. Such a case is designated as a "probable case."

Confirmatory case definitions

United States

A patient must have had paralysis clinically and epidemiologically compatible with poliomyelitis and, at 60 days after onset of symptoms, had a residual neurological deficit, had died, or had no information available on neurologic residue*.

World Health Organization

Probable case (see under screening definition) by history and physical examination, and any one of the following: 1) positive virus culture for poliovirus; 2) positive serology; 3) epidemiologic linkage to another suspected or confirmed case; 4) residual paralysis 60 days after onset; 5) death of a suspected case; or 6) lack of followup of a suspected case.

---

*This classification was formerly known as the Best Available Paralytic Poliomyelitis Case Count (BAPPCC) case definition

developed to distinguish between suspected and confirmed cases [12]. For poliomyelitis reporting purposes, screening case definitions are used with the objective of maximizing the sensitivity of surveillance (i.e., including all true cases of poliomyelitis). After intensive investigation, confirmatory case definitions are then applied to individual suspected cases to maximize the specificity of a system (i.e., to exclude all non-poliomyelitis cases). The United States and WHO both use sensitive screening case definitions for reporting suspected cases of poliomyelitis (Table 1). However, the major difference is that WHO screening case definition encompasses all cases of acute flaccid paralysis including cases of physician diagnosed Guillain-Barré syndrome [10,11] (Table 1).

It is likely that the screening case definition recommended by WHO is more sensitive and less specific than the one used in the United States. However, each of these case definitions responds to different needs. In the United States, wild poliovirus circulation probably has been eliminated [1], and the last endemic or epidemic poliomyelitis cases due to wild poliovirus were reported in 1979. Thus, the major reason for residual paralysis is due to background illness that may be misclassified as vaccine-associated paralytic poliomyelitis. On the other hand, in developing countries with endemic polio-myelitis, there is a need to emphasize sensitivity to detect virtually all cases of poliomyelitis. The confirmatory case definitions are similar for the two set-tings, with the only differences being that the U.S. definition requires paralytic residua, while the WHO definition includes poliomyelitis cases without

Table 2. Alternative case definitions for screening purposes

Study	Alternative case definition(s)	Sensitivity (95% CI)*	Specificity (95% CI)*
Andrus [14]	Age < 6 years *and* fever at paralysis onset *and* ≤ 4 days for paralysis to develop completely	64% (47%, 82%)	82% (80%, 84%)
Biellik [13]	Case reported from same State[†] *and* absence of involvement of four limbs *and* fever	85% (72%, 99%)	58% (53%, 63%)
Dietz [15]	1) Proximal paralysis *and* fever at onset and progression < 4 days;	16% (12%, 21%)	99% (96%, 99%)
	2) Proximal *and* unilateral paralysis *and* fever at onset;	16% (12%, 21%)	99% (96%, 99%)
	3) Proximal *and* unilateral paralysis *and* progression < 4 days;	16% (12%, 21%)	99% (96%, 99%)

*95% confidence interval
[†]Same State of Brazil

residual neurological deficits if poliovirus is isolated. Nonetheless, non-paralytic cases with wild poliovirus isolates would also be aggressively followed up and investigated in the U.S.

Although the sensitivity and specificity of the U.S. screening case definition is not known, several studies have been carried out in the Americas to evaluate different combinations of signs and symptoms for poliomyelitis to arrive at screening case definitions with improved specificity, without substantial scarifice of sensitivity [13,14,15]. These studies reported similar findings, and suggested that age (< 6 years), fever at onset, and rapid progression toward maximum extent of paralysis are important predictors of eventual case confirmation by virus culture (Table 2). Specific pattern of paralysis (proximal, unilateral, or absence of paralysis in all four limbs) also increased the specificity, with varying degrees of loss in sensitivity [13,14,15].

Modifying case definitions to address changing needs

The findings of these case-definition studies, and the observation that the age distribution of reported poliomyelitis cases from poliomyelitis-endemic areas concentrated in the first 2–4 years of life, recently prompted WHO to consider a change in the standard screening case definition to: "A case of polio-myelitis is defined as any patient with acute flaccid paralysis (including any child less than 6 years of age diagnosed to have Guillain-Barré syndrome) for which no other cause can be identified [Ward N: Personal communication 1992]."

The WHO region of the Americas was faced with a different problem as the eradication objective drew near, that is, how to achieve the eradication goal using a confirmatory case definition (that of WHO) that lacked high specificity. Consequently, this region adopted a more specific confirmatory case definition* beginning 1990, that required wild-type poliovirus isolation from stool [16,17].

Adequacy of surveillance

Guidelines for evaluating surveillance systems have been published for the United States [18], and specific elements recommended for poliomyelitis surveillance have been proposed [11,19]. At a minimum, surveillance systems for poliomyelitis eradication require, (1) passive reporting of individual cases from health personnel; and (2) active reporting, including the active search for the occurrence or the absence of cases from curative and preventive institutions. While surveillance for many infectious causes is in place in many developing countries, the demands for poliomyelitis eradication are likely to exceed the scope of these current systems. To prioritize limited resources, WHO has proposed staging criteria to better target resources depending on the epidemiologic situation in each country [11,20]. Thus, rigorous poliomyelitis case classification systems are not likely to be valuable in countries where poliomyelitis is highly endemic (i.e., the vast majority of poliomyelitis cases are due to wild poliovirus), and where increases in vaccination coverage deserve the highest priority. However, countries that have achieved a substantial degree of control of poliomyelitis and in which further poliomyelitis control is a recognized priority, should consider surveillance systems that include the following elements: 1) immediate reporting of suspected poliomyelitis cases to designated public health officials; 2) conducting individual case investigation promptly (i.e. initiating investigation within 48 hours after notification); and 3) collecting appropriate clinical, epidemiological, and laboratory information using a standardized case investigation form. In addition, curative and preventive institutions should be required on a weekly basis to report whether they had encountered a patient meeting the screening case definition for poliomyelitis. Recently, this aspect (i.e., active reporting from institutions) was evaluated in Venezuela [21].

The Pan American Health Organization (PAHO) has developed a series of surveillance indicators to monitor poliomyelitis surveillance for individual countries [19], and to target limited resources appropriately. Since poliomyelitis surveillance starts with surveillance of acute flaccid paralysis and because countries that were thought to have good surveillance found a rate of AFP of approximately 1 case per 100 000 population < 15 years of age, PAHO established that surveillance systems should detect AFP at a rate of at least 1 case per 100 000 population even in the absence of poliomyelitis.

---

*A **confirmed** case of poliomyelitis required AFP *and* wild-type poliovirus isolation; a **compatible** case required AFP *and* residual paralysis, death, or lack of follow-up. Compatible cases may be discarded if two adequate stool specimens tested negative in three different laboratories [16].

Countries reporting rates of AFP < 1/100 000 may not have sensitive surveillance systems [16,17].

More recently, PAHO's International Certification Commission has proposed the following criteria for evaluating whether a country should be considered as free of poliomyelitis: 1) absence of wild-type poliovirus isolate for at least three years in a region; and 2) all AFP cases investigated with 24 hours of notification, with at least two stool specimens collected within 15 days of onset of paralysis [16]. Thus, demonstration of adequate surveillance will be of critical importance for poliomyelitis eradication certification purposes.

Requirements for laboratory support

In addition to establishing effective surveillance for poliomyelitis, laboratory support for the isolation and characterization of polioviruses from stool specimens should be organized in parallel [22,23]. Thus, virology laboratories with experience in viral isolation and which work with appropriate cell lines and tissue cultures need to be recruited into the eradication program, to ensure that countries that are approaching elimination of poliomyelitis have access to adequate laboratory support (including isolation, serotyping and intratypic differentiation of poliovirus). PAHO has implemented a regional laboratory network that provides this level of laboratory support. In addition, PAHO has developed specific laboratory indicators to monitor the functioning of the network laboratories. These indicators include: 1) the number of stool specimens collected from individual cases; 2) the condition of the stool specimens upon arrival in the laboratory (under adequate cold chain, packing adequate, etc.); 3) the processing time (i.e., the interval from receiving a specimen to reporting results); and 4) the proportion of stool specimens in which non-polio enteroviruses are isolated [16,17,19,24,25].

Expert committee for review of suspected cases

Poliovirus is not the only cause of poliomyelitis-like paralytic disease. Because a number of infectious and non-infectious causes can produce a disease mimicking paralytic poliomyelitis, many industrialized countries have made use of expert review committees to evaluate each suspected case of poliomyelitis individually, and determine whether individual cases meet the confirmatory case definition for poliomyelitis [10]. Similarly, WHO recommends the use of expert review committees to ensure a uniform review mechanism for suspected cases [11]. Because the differential diagnosis for paralytic poliomyelitis is extensive [26,27], expert review committees are usually composed of pediatricians (including pediatric infectious disease specialists and pediatric neurologists), epidemiologists, and virologists, to allow the best possible interpretation of available clinical, epidemiologic and laboratory information on each suspected poliomyelitis case, and to determine the appropriateness of control measures taken after notification of each case.

Classification of poliomyelitis cases

United States

The classification system for reported poliomyelitis cases in the United States (known as the Epidemiologic and Laboratory Classification System for Paralytic Poliomyelitis Cases [ELCPPC]) integrates epidemiologic criteria, virus isolation, and strain characterization results [10]. The classification of paralytic poliomyelitis cases in the United States for the period 1980–89 is provided in Table 3. This system (ELCPPC) replaced an earlier classification system initiated in 1976 that was known as the "epidemiologic classification of paralytic poliomyelitis cases (ECPPC)." The ECPPC assigned cases into one of four categories: epidemic, endemic, imported or immune deficient and used both epidemiologic and laboratory information to more definitively classify confirmed cases of paralytic poliomyelitis [28]. Because both the last epidemic of poliomyelitis and the last known case of endemic paralytic poliomyelitis due to wild poliovirus in the United States occurred in 1979 [29] – and approximately one case of imported paralytic poliomyelitis is reported annually – the small number of cases of vaccine-associated paralytic poliomyelitis (VAPP) annually among OPV vaccine recipients and their susceptible contacts [28,30] have become the predominant form of paralytic poliomyelitis in the United States.

The classification system adopted in the United States in 1985 (the ELCPPC) and applied retrospectively to cases since 1980 was directly related to improvements in laboratory methods. Poliovirus isolates were first reliably identified at the Centers for Disease Control and Prevention (CDCP) by their antigenic properties, using specific polyclonal antibodies [31], an approach further refined by the use of cross-adsorbed antisera described by van Wezel [32]. While the specificity of antigenic characterization has improved with the development of panels of neutralizing monoclonal antibodies [33,34,35,36], polioviruses are now most definitively identified by molecular methods such as oligonucleotide fingerprinting or partial genomic sequencing [37,38,39], or polymerase chain reaction [40], which measure or use the extent of genetic sequence relationship among isolates.

World Health Organization

Following notification of a suspected case of poliomyelitis based on the WHO screening case definition, an immediate case investigation should be initiated (recommended to begin within the first 48 hours after notification). The principal purposes of the case investigation are to collect detailed clinical and epidemiological data from a suspected case, and to arrange for appro-

Table 3. Epidemiologic and Laboratory Classification of Paralytic Poliomyelitis Cases (ELCPPC): United States, 1980 Through 1989*

Categories	Number of Cases
I. *Sporadic* A case of paralytic poliomyelitis not linked epidemiologically to another case of paralytic poliomyelitis.	
A. Wild virus poliomyelitis – Virus characterized as wild virus Vaccine-associated poliomyelitis	0
B. 1. Recipient – OPV was received 4 to 30 days before onset of illness	
a. Virus characterized as vaccine-related	23
b. Virus not isolated or not characterized	8
2. Contact – Illness onset was 4 to 75 days after OPV was fed to a recipient in contact with patient and contact occurred within 30 days before onset of illness	
a. Household – vaccinee and patient regularly share the same household for sleeping	
i. Virus characterized as vaccine-related	11
ii. Virus not isolated or not characterized	6
b. Non-household	
i. Virus characterized as vaccine-related	9
ii. Virus not isolated or not characterized	3
3. Community – No history of receiving OPV or of contact with an OPV recipient, as defined in 1 and 2, and virus isolated and characterized as vaccine-related	4
C. Poliomyelitis with no history of receiving OPV or of contact with an OPV recipient, as defined in B1 and B2, and virus not isolated or not characterized	1
II. *Epidemic* A case of paralytic poliomyelitis linked epidemiologically to another case of paralytic poliomyelitis.	
A. Not a recipient of OPV	
1. Virus characterized as wild virus	0
2. Virus characterized as vaccine-related	0
3. Virus not isolated or not characterized	0
B. OPV recipient – OPV received 4 to 30 days before onset of illness	
1. Virus characterized as wild virus	0
2. Virus characterized as vaccine-related	0
3. Virus not isolated or not characterized	0
III. *Immunologically abnormal* Proven or presumed	
A. Wild virus poliomyelitis – Virus characterized as wild virus	0
B. Vaccine-associated poliomyelitis	
1. Recipient – OPV was received 4 to 30 days before onset of illness[†]	
a. Virus characterized as vaccine-related	9
b. Virus not isolated or not characterized	2
2. Contact – Illness onset was 4 to 75 days after OPV was fed to a recipient in contact with patient and contact occurred within 30 days before onset of illness[†]	

Table 3. (*Contd.*)

Categories	Number of Cases
a. Household – vaccinee and patient regularly share the same household for sleeping	
i. Virus characterized as vaccine-related	2
ii. Virus not isolated or not characterized	1
b. Non-household	
i. Virus characterized as vaccine-related	1
ii. Virus not isolated or not characterized	2
3. Community – No history of receiving OPV or of contact with an OPV recipient, as defined in 1 and 2, and virus isolated and characterized as vaccine-related	1
C. Poliomyelitis with no history of receiving OPV or of contact with an OPV recipient, as defined in B1 and B2, and virus not isolated or not characterized	0
IV. *Imported*	
Poliomyelitis in a person (U.S. resident or other) who has entered the United States.	
A. Virus characterized as wild virus	
1. Onset of illness within 30 days before entry	2
2. Onset of illness within 30 days after entry	1
B. Virus characterized as vaccine-related	
1. Onset of illness within 30 days before entry	0
2. Onset of illness within 30 days after entry	0
C. Indeterminate – Virus not isolated or characterized	
1. Onset of illness within 30 days before entry	1
2. Onset of illness within 30 days after entry	1
Total	88

*From Strebel et al. [1] and CDC unpublished data
†Receipt of OPVmore than 30 days prior to onset or contact more than 75 days before onset does not preclude classification in these categories.

priate stool specimens (i.e., two stool specimens collected on two separate days within the first 15 days after onset of paralytic illness). Parallel to the individual case investigation, active surveillance for other potential cases needs to be instituted, and appropriate control activities considered [11]. After all information, including laboratory data, is available, cases are examined by an expert review committee. This committee then classifies confirmed cases into four categories: 1) vaccine-associated; 2) wild poliovirus/imported; 3) wild poliovirus/indigenous; and 4) unknown/other.

In the WHO region of the Americas, adoption of a revised confirmatory case definition required a change in the classification system. In the new system, three categories are currently used to finally classify suspected cases of poliomyelitis: 1) confirmed cases (require wild-type poliovirus isolation); 2) compatible cases (residual paralysis, death, or lost to followup); and 3) discarded cases. Suspected cases in the second category may be discarded if two adequate stool specimens – that is, collected within 15 days after onset of paralysis and shipped under the reverse cold chain – are negative for virus isolation in three different laboratories [16].

Key differences between U.S. and WHO systems

The U.S. classification system includes additional classification categories and a series of specific definitions [10], including: (1) "recipient case" of VAPP must have received OPV 4–30 days prior to onset of illness; (2) "contact case" must have had onset of illness 4–75 days after OPV was fed to a recipient in contact with the case; (3) "immunologically abnormal" includes cases in persons with proven or presumed immunological abnormalities, regardless of the etiology of their abnormality (primary immunodeficiency, drug induced suppression, etc.), and (4) "imported" includes cases in persons (U.S. residents or others) with onset within 30 days before or after entry into the United States.

Limitations of current classification systems

The usefulness of classification systems for poliomyelitis cases depends on the quality of individual investigations of suspected poliomyelitis cases. Detailed medical, epidemiological, and laboratory data are required in order to confirm poliomyelitis cases, and to assign appropriate classification categories. Both the classification system used in the United States, and the system recommended by WHO, rely on sophisticated laboratory methods that distinguish between vaccine-related and wild-type strains of polioviruses. However, isolation of vaccine-related poliovirus in stool specimens of patients with suspected poliomyelitis strengthens the diagnosis, but, by itself, does not constitute proof of a causal association of such viruses with paralytic poliomyelitis. Other non-polio enteroviruses (particularly enterovirus 71) may cause a polio-like illness that may result in death (bulbar paralysis) or residual paralysis indistinguishable from actual poliomyelitis [26,27,41]. Thus, to monitor progress toward poliomyelitis eradication, and to obtain the most useful information from classification systems, high quality laboratory support for developing country settings is required [16,17,23].

Future directions

Classification systems for poliomyelitis are an integral part of the investigation, followup and confirmation process. Accurate determination of the final disposition of cases of acute flaccid paralysis (AFP) permits the monitoring of progress toward poliomyelitis eradication; the identification of risk groups in need of additional attention; and eventually the certification of a country as free of poliomyelitis. As an increasing number of countries approach the eradication goal, the remaining poliomyelitis case burden will likely be due predominantly to OPV, parallel to the experience in industrialized countries with vaccine-associated paralytic poliomyelitis. Thus, a more extensive classification system may be needed in developing countries once greater control of poliomyelitis has been achieved.

Accurate determinations of suspected cases of poliomyelitis increasingly requires sophisticated laboratory support that is not usually available in most developing countries. This need was first recognized by PAHO, and resulted in a comprehensive laboratory network for the Americas. WHO is currently establishing a global laboratory network to support the poliomyelitis eradication initiative. These laboratory networks will require extensive transfer of technology and training programs to permit differentiation of vaccine-related and wild-type polioviruses. Such transfer of technology has been achieved to date in the WHO region of the Americas. In the future, additional demands on laboratory support may include: 1) typing of non-polio enteroviruses isolated from AFP cases; and 2) serological tests using an IgM assay for polioviruses which would permit rapid diagnosis from a single serum specimen.

While the U.S. classification system is complex, the greater descriptive accuracy of each category and subcategory provides more specific and useful information, particularly concerning vaccine-associated disease. Information on the relative occurrence of wild-virus and vaccine-associated cases within each epidemiologic category can be obtained rapidly from this tabular format. This classification system has proved useful in the United States for assessing the risk of vaccine-associated paralytic poliomyelitis (VAPP), identifying special groups at higher risk of VAPP, and stimulating discussion regarding the optimal vaccination policy in the U.S. For example, the risk of VAPP among recipient of OPV for the period 1980–1989 was estimated as one case of paralytic poliomyelitis per 700 000 doses of OPV distributed for those receiving their first dose of OPV, while the risk of subsequent OPV doses was approximately ten-fold lower [1].

As current worldwide efforts to immunize children against poliomyelitis are expanded and the effectiveness of these efforts is documented by a sustained decrease in poliomyelitis cases, classification systems will have to become more sophisticated to be useful as standardized tools to track the success of these programs and, eventually, to help in certifying countries or regions as free of wild poliomyelitis.

References

1. Strebel PM, Sutter RW, Cochi SL, Biellik RJ, Brink EW, Kew OM, Pallansch MA, Orenstein WA, Hinman AR (1992) Epidemiology of poliomyelitis in the United States one decade after the last reported case of indigenous wild virus-associated disease. Clin Infect Dis 14: 568–579
2. Kim-Farley RJ, Bart KJ, Schonberger LB, Orenstein WA, Nkowane BM, Hinman AR, Kew OM, Hatch MH, Kaplan JE (1984) Poliomyelitis in the U.S.A.: Virtual elimination of disease caused by wild virus. Lancet ii: 1315–1317

3. Expanded Programme on Immunization (1992) Global overview of poliomyelitis. Geneva, Switzerland: World Health Organization, 1992 (Technical Commission Document 56.1)

4. Robertson SE, Chan C, Kim-Farley R, Ward N (1990) Worldwide status of poliomyelitis in 1986, 1987 and 1988, and plans for its global eradication by the year 2000. World Health Stat Q 43: 81–90

5. Horstmann DM, Quinn TC, Robbins FC (eds) (1984) International Symposium on Poliomyelitis Control. Rev Inf Dis 6 [Suppl]: 301–601

6. Nathanson N (1982) Eradication of poliomyelitis in the United States. Rev Inf Dis 4: 940–945

7. Resolution of the 41st World Health Assembly (1988) Global Eradication of Poliomyelitis by the Year 2000. Geneva, Switzerland: World Health Organization (WHA 41.28)

8. Expanded Programme on Immunization (1991) EPI global overview. Geneva, Switzerland: World Health Organization (EPI/GAG/WP.1)

9. Hinman AR, Foege WH, de Quadros CA, Patriarca PA, Orenstein WA, Brink EW (1987) The case for global eradication of poliomyelitis. Bull WHO 65: 835–840

10. Sutter RW, Brink EW, Cochi SL, Kew OM, Orenstein WA, Biellik RJ, Hinman AR (1989) A new epidemiologic and laboratory classification system for paralytic poliomyelitis cases. Am J Public Health 79: 495–498

11. Expanded Programme on Immunization (1989) Global poliomyelitis eradication by the year 2000. Manual for managers of immunization programmes. Geneva, Switzerland: World Health Organization (WHO/EPI/POLIO/89.1)

12. Centers for Disease Control (1990) Case definitions for public health surveillance. MMWR 39 (RR-13): 1–43

13. Biellik RJ, Bueno H, Olivé JM, de Quadros C (1992) Poliomyelitis case confirmation: characteristics for use by national eradication programmes. Bull WHO 70: 79–84

14. Andrus JK, de Quadros C, Olivé JM, Hull HF (1992) Screening for acute flaccid paralysis for polio eradication: Ways to improve specificity. Bull WHO 70: 591–596

15. Dietz V, Lezana M, Sancho CG, Montesano R (1992) Predictors of poliomyelitis case confirmation at initial clinical evaluation: Implications for poliomyelitis eradication in the Americas. Int J Epidemiol 21: 800–806

16. Andrus JK, de Quadros CA, Olivé JM (1992) The surveillance challenge: Final stages of eradication of poliomyelitis in the Americas. MMWR 41: 21–26

17. de Quadros CA, Andrus JK, Olivé JM, de Macedo CG (1992) Polio eradication from the Western hemisphere. Ann Rev Publ Health 13: 239–252

18. Centers for Disease Control (1988) Guidelines for evaluating surveillance systems. MMWR 37: 1–18

19. Expanded Program on Immunization (1988) Polio eradication field guide. 2nd edition, Washington, DC: Pan American Health Organization (ISBN 92 75 13006X)

20. Wharton M (1992) Evaluation of weekly negative reporting of acute flaccid paralysis, Venezuela, 1991. PAHO Bull 14: 4–5

21. Wright PF, Kim-Farley RJ, Robertson SE, Scott RM, Ward NA, Henderson RH (1991) Strategies for the global eradication of poliomyelitis by the year 2000. N Engl J Med 325: 1174–1179

22. World Health Organization (1989) Laboratory support for poliomyelitis eradication: Memorandum from a WHO meeting. Bull WHO 67: 365–367

23. Expanded Programme on Immunization (1990) Global poliomyelitis eradication by the year 2000. Manual for the virological investigation of poliomyelitis. Geneva, Switzerland: World Health Organization (WHO/EPI/CDS/POLIO/90.1.)

24. Centers for Disease Control (1990) Update: Progress toward eradicating poliomyelitis from the Americas. MMWR 39: 557–561

25. de Quadros CA, Andrus JK, Olivé JM, da Silveira C, Eikhof RM, Carrasco P, Fitzsimmons JW, Pinheiro FP (1991) Eradication of poliomyelitis: progress in the Americas. Pediatr Infect Dis J 10: 222–229

26. Gear JHS (1984) Nonpolio causes of polio-like paralytic syndromes. Rev Infect Dis 6 [Suppl]: 379–384
27. Melnick JL (1984) Enterovirus type 71 infections: A varied clinical pattern sometimes mimicking paralytic poliomyelitis. Rev Infect Dis 6 [Suppl]: 387–390
28. Nkowane BM, Wassilak SG, Orenstein WA, Bart KJ, Schonberger LB, Hinman AR, Kew OM (1987) Vaccine-associated paralytic poliomyelitis. United States: 1973 through 1984. JAMA 257: 1335–1340
29. Poliomyelitis Surveillance Summary: 1979. Atlanta, Centers for Disease Control, April 1981.
30. Schonberger LB, McGowan JE, Gregg MB (1976) Vaccine-associated poliomyelitis in the United States: 1961–1972. Am J Epidemiol 104: 202–211
31. Nakano JH, Hatch MH, Thieme ML, Nottay B (1978) Parameters for differentiating vaccine-derived and wild poliovirus strains. Prog Med Virol 24: 178–206
32. van Wezel AL, Hazendonk AG (1979) Intratypic serodifferentiation of poliomyelitis virus by strain-specific antisera. Intervirology 11: 2–8
33. Humphrey DD, Kew OM, Feorino PM (1982) Monoclonal antibodies of four different specificities for neutralization of type 1 polioviruses. Infect Immun 36: 841–843
34. Minor PD, Schild GC, Ferguson M, Mackay A, Magrath DI, John A, Yates PJ, Spitz M (1982) Genetic and antigenic variation in type 3 polioviruses: Characterization of strains by monoclonal antibodies and T1 oligonucleotide mapping. J Gen Virol 61: 167–176
35. Crainic R, Couillin P, Blondel B, Cabau N, Boue A, Horodniceanu F (1983) Natural variation of poliovirus neutralization epitopes. Infect Immun 41: 1217–1225
36. Osterhaus ADME, van Wezel AL, Hazendonk TG, Uytdehaag FGCM, van Asten JAAM, van Steenis B (1983) Monoclonal antibodies to polioviruses. Comparison of intratypic strain differentiation of poliovirus type 1 using monoclonal antibodies versus crossadsorbed antisera. Intervirology 20: 129–136
37. Kew OM, Nottay BK, Obijeski JF (1984) Applications of oligonucleotide fingerprinting to the identification of viruses. In: K. Maramorosch and K. Koprowski (eds). Methods in Virology. Academic Press, New York, 8: 41–84
38. Rico-Hesse R, Pallansch MA, Nottay BK, Kew OM (1987) Geographic distribution of wild poliovirus type 1 genotypes. Virology 160: 311–322
39. da Silva EE, Pallansch MA, Holloway BP, Cuoto Oliveira MJ, Schatzmayr HG, Kew OM (1991) Oligonucleotide probes for the detection of the wild poliovirus types 1 and 3 endemic to Brazil. Intervirology 32: 149–159
40. Yang Chen-Fu, De L, Holloway BP, Pallansch MA, Kew OM (1991) Detection and identification of vaccine-related polioviruses by the polymerase chain reaction. Virus Res 20: 159–179
41. Hayward JC, Gillespie SM, Kaplan KM, Packer R, Pallansch M, Plotkin S, Schonberger LB (1989) Outbreak of poliomyelitis-like paralysis associated with enterovirus 71. Pediatr Infect Dis J 8: 611–616

19

Monitoring of the environment as a means of poliovirus surveillance

Y. Ghendon

World Health Organization, Geneva, Switzerland

Summary

For certification that the eradication of poliomyelitis has been achieved it is essential to be sure that no wild polioviruses are circulated in the population, and no such viruses can be isolated from healthy people or from environmental samples. The central problem is how to detect wild poliovirus rapidly and reliably and technical difficulties arise due to the circulation of live polio vaccine viruses and other enteroviruses. New sensitive and more specific methods that allow detection of wild poliovirus in environmental specimens is urgently required. The new approach will include the use of the PCR in combination with other techniques. One of the important questions of environmental surveillance of wild polioviruses is the duration of such surveillance.

1. Introduction

The elimination of wild poliovirus circulation in mankind and the environment is one of the main objectives of the World Health Organization (WHO) poliomyelitis eradication initiative.

Eradication of any disease is not only elimination of the clinical form of the disease, but also eradication of the agent causing the disease.

Eradication of poliomyelitis is more difficult than eradication of smallpox. In contrast to smallpox virus, poliovirus infects most people without causing disease and is excreted by those infected for a considerable period of time. The virus may find its way into the environment via faecal excretion, may survive in waste water for several months and may thus be the source of new infections. It should be noted that wild poliovirus has been detected in sewage or river water even in those countries where no cases of poliomyelitis associated with wild polioviruses have occurred for several years [10].

For example, in the Netherlands where during the last decade there have been no cases of paralytical poliomyelitis, the wild polioviruses can be isolated from river water, and in one of the rivers (Meuse) during 1982–1985 fifteen strains of wild poliovirus types 1 and 2 were isolated [9].

For certification that eradication of poliomyelitis has been achieved it is essential to be sure that no wild polioviruses are circulated in the population and no such viruses can be isolated from healthy people or from environmental samples. Such information becomes extremely important because the eradication of poliomyelitis leads to the cessation of vaccination, as happened after the eradication of smallpox, and if wild poliovirus still circulate cessation of vaccination can lead to large outbreaks of poliomyelitis.

2. Environmental surveillance

Environmental surveillance will, therefore, be crucial to the success of the eradication programme and the absence of circulating wild poliovirus will be one of the criteria to declare countries free of wild poliovirus transmission.

Although several parts of the world have eliminated indigenous polioviruses and are close to becoming free of wild polioviruses transmission it is necessary to develop means of monitoring the environment and especially sewage to detect the importation of wild polioviruses from remaining endemic areas.

The greater part of the virus population of sewage consists of viruses of human origin and particularly those which are excreted in large quantities with faeces. These viruses are referred to as enteric viruses and include polioviruses, coxsackieviruses, echoviruses, and also adenoviruses, reoviruses, rotaviruses, and the Norwalk-type agents.

It is very important to note that polioviruses may survive in the environment for several months with the survival time being affected by temperature, water flux and several other factors.

It has been known for over 30 years that a definite correlation exists between the isolation of polioviruses and the number of clinical cases within a community. Polioviruses can be demonstrated in sewage using only grab sample techniques (i.e. no concentration when only 3–4 persons per thousand are excreting the virus). Using modern concentrating and detecting techniques, it may be possible to detect wild poliovirus in sewage if only one person in 10–50 000 is excreting the virus [2].

3. Samples examination

The examination of environmental samples for the presence of wild polioviruses requires three general steps: (A) collecting a representative sample, (B) concentrating the viruses in the sample, and (C) identifying and estimating quantities of the concentrated viruses. All these three steps need to be improved.

The detection of viruses in different kinds of water and sewage has been proffered for a number of methodologies but up to the present time there is no optimals and standard methodology for sampling, recovering, concentrating and isolating viruses in these samples.

Data from different countries showed that environmental samples were collected from varying sampling sites. However, most countries focused on sampling from sewage treatment plans, either at the intake station or from the activated sludge – which sampling is better should be investigated. Several sampling methods are in use and fall into one of two categories:

(a) "continuous sampling" (i.e. suspension for a certain period of time of gauze pads or macroporous glass beads packed in small bags in sewage streams at the inlet of sewage treatment plants. Viruses in the sewage absorb onto the pads or beads and are eluted thereof for inoculation on susceptible cells; one of the interesting approaches is using beads to which specific antibodies were bound (antigen-capture sampling).

(b) "grab sampling", i.e. taking a single sample of a certain volume from the sewage system, either at the inlet of sewage treatment plants or from the activated sludge. After treatment to remove toxic or interfering factors, the sample is inoculated on susceptible cells.

The optimal sampling volume or time depends on the sampling method, the nature of the sample, the application of concentration techniques and on the virus detection method used. These have not (yet) been established.

For raw sewage "grab sampling", samples from 1 to 5 litres have usually been sufficient for the demonstration of viruses, but it is not clear if this amount will be appropriate for countries which are close to becoming free of clinical cases of poliomyelitis in which a low quantity of wild polioviruses may be present in sewage.

Regarding sampling volume, it should be noted that an increase in the number of sewage plants sampled is preferable to an increase of the volume sampled at one single sewage plant.

Poliovirus is most readily detected in urban sewage. Dr J. John, India (personal communication), has found that among healthy under-five-year-olds from whom stool specimens were collected for viral isolation, 2% in urban areas were generally excreted for longer periods of time by young children.

In view of these factors, it seems that sampling sites, as a priority, should include urban sewage, particularly from sites near hospitals, schools, pre-schools, railway stations, and sewage treatment intakes. Where sewage systems exist in rural areas, similar sites could be monitored. In rural areas where no sewage systems are present, random testing of faeces from individuals in high-risk areas would be a priority above environmental monitoring.

Sampling size also relates to the proportion of the population monitored in the surveillance programme. A substantial percentage of the population in a country should be covered before a decision can be made regarding the absence of wild poliovirus circulation in a given area is not yet clear and should be the subject of special studies.

It is also very important to determine optimal frequency and timing of sampling. Although a certain seasonality of poliovirus excretion is well-documented, environmental sampling should be carried out throughout the year. In view of the limitations due to the present, rather laborious and time-consuming methods for environmental monitoring of poliovirus, it can be recommended that sampling at sewage treatment plants should take place at least monthly. Later on, when improved, more efficient virus detection methods are available, the frequency may be increased.

Regarding surveillance of wild poliovirus in healthy people, the main problems are related to estimation of number of people (children) for sampling, the frequency of sampling and the duration of such surveillance. If a very sensitive method for detection of wild polioviruses in faeces without cultivation of viruses is developed, it would be possible to analyse a mixture of 100–1000 samples.

As was noted above, in the analysis of sewage it is possible to detect one virus excreter among 10 000–50 000 non-excreters, but the main problem of environmental surveillance is posed by the difficulty in detecting wild-polioviruses in the widespread presence of live poliovirus vaccine strains and other enteroviruses using ordinary methods of poliovirus isolation. New, sensitive and more specific methods that allow detection of wild poliovirus in environmental specimens is urgently needed.

This problem was discussed at the WHO consultation of Methods for Detection of the Wild Polioviruses in Environmental and Clinical Samples [12].

4. Methods for detection of virus samples

So far, the multiplication of polioviruses in cell cultures in vitro has been the most successful method for their detection. Cell culture multiplication has excellent sensitivity because it can detect one infectious unit of virus. Specificity, however, is limited since most cells also permit replication of other enteroviruses. This lack of specificity necessitates the use of neutralizing antibodies for further identification and characterization of viral isolates. This is one of the reasons why the cell culture method is laborious and time-consuming and, therefore, less suitable for large scale application in surveillance programmes for the detection of wild poliovirus circulation.

After isolation in cell culture polioviruses are identified and serotyped by neutralization with type-specific antisera. Thereafter cross-absorbed antisera or monoclonal antibody panels are used for intratypic differentiation. In this way the great majority of isolates can readily be identified as wild or vaccine-derived viruses.

In general, the combination of virus isolation in cell culture and neutralization with specific antibodies works well in examination of patients' specimens. Major difficulties, however, are encountered when environmental samples have to be examined for the presence of wild polioviruses. The main reason for this is that environmental samples will nearly always contain non-polio enteroviruses and, in regions that use the live poliovirus vaccine, also very high levels of vaccine-derived polioviruses. Attempts have been made to circumvent this problem for example by including non-polio enterovirus neutralizing antibodies in the isolation assay. However, no solution has been found that allows large-scale, routine application.

Recently poliovirus-sensitive transgenic mice were produced by introducing the human gene encoding cellular receptors for poliovirus into the mouse genome (6,11). At the above WHO consultation the experts proposed that the development of mouse cell lines expressing the human receptor for polioviruses may offer new possibilities to use cell culture techniques for the selective multiplication of polioviruses from environmental samples.

However, it will still be necessary: to differentiate between wild and vaccine-derived polioviruses isolated on these cells; to evaluate whether other human enteroviruses replicate in these cells; to beware of animal viruses present in the environment that may also replicate in the mouse cell lines.

To improve selective replication of wild polioviruses, combinations of Sabin virus – specific monoclonal antibodies may be used to neutralise possible vaccine-derived viruses in the sample. Thereafter, viruses that were selected by this process can be identified and characterized as wild polioviruses using a combination of serological and molecular methods.

The new approach will include the use of the polymerase chain reaction (PCR) in combination with other techniques. PCR is a novel technique recently developed for the in vitro amplification of DNA. It is a cyclical procedure which directs the exponential amplification of a complementary copy DNA by using reverse transcriptase. Each PCR cycle usually takes approximately 3–5 minutes and is repeated 20–40 times. After 30 cycles a single copy of DNA can be multiplied by up to 10^9-fold.

PCR has exquisite analytical sensitivity and under the best conditions the procedure allows the detection of a single copy of a viral genome even if present in an excess of cellular or other virus DNA or RNA.

For polio the PCR technique can be implemented for rapid analysis of clinical samples without cultivation of virus. An even more important application will be its use to detect the silent transmission of wild polioviruses. Because of its high specificity and sensitivity, PCR is capable of identifying less than 2.5 wild poliovirus genomes in the presence of a large excess of Sabin attenuated polio virus [5].

Recently, primers pairs have been designed that selectively amplify the sequences of each of three Sabin attenuated poliovirus strains [7,15]. From the 5' end of the VP1 gene, specific PCR primer pairs and hybridization probes were prepared for the attenuated strains type 1, 2 and 3, yielding amplificates of 97 bp, 71 bp and 44 bp, respectively. With these primers it was possible to identify appropriate serotype of Sabin strains in a mixture of 1 + 2, 1 + 3, 2 + 3 and 1 + 2 + 3 serotypes of Sabin strains. The detection limit was approximately 10^{-5} pg of viral RNA. It was shown that when PCR was used in combination with poliovirus strain-specific ^{32}P labelled oligonucleotic probes, the limit of detection was ≤ 2.5 poliovirus genomes, exceeding the sensitivity of poliovirus isolation in cell culture by at least one hundred fold. In studies with sewage artificially contaminated by polioviruses type 1, it was found that wild poliovirus sequences can be selectively amplified in the presence of a large stoichiometric excess ($\geq 10^6$-fold) of Sabin virus genomes [5].

The requirements for identification of wild polioviruses in environmental samples by PCR differ from those appropriate for the analysis of clinical specimens from individual patients. There are two major problems: differentiation of polioviruses from non-polio enteroviruses (or other picornaviruses), and identification of poliovirus genotypes to detect wild poliovirus.

Two kinds of PCR primers have so far been used for clinical specimens.

The first-primers with broad specificities. These inevitably also amplify non-polio enteroviruses. Several sets of primers matching conserved sequences in the 5' noncoding region have been developed that recognize nearly all enteroviruses [3,8,14]. These can be used to screen the presence of enteroviruses in samples.

The second – primers which are genotype-specific. The design of these requires prior knowledge of the sequences of the virus under investigation. As it was noted above such primers have been developed specifically for oral polioviruses and for several different genotypes of wild polioviruses [7,15].

These approaches are quite reliable for the identification of polioviruses in clinical specimens.

Detection of wild polioviruses in specimens containing mixtures of viruses, such as environmental samples (sewage or waste water) or pooled clinical specimens, requires the use of methods for enrichment or amplification of wild virus sequences.

It was demonstrated earlier that the wild polioviruses circulating in different regions of the world have some differences in the VP.1-2A region of viral genome [13]. The genotype of polioviruses was defined by these authors as a group of polioviruses having no more than 15% genomic divergence within this 150-nucleotide VP1-2A interval. It was found that genetically related polioviruses cluster geographically.

Recently it was found that for identification of wild poliovirus in clinical or environmental samples by PCR it is necessary to use genotype-specific primers for the indigenous wild polioviruses circulating in the appropriate geographical region and that direct amplification by PCR does work under field conditions if the sequences of the indigenous wild genotype are known and primers designed accordingly [7]. However, for non-endemic regions, the genomic sequence heterogeneity of the poliovirus genotypes potentially imported is very large.

Under such circumstances, the use of genotype – specific primers becomes technically unreliable and carries the risk that wild genotypes may be present but not detected because the appropriate primer sets were not used. Thus, new approaches based on a combination of techniques must be developed.

There is already a large body of evidence that recombination among poliovirus (and other enteroviruses) genomes occurs frequently in natural infection and upon immunization with OPV. Thus, different genomic regions may derive from quite independent poliovirus (and other enterovirus) lineages. Because the poliovirus capsid contains the sequences necessary for binding to the cellular receptor as well as the determinants of viral serotype, primers for environmental surveillance should be directed to capsid-encoding regions of the genome. Genotype-specific primers can be systematically designed from these sequences of known poliovirus genomes. However, it appears unlikely from the available sequence data, that poliovirus-specific or serotype-specific PCR primers can be developed.

The central problem is how to detect wild polioviruses rapidly and reliably and technical difficulties arise because of the circulation of OPV-derived viruses and other enteroviruses. The most promising approach appears to be the use of generic primers in combination with other methods. It is desirable to apply selective conditions prior to the PCR step and other analytic methods following PCR. A good example is so called "antigen-capture PCR" as described for detection of hepatitis A virus [4].

Because of the limitations of the established cell culture and PCR methods as applied individually it is likely that a combination of neutralization, cell culture, PCR and genome sequencing will be necessary to detect the presence of wild poliovirus circulation.

One possible approach can be outlined by WHO experts as follows [12].

(a) Environmental samples should be treated by standardized methods and then neutralized using a combination of polyclonal antibodies against two of the poliovirus serotypes $(1 + 2, 2 + 3, 1 + 3)$ and a pool of neutralising monoclonal antibodies against the Sabin strain of the remaining serotype.

(b) The resulting suspension should be propagated on cells permissive for polioviruses but restrictive for the growth of other enteroviruses e.g. mouse cells carrying the human receptor for polioviruses.

(c) After propagation, reverse transcription and PCR amplification should be attempted using generic primers.

(d) The resulting PCR product should be further characterized for example by methods such as hybridization, restriction enzyme digestion to identify restriction fragment length polymorphism profiles [1], second cycle PCR or nucleotide sequencing.

It is planned that a collaboration will be established by WHO between several laboratories with the necessary expertise to investigate the above methods and establish when they could be used reliably to monitor environmental samples.

Finally, it should be stressed that one of the important questions of environmental surveillance of wild polioviruses is the duration of such surveillance. It is to be expected that environmental surveillance will have to be continued for several years after global disappearance of clinical cases of poliomyelitis. The absence of the wild virus from the environment will have to be demonstrated for several years before a decision on stopping vaccination can be made. How long this will take is not clear yet and should be the object of additional discussions.

5. Conclusion

Poliomyelitis still causes thousands of cases of paralytical disease inspite of the availability of safe and efficacious vaccines. Improvement of the vaccination programme and correct implementation of vaccines could not only prevent, but even eradicate poliomyelitis. Laboratory diagnosis of poliomyelitis and environmental surveillance will be an important key to the success of WHO's goal of global eradication of poliomyelitis by the year 2000.

References

1. Ballant J, Guillot S, Candrea A, Delpeyroux F, Crainic R (1991) The natural genomic variability of poliovirus analyzed by a restriction fragment length polymorphism assay. Virology 184: 645–654
2. Hovi T (1991) Quantitative aspects of poliovirus excretion and survival in the environment. Working paper at WHO/NPHI meeting on Environmental Surveillance of Wild Polioviruses Circulation in Europe, 5–6 April, Helsinki, Finland
3. Hypiä T, Auvinen P, Maaronen N (1989) Polymerase chain reaction for human picornaviruses. J Gen Virol 70: 3261–3268
4. Jansen R, Siegl G, Lemon S (1990) Molecular epidemiology of human hepatitis A virus defined by an antigen-capture polymerase chain reaction method. Proc Natl Acad Sci USA 87: 2867–2871
5. Kew O, Yang C-F (1990) PCR in the laboratory diagnosis of poliomyelitis. Working paper of WHO workshop on the use of PCR in the laboratory diagnosis of poliomyelitis, 29 October–1 November, Bilthoven, Netherlands
6. Koike S, Taya C, Kurata T, Abe S, Ise I, Yonekawa H, Nomoto A (1991) Transgenic mice susceptible to poliovirus. Proc Natl Acad Sci USA 88: 951–955
7. New approaches to poliovirus diagnosis using laboratory techniques: Memorandum from a WHO meeting (1992) Bull WHO 70: 27–33
8. Olive D, Al-Mufti S, Al-Mulla W, Khan M, Pasca A, Stanway G, Al-Nakib W (1990) Detection and differentiation of picornavirus in clinical samples following genomic amplification. J Gen Virol 71: 2141–2147
9. van Olphen N, van Loon A (1991) Detection of poliovirus in river water in the Netherlands (1979–1989) Working paper of WHO/NPHI meeting on Environmental Surveillance of Wild Polioviruses Circulation in Europe, 5–6 April, Helsinki, Finland
10. Pövry T, Stenvik M, Hovi T (1988) Virus in sewage waters during and after a poliomyelitis outbreak and subsequent nationwide oral poliovirus vaccination campaign in Finland. Appl Environ Microbiol 54: 371–374
11. Ren R, Constantini F, Gorgacz E, Lee J, Racaniello V (1990) Transgenic mice expressing a human poliovirus receptor: a new model for poliomyelitis. Cell 63: 353–362
12. Report of WHO Informal Consultation on Primers and Methods for Detection of the Wild Polioviruses in Environmental and Clinical Samples with PCR, (1992) Geneva, 26–27 March, WHO unpublished document
13. Rico-Hesse R, Pallansch N, Nottay B, Kew O (1987) Geographic distribution of wild poliovirus type 1 genotypes. Virology 160: 311–322
14. Rotbart H (1990) Enzymatic RNA amplification of the enteroviruses. J Clin Microbiol 28: 438–422
15. Yang C-F, De L, Holloway B, Pallansch U, Kew O (1991) Detection and identification of vaccine-related polioviruses by the polymerase chain reaction. Virus Res 10: 159–179

20

Criteria for the certification of poliomyelitis eradication

I. Arita

Agency for Cooperation in International Health, Kumamoto City, Japan

Summary

In 1988 the World Health Organization launched the global programme for eradication of poliomyelitis by the year 2000. Substantial progress was made in Latin America, where eradication is imminent.

In this paper, discussed are criteria of certification of eradication of poliomyelitis in Latin America. In addition to the formation of commission for certification, the importance is stressed on the continuing epidemiological surveillance based on acute flaccid paralysis, laboratory surveillance of wild polio virus and maintenance of a good immunity level. The concept for certification as described here would be useful in the future in other continents where polio eradication programmes are now in progress.

I. Introduction

In 1988, the World Health Assembly (WHA) resolved to launch the global programme for eradication of poliomyelitis with the year of 2000 as the target date for its completion. The programme has been organized with the collaboration of WHO, UNICEF, UNDP, the World Bank, bilateral agencies, Rotary International, and other NGOs. Since its inception, the reported incidence of poliomyelitis has declined, to the extent in 1991, 12,247 cases were reported, which constituted only 30% of the total reported in 1988 (Fig. 1).

The PAHO (Pan American Health Organization)/WHO Regional Office for Americas made substantial progress [1], such that the incidence in the Americas rapidly declined, what was apparently the last case of laboratory confirmed poliomyelitis, occurring in Peru in August, 1991. As of May 1992, despite ongoing intensive surveillance throughout the Region, no case of poliomyelitis had been reported over a period of nine months.

Since the interruption of poliomyelitis transmission was in sight, the PAHO organized meetings to discuss the criteria for the certification of polio-

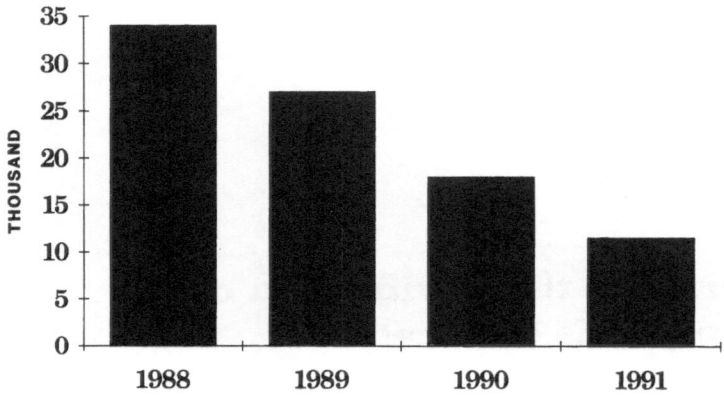

Fig. 1. Reported incidence of poliomyelitis, 1988–1991

myelitis eradication in Washington in July, 1990 and in Rio de Janeiro in March 1991.

The progress of poliomyelitis eradication efforts has been much slower in the other five WHO Regions (Europe, Western Pacific, South East Asia, Eastern Mediterranean and Africa), and certification activities will not be embarked upon for some years.

In this chapter, the criteria for certification of poliomyelitis eradication are discussed with reference to the situation in the Americas. Subsequently, the discussion is extended to anticipated certification activities in the other WHO regions and to the final target of the global certification of poliomyelitis eradication. Experience with the certification of smallpox eradication, which was completed in 1980, provides a model for the certification of poliomyelitis eradication.

II. Definition of eradication

Poliomyelitis eradication is defined in this chapter as the complete interruption of the transmission of wild polioviruses in human populations throughout the world.

Since there is no animal reservoir, and no evidence that infected persons maintain the virus for extended periods of time, eradication is technically feasible. However, in poliomyelitis there are a large number of subclinical infections (100 to 1000 per clinically manifested case), so that the absence of the clinical illness in the human population is not sufficient to ensure that the circulation of wild viruses has ceased. In this respect, poliomyelitis differs for smallpox, where the absence of clinical disease in the human population was a reliable index of the interruption of transmission. In fact, in 1980 when the global eradication of smallpox was certified by the Global Commission for Certification of Smallpox Eradication, the absence of subclinical infection was an important characteristic of smallpox on which analysis of the data by the Commission depended [2].

Once wild polioviruses disappear from the human population, there are theoretically two possibilities of the virus return. Firstly, strains of the wild polioviruses might be retained in a number of laboratories, and could escape from the laboratories and cause human infections. Although the risk is small, this possibility will have to be evaluated when the certification is concluded. In the smallpox eradication programme, WHO developed a system for the international registration of laboratories retaining smallpox virus, and eventually most of the laboratories destroyed their stocks of virus. However, virus stocks were retained in high security facilities in the WHO Collaborating Laboratories in Moscow and Atlanta. These stocks will be destroyed in 1993, when the studies on smallpox virus genome are completed.

Another theoretical risk for the reintroduction of wild virus is that of mutation of vaccine strains of virus to full virulence and transmissibility. However, in spite of the use of vaccine strains on a vast scale for many years, there is no evidence of any such change in any strains, and this risk can be ignored.

III. Criteria for poliomyelitis eradication

Based on the definition just outlined, what are the criteria for which eradication of poliomyelitis can be certified in an area? Although such criteria may vary according to the epidemiological situation of poliomyelitis and development of health services in the area, there are essentially three major criteria for the purpose, [1] surveillance of acute flaccid paralysis (AFP), [2] surveillance of wild polioviruses and [3] poliomyelitis immunization programme. If these three measures are effectively carried out, health services in each country will have the means to discover whether transmission of wild polioviruses is continuing within their jurisdictions.

1. Surveillance of acute flaccid paralysis

In PAHO, the current poliomyelitis eradication programme was started in 1985, three years before the World Health Assembly's resolution in 1988. As the routine vaccination as well as supplemental vaccination reduced the number of susceptible in the Latin America, the incidence has declined rapidly. Of 23 countries in the Region, only five countries reported cases in 1990 and only two countries in 1991. The last case occurred in August 1991 in Peru.

As the incidence of poliomyelitis declined, PAHO initiated surveillance on acute flaccid paralysis (AFP). Each year more than 2000 cases of AFP have been reported. Two stool specimens were taken from each AFP case and from each of its contacts, and laboratory tests were carried out to ensure that such cases were not due to infections with polioviruses (Figs. 2, 3).

Fig. 2. Confirmed cases of poliomyelitis. Region of the Americas, 1991, Last dates of onset: Colombia: 24 May 1991, Peru: 23 August 1991

Because with poliomyelitis the absence of clinical disease does not necessary mean that transmission has ceased, certification of eradication requires this type of intensive and careful surveillance.

Further, in PAHO, it was found that to be effective, the implementation of AFP surveillance should meet the following requirements:

(a) Weekly negative reporting of AFP should be implemented at all sentinel health units and at least 80% should report weekly.

(b) All (vs. 80%) cases of AFP should be investigated within 48 hours of reporting. Two stools should be collected from each case at this first encounter (Fig. 4) (Table 1).

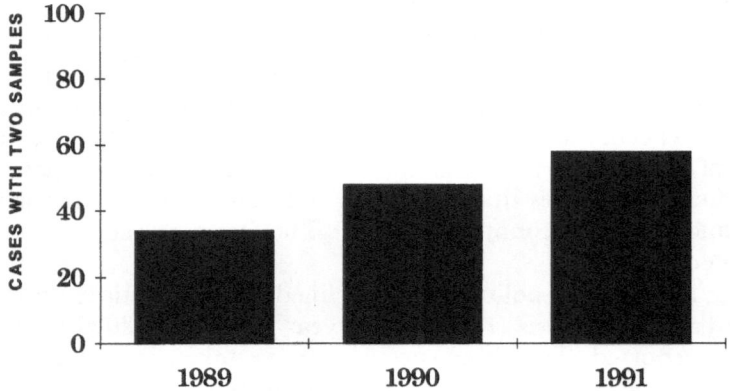

Fig. 3. Proportion of AFP cases with two adequate stool samples taken. Region of the Americas, 1989–1991

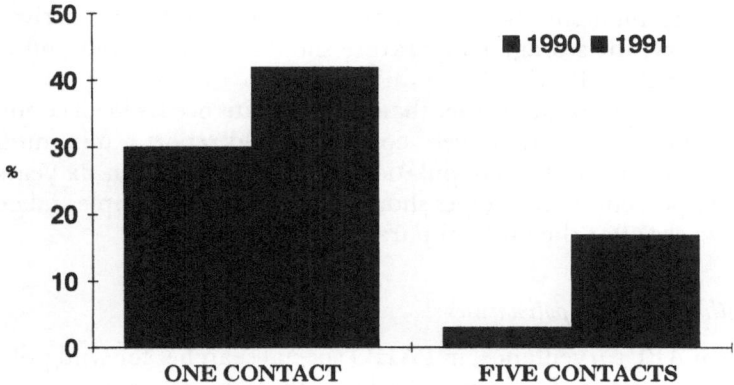

Fig. 4. Proportion AFP cases with stool samples taken from at least one and 5 contacts. Latin America, 1990–1991

Table 1. Classification of AFP cases. Region of the Americas, 1989–1990

Category	1989 (%)	1990 (%)
Wild (confirmed)	24 (19)	18 (17)
Compatible		
Deaths	18 (14)	13 (12)
Sequelae	59 (46)	41 (39)
Lost	20 (16)	21 (20)
Vaccine-associated	7 (5)	13 (12)
Sub-total	128 (100)	106 (100)
Discarded	1802	2297
Total AFP cases	1930	2403

Source: PAHO by Dr. J. K. Andrus, Dr. C. de Quadros, 1992

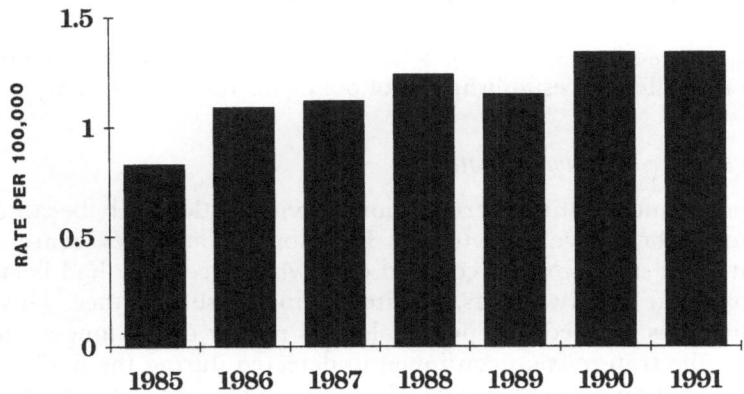

Fig. 5. Rate of AFP per 100 000 persons per year below 15 years. Latin America

(c) From the moment the stool sample is taken until the arrival at the laboratory, the storage temperature should be monitored and recorded to be less than 10°C.

(d) Since AFP due to cause other than poliomyelitis occurs in all communities, it is to be expected that every country should report a minimum rate of AFP of 1.0 per 100 000 population in children less than six years of age. Eighty percent of these cases should have two stool samples taken within two weeks after the onset of paralysis (Fig. 5).

2. Surveillance of wild polioviruses

Apart from AFP surveillance, in PAHO special searches for wild polioviruses are being conducted. This consists of [1] collection and examination of stool specimens taken from the high risk communities, such as a community where a "compatible case" has occurred, ("Compatible cases" are those reported to be poliomyelitis suspect, but who either died or could not be traced). [2] The second method, namely sewage sampling, has been developed in recent years and has successfully demonstrated the presence of wild poliovirus in sewage specimens. However, further studies will be required to increase its sensitivity with reference to site, sample size, method of concentration, etc. A manual detailing reliable methods of sewage sampling needs to be developed.

3. Immunization coverage

This criterium is not directly related to the absence of poliomyelitis due to wild polioviruses, but a high level of immunization coverage is a further assurance that high immunity in the community will act as a barrier to the spread of polioviruses imported from other endemic areas. Current experience in PAHO indicates that a rate of 80–85% is an attainable level and should be maintained in each country.

IV. Execution of certification activities

Certification activities meeting with the technical criteria just described, require the following establishment of policy for their execution.

1. Length of certification activities

Countries cannot continue certification activities indefinitely because of lack of resources, hence we need to ask: How long should they continue? The eradication of smallpox was certified only when a country had been free of smallpox for at least two years, despite continuing surveillance. This period, two years, was decided on epidemiological review of the longest period of time that the transmission continued undetected during the field operation of the eradication programme, which was eight months. To provide a major of safety, this period was multiplied by three.

In poliomyelitis eradication programme, PAHO has recommended that the surveillance and immunization activities should continue for at least three years after the last case in order to certify that poliomyelitis has been eradicated. The decision is arbitrary, but it is conceived that three years would be sufficient length of time to discover the hidden circulation of wild polioviruses by the recommended surveillance measures.

2. Commission for certification

The decision as to whether a country or region is free of wild polioviruses should be made by an authoritative and independent group, namely a commission, since such decision may influence the measures to be taken by national health authorities during the post eradication era in terms of possible modification of surveillance method and immunization programme or its final stoppage. In 1990, the PAHO formed a commission entitled the International Certification Commission on Poliomyelitis Eradication (ICCPE). Assuming that the Peruvian case is the last polio case in the Region, ICCPE will start to review for each recently endemic country data related to the criteria. Such data will be prepared by the national polio eradication programme.

As the global programme for eradication of poliomyelitis progresses, commissions similar to ICCPE will be created in each WHO Region, and eventually a Global Commission for Certification of Poliomyelitis Eradication will be formed, to certify whether the eradication has globally been achieved. Much greater efforts and national commitment, however, are required to make the global programme successful during the next eight years, if global eradication is to be achieved by the year 2000.

3. The epidemiological situation in different countries in the PAHO region

The epidemiological situation of poliomyelitis varies in different parts of the Americas. Poliomyelitis has been endemic in many Latin American countries during the past 3–4 years. However, more than 10 years have elapsed in North America since the last wild poliovirus was isolated; more than nine years in the English speaking Caribbean countries; and more than five years in the countries of the Southern Cone of the South America.

Thus, although poliomyelitis was endemic in the American region as a whole, some countries have been free of the infection for a long time. The question as to whether criteria for certification should be fully applied to all these countries will have to be examined by the ICCPE.

V. Certification in the other WHO regions

In Western Pacific Region, the second Technical Advisory Group on Poliomyelitis Eradication meeting held in Sebu, Philippines in 1991 confirmed

Table 2. Number of submitting zero poliomyelitis report to WHO

WHO Region	Number of countries	Number of countries reporting zero poliomyelitis incidence		
		1989	1990	1991
Africa	47	10	9	7
America	23	12	18	21
East Mediterranean	23	8	9	8
Europe	46	21	25	26
South East Asia	11	3	4	2
Western Pacific	35	29	29	30
	185	83	94	94

Source: WHO, EPI
The balance of countries is either reporting poliomyelitis or sending no report.
For example, of a total number of 185 countries, 94 reported zero incidence,
and the balance of 91 reported cases or did not send any report in 1991.

the previous commitment that the programme should eradicate the disease in the Region by 1995. Of 35 countries, five reported a total of 2615 cases in 1991, China reporting 70% of the regional incidence. If the current problem of shortage of the vaccine is rapidly solved, the Region is likely to meet the 1995 target for eradication, and the preparation for certification should start sometime in 1994 (Table 2).

In 1991, of 46 countries in the European Region, 15 countries reported a total of 313 cases. As the economic situation deteriorates in Eastern European countries as well as Confederation of Independence State (CIS), the risk of increased transmission is becoming greater. There is also a shortage of vaccine in CIS. Certification of eradication in Europe will have to wait for a few years until these situations have hopefully improved.

In the South East Asia and Eastern Mediterranean Regions, of a total of 34 countries, more than 20 countries are currently reporting cases, (8364 cases in 1991). The surveillance as well as immunization programmes in both regions need to be improved considerably before certification activities can be discussed.

Of 47 countries in the African Region, almost all either reported polio cases in 1991 or have not submitted a report to WHO. Although some country programmes in Africa might achieve interruption of polio transmission somewhat more easily than countries in other Regions, because of the rather low population density, greater efforts much be exercised to achieve the eradication by 2000. Further studies on surveillance methodology as well as on heat-stable vaccine are needed to promote progress in Africa. Certification may come last in this Region, but by then experience in the other Regions may become available to promote eradication and certification activities.

VI. Global certification

Assuming that there is no unexpected setback, the global certification of poliomyelitis eradication could be achieved three years after the last case of naturally occurring poliomyelitis is discovered, anywhere in the world. Until the global certification of poliomyelitis eradication, all nations throughout the world should maintain effective surveillance, capable of discovering importations or any ongoing circulation of wild poliovirus, and should continue immunization so as to maintain a reasonable level of the immunity in the population.

An immediate concern is that when the Americas are certified free of poliomyelitis, other Regions, especially Africa, are still likely to have a relating high incidence of poliomyelitis, causing the risk of importation from Africa to Americas, although such a risk might be remote. The Americas will therefore have to maintain a good surveillance and immunization programme after certification. Such efforts, however, may be more readily sustained because PAHO has now initiated programmes to strengthen special immunization and surveillance campaigns, such as measles death reduction and neonatal tetanus elimination, and a poliomyelitis component can be combined with these efforts.

At the same time, it is extremely important for the other WHO Regions to expedite the progress of poliomyelitis eradication. In the eradication of smallpox, its cost benefit was estimated as follows: the world community spent about $300 million for the intensified smallpox eradication programme from 1967 to 1980, but the discontinuation of vaccination as well as all related measures for the prevention of smallpox, when the global certification was made in 1980, resulted in an annual saving of approximately $1000 million. In case of poliomyelitis eradication, after the global certification, preventive measures against poliomyelitis infection including immunization could be discontinued. The resulting saving would be much greater than with small-pox eradication, since poliomyelitis immunization is much more expensive than smallpox immunization. It follows that the shorter the eradication period the greater the benefit from the achievement.

In the post eradication era, certain measures would be still kept in place as an insurance policy to cope with unexpected situations. Such measures include the maintenance of international registration of suspected case of poliomyelitis, stocks of certain amount of polio vaccine (both oral and in-activated vaccine), and the supervision or destruction of wild polioviruses being retained in laboratories. Similar measures were taken during the post smallpox eradication era [3].

VII. Conclusion

Since the inception of the global eradication of poliomyelitis in 1988, the Regional programme in the Americas has achieved great progress, and eradication is imminent. In 1990–1991, the PAHO formed the regional International Certification Commission on Poliomyelitis Eradication which will proceed with certification activities for the next three years. The maintenance of sensitive surveillance on both AFP and wild polioviruses and high level immunization coverage are major criteria for certification.

The progress in the other WHO Regions varies. It would be premature to initiate the certification activities in any Region except for PAHO, but all the Regions should pay attention to the methods having adopted for certification being applied in PAHO Region as a model for their eventual certification work.

The global certification of poliomyelitis is envisaged sometime early 2000s if the programme progresses as planned, but at this moment it is most important for all Regions other than the Region of the Americas to strengthen their current efforts so that circulation of wild polioviruses in respective Regional areas can be interrupted as soon as possible.

Acknowledgement

All the figures and tables presented in this chapter are taken from PAHO and WHO resource materials. I am most grateful both to Dr. Frank Fenner, the chairman of the WHO Global Commission for Certification of Smallpox Eradication (1978–1980) and to Dr. Ciro de Quadros, Senior Advisor on Immunization, PAHO for critical review and advise on the manuscript.

References

1. de Quadros CA, Andrus JK, Olive JM, de Marcedo CG, Henderson DA (1992) Polio Eradication: The Experience in the Americas. Ann Rev Publ Health 13: 239–252
2. Arita I (1979) Virological evidence for the success of the smallpox eradication programme. Nature 279: 293–298
3. Fenner F, Henderson DA, Arita I, Jezek Z and Ladnji ID (1988) Smallpox and its Eradication, World Health Organization, Geneva

21

Poliovirus vaccine formulations

P. A. Patriarca, R. W. Linkins, and **R. W. Sutter**

Division of Immunization, Centers for Disease Control and Prevention, Atlanta, Georgia, USA

Summary

Most of the early trials of Sabin-derived, live attenuated oral poliovirus vaccine (OPV) involved sequential administration of a single dose of each poliovirus type in monovalent form. Replication of each Sabin strain in the gastrointestinal tract was usually demonstrated in 90–100% of seronegative subjects at dosage levels of $\geq 10^5$ $TCID_{50}$. However, when the same dosage of each Sabin type were mixed together and administered as a trivalent preparation (TOPV), intestinal excretion and subsequent antibody production were consistently lower for types 1 (50%–70%) and 3 (55%–75%), with no apparent effect on type 2 (96%–100%). In an effect to compensate for these deficiencies, which were largely attributed to interference from type 2, "balanced" formulations were developed and tested in Canada in the early 1960s in which higher relative potencies of types 1 (10^6 $TCID_{50}$) and 3 ($10^{5.5}$ [300 000] $TCID_{50}$) were associated with coresponding increases in seroconversion rates. Limited studies in developing countries have revealed similar findings, although antibody responses to types 1 and 3 after three doses of the currently recommended formulation of TOPV have often remained suboptional. In view of the limited resources currently available for the poliomyelitis eradication initiative and the need to maximize the efficiency of vaccine delivery by achieving higher rates of seroconversion early in life, the development of more immunogenic formulations of TOPV remains a high priority.

Introduction

Widespread use of trivalent oral poliovirus vaccine (OPV) has been associated with the virtual elimination of paralytic poliomyelitis in many industrialized countries, as well as substantial declines in the incidence of the disease throughout much of the developing world [1,2]. Nevertheless, an estimated 150 000 paralytic cases continue to occur each year, primarily in tropical and subtropical areas [3,4]. Although much of the remaining burden of polio-myelitis can still be attributed to cases occurring in unvaccinated or partially vaccinated children, vaccine failure remains an important problem, partic-ularly with respect to poliovirus types 1 and 3 [5–8]. Factors contributing to OPV failure are complex, and appear to include high levels of residual maternal antibody, high levels of secretory antibody present in colostrum and breast milk, malnutrition, a high risk of concurrent infection with non-polio enteroviruses and other enteric pathogens, and other host factors which have not been fully elucidated [7,8]. Another factor which may also influence the immune response to vaccination is the formulation of the vaccine itself, which can be defined in terms of both absolute and relative potency of each of the three Sabin strains [8]. Careful evaluation of vaccine formulation and its effect on neutralizing antibody responses assumes great importance for several reasons: first, vaccine formulation represents one of the few factors that can be easily modified; second, no other modifications in the routine vaccination schedule or method of delivery would theoretically be required; and third, selection of optimal formulations could achieve higher rates of seroconversion at each visit, thereby improving the efficiency of vaccination and the likelihood of protection against all three serotypes earlier in life. In this chapter, we will provide an historical review of the development of live, attenuated poliovirus vaccines; review available data regarding the effect of vaccine formulation on serologic responses to OPV; and suggest potential avenues for additional research in this area.

Development and use of monovalent and trivalent oral poliovirus vaccines

The development of live, attenuated poliovirus vaccines was one of the most important events in the history of poliomyelitis control. Years of painstaking research to develop suitable candidate strains was carried out at three different institutions (Children's Hospital of Cincinnati [9], Lederle Laboratories [10], and the Wistar Institute [11]), and led to the production of nine different monovalent preparations that were eventually tested in humans [12]. Because of their superiority in inducing high levels of detectable antibody in susceptible (seronegative) recipients, low degrees of neurovirulence in monkeys, and genetic stability following replication in the human host, the three candidate strains (for types 1, 2, and 3) developed by Sabin were eventually chosen for licensure and mass production throughout much of the world [12].

Most of the early trials of Sabin-derived OPV involved sequential administration of a single dose of each poliovirus type in monovalent form. Replication in the gastrointestinal tract was usually demonstrated in 90–100% of seronegative subjects at dosage levels of 10^5 $TCID_{50}$, although most vaccine preparations were evaluated at somewhat higher dosages [13–17]. However, when the same dosage of each Sabin type were mixed together and administered as a trivalent preparation, intestinal replication and subsequent antibody production were consistently lower for types 1 (50%–70%) and 3 (55%–75%), with virtually no effect on type 2 (96%–100%) [18–22]. These observations set the stage for evaluating the effect of altering the dosages of

Table 1. Virus excretion and antibody responses to poliovirus type 1 in infants after administration of one dose of trivalent oral poliovirus vaccines of differing composition, Montreal, Canada, 1966 (adapted from Ref. [24])

Vaccine formulation (in $\log_{10}$ TCID$_{50}$ of Sabin type:			Overall ratio*	No. subjects	% with excretion	Sero-conversion rate (%)	GMT
1	2	3					
6.0	5.0	5.5	10:1:3	19	68	32	5
5.7	3.4	4.7	20:1:2	16	81	50	13
6.0	3.0	4.7	1000:1:50	14	93	56	21
5.7	–	4.7	10:–:1	10	90	60	35

*Overall ratio of the three Sabin types (1, 2, and 3) in each formulation, respectively, with lowest dosage designated as "1"

the three types to compensate for differences in their relative rates of replication in vivo and in vitro [8,12].

In an effect to achieve high rates of seroconversion to all three serotypes of poliovirus with the fewest number of doses possible, investigators at Connaught Medical Research Laboratories (Ontario, Canada) were among the first to experiment with "balanced" formulations of TOPV. A study carried out in the city of Prince Albert, Saskatchewan, in 1961 was the first time in which the current and most widely used formulation of TOPV (containing $10^6, 10^5$, and $10^{5.5}$ (300 000) TCID$_{50}$ of Sabin types 1, 2, and 3, respectively) was evaluated on a large scale [23]. Of the 23 000 persons vaccinated, paired sera (for neutralizing antibody studies) and stool specimens (for virus excretion studies) were obtained from a sample of ~600 persons, approximately two-thirds of whom were infants and young children. Following a single dose of this balanced "10:1:3" formulation, nearly 100% of these persons had detectable antibody to all three serotypes, including 103 (97%) of the 106 previously seronegative subjects [23]. Excretion of vaccine virus was also documented in a high proportion of vaccinees, particularly in those with absent or low levels of pre-existing antibody. On the basis of this study, the "10:1:3" formulation of TOPV was licensed for use in Canada in 1962, and in the United States in 1963. Although additional studies carried out by Connaught suggested that seroconversion rates to types 1 and 3 could be increased further when the dosage of type 2 was lowered to $< 10^4$ TCID$_{50}$ (Table 1), the "10:1:3" vaccine remained the standard recommended formulation throughout much of the world [24].

Experiences with 10:1:3 and other "balanced" formulations in the developing world

Most of the early trials of OPV conducted in tropical and subtropical areas of Latin America and Africa [13,21,22,25,26] involved the administration of candidate strains developed by Lederle Laboratories [10] and the Wistar

Institute [11]. In general, these trials involved sequential administration of monovalent preparations, and showed little, if any, differences in serologic responses compared with earlier studies in industrialized countries [12,27]. Thus, when the "balanced" trivalent formulation developed in Canada was shown to be effective, most public health authorities believed that it was not necessary to test this formulation or consider alternative formulations in developing countries.

Over the past 25 years, however, considerable data have accumulated regarding the use and immunogenicity of Sabin-derived OPV in the developing world, and suggest that the $10:1:3$ formulation may not provide optimal protection against all three poliovirus serotypes. A recent review of more than 50 of these studies [8] emphasizes the wide variation in reported rates of seroconversion among different countries, with the overall experience indicating that neutralizing antibody responses to types 1 and 3 are often suboptimal: whereas a weighted average of 90% of children seroconverted to type 2 after three doses of the balanced "$10:1:3$" formulation, only 73% and 70% of these children seroconverted to types 1 and 3, respectively. Three of these studies compared antibody responses to different formulations of OPV in the same or similar populations, and are reviewed in more detail below.

The first of these studies took place in Thailand in 1967, and compared antibody responses in two groups of triple-seronegative children who received three doses of either a "balanced" formulation of TOPV (containing $500\,000$, $100\,000$, and $200\,000$ $TCID_{50}$ of Sabin types 1, 2, and 3, respectively) or a formulation containing an equivalent dosage ($200\,000$ $TCID_{50}$) of all three serotypes [28]. As shown in Table 2, overall seroconversion rates to types 1 and 3 (after both two and three doses of TOPV) were significantly higher among children who received the "balanced" formulation. More pronounced differences in children ≥ 12 months of age (Table 2) were most likely due to a high incidence of co-infection with non-polio enteroviruses (50%) in younger children [28]. Because responses to type 2 were similar in both vaccine groups, regardless of age, differences for types 1 and 3 were again attributed to reductions in the relative amount of type 2 (and consequent interference) in the "balanced" formulation.

The second study took place in India in 1972 [29], and compared antibody responses between three groups of unvaccinated infants who were given one dose of a high-potency, monovalent vaccine containing either type 1 (10^7 $TCID_{50}$), type 2 (10^6 $TCID_{50}$), or type 3 ($10^{6.5}$ [$300\,000$] $TCID_{50}$), and one group of infants who received the same dosage of each type in trivalent form. Seroconversion rates to types 1, 2, and 3 in the monovalent groups were 98%, 93%, and 76%, respectively, compared with only 42% ($p<.01$), 85% ($p=NS$), and 31% ($p<.01$) in the TOPV group. An earlier study of the standard "$10:1:3$" vaccine carried out by the same investigators [30] demonstrated seroconversion rates of 29%, 67%, and 31% to types 1, 2, and 3, respectively (after a single dose), among infants of similar age, neighborhood, and social class as those included in the study in 1972.

Table 2. Antibody responses in triple-seronegative children following two and three doses of trivalent oral poliovirus vaccine, Bangkok, Thailand, 1967 (adapted from Ref. [28])

Time after vaccination	Age (in years)	Formulation[a] (no. tested)	Seroconversion rate (in %)		
			Type 1	Type 2	Type 3
6 wks	< 1	5:1:2 (n = 12)	58	100	100[b]
after	< 1	1:1:1 (n = 66)	53	86	52
2nd	1	5:1:2 (n = 41)	90[b]	100	98[b]
dose	1	1:1:1 (n = 17)	53	94	53
	2	5:1:2 (n = 26)	92[b]	100	100[b]
	2	1:1:1 (n = 10)	60	90	60
	Total	5:1:2 (n = 81)	86[b]	100	99[b]
	Total	1:1:1 (n = 93)	57	89	56
4 wks	< 1	5:1:2 (n = 12)	58	100	100[b]
after	< 1	1:1:1 (n = 66)	56	86	53
3rd	1	5:1:2 (n = 41)	95[b]	100	100[b]
dose	1	1:1:1 (n = 16)	69	94	56
	2	5:1:2 (n = 25)	100[b]	100	100[b]
	2	1:1:1 (n = 10)	60	90	70
	Total	5:1:2 (n = 78)	90[b]	100	100[b]
	Total	1:1:1 (n = 92)	69	95	69

[a]"5:1:2" = TOPV formulation containing 500 000, 100 000, and 200 000 $TCID_{50}$ of Sabin types 1, 2, and 3, respectively). "1:1:1" = TOPV formulation containing 200 000 $TCID_{50}$ for all three Sabin types
[b]Significantly higher vs. 1:1:1 formulation, $p < .05$, chi-square test

The difference in antibody response to type 1 observed between the two TOPV studies (42% vs. 29%, $p < .05$) – in spite of the same relative ratio between the three vaccine components (10:1:3) – was among the first to suggest that the absolute dosage of type 1 may also play an important role in seroconversion, an observation that will be discussed in more detail later in this chapter.

The third study examining responses to "balanced" formulations in developing countries was carried out in Brazil in 1986, in the midst of one of the largest type 3 outbreaks ever reported [31]. Following an epidemiologic investigation which documented selective failure of the type 3 component in the standard (10:1:3) formulation, the Brazilian Ministry of Health attempted to control the outbreak by manufacturing a new vaccine containing twice the amount of type 3 (i.e., $10^{5.8}$ [600 000] $TCID_{50}$). To evaluate the impact of this change on seroconversion rates to type 3 441 children < 5 years of age were randomly assigned to receive a single dose of either standard (10: 1: 3) TOPV, the new, higher-potency TOPV (10: 1: 6), or a monovalent vaccine containing 300 000 $TCID_{50}$ of type 3; 235 (53%) of the 441 children had no detectable antibody to type 3 at the time of vaccination. Although rates of seroconversion to types 1 and 2 were equivalent following vaccination

Table 3. Rates of seroconversion to type 3 following one dose of trivalent or monovalent oral poliovirus vaccine, Pernambuco, Brazil, 1986 (adapted from reference [31]).

Vaccine Formulation[a]	No. subjects	Seroconversion rate to type 3 (in %) [95% CI[b]]	Response ratio [95% CI] (compared with 10:1:3 vaccine)
Seronegative to type 3:			
10:1:3	83	16% [8, 23]	1.0 (referent)
10:1:6	73	42% [31, 54]	2.7 [1.6, 4.7], p < .0005
−:−:3	79	52% [41, 63]	3.3 [2.0, 5.4], p < .00003
Triple-seronegative:			
10:1:3	10	10% [0, 29]	1.0 (referent)
10:1:6	13	54% [27, 71]	5.4 [0.8, 35.6], p < 0.081
−:−:3	15	53% [28, 79]	5.3 [0.9, 33.5], p < 0.074

[a]Formulation expressed in $TCID_{50} \times 10^5$ of Sabin types 1, 2, and 3, respectively
[b]CI = confidence interval

with either formulation of TOPV. seronegative children who received the new formulation were 2.7 times more likely to seoconvert to type 3 (p < .005, Table 3). Similar differences for type 3 were observed when monovalent vaccine was compared with standard TOPV, even though both groups had received the same dose (300 000 $TCID_{50}$) of type 3 antigen (Table 3). The low rate of seroconversion to type 3 in the standard TOPV group was associated with a higher rate of reinfection with type 2 vaccine virus, which also appeared to interfere with seroconversion to type 1 [31]. These findings extend earlier observations that interference from Sabin type 2 may substantially contribute to type-specific vaccine failure, and provided additional evidence that such interference may be overcome by changing the relative dosages of the three Sabin strains.

In view of the results from the study in Brazil, and the continued transmission of wild type 3 polioviruses throughout much of Latin America in 1986 and 1987, the Pan American Health Organization (PAHO) recommended that higher-potency type 3 vaccines be used throughout the Region beginning in 1987. Aside from a type 3 outbreak in Mexico in 1990 that was linked to the use of vaccine of substandard potency produced by the Mexican Ministry of Health, only sporadic type 3 cases were documented in Latin America after 1987, with none being reported since 1990 [32]. In view of this success, the Global Advisory Group of the Expanded Programme on Immunization (EPI) recommended in 1990 that vaccines containing 10^6, 10^5, and $10^{5.8}$ infectious units per dose for types 1, 2 and 3, respectively, with a balance of 10:1:6, be considered the minimum standard for TOPV [33].

Conclusions and directions for further research

Although the use of the new, higher-potency formulation of TOPV may lead to improved control of type 3 poliomyelitis in the developing world, continued evidence of relatively poor seroconversion rates to type 3 [8,34] underscores the need for further research to ensure that the goal of global eradication of poliomyelitis can be achieved by the year 2000 [2]. Whether further alterations in formulation would lead to improved antibody responses in uncertain, but results from an unpublished study conducted in Guam in 1962 found a striking dose-response relationship between type 3 potency and the ultimate seroconversion rate to type 3 after three doses of TOPV (U.S. Food and Drug Administration; unpublished data). As shown in Fig. 1, increases in type 3 potency above $10^{5.8}$ (600 000) $TCID_{50}$ may have the potential for increasing seroconversion rates well beyond 90%. Recent potency tests of lots of TOPV produced in the United States by Lederle Laboratories found an average of $10^{6.3}$ (2.2 million) $TCID_{50}$ of type 3 [8]. The absence of an increased risk of vaccine-associated paralysis with the Lederle vaccine [1,35] suggests that there is considerable room for safely increasing the potency of the type 3 component. Increasing type 3 potency may also reduce the effect of interference from non-polio enteroviruses, which have been shown to influence antibody responses to type 3 more so than types 1 or 2 [6,8,36]. Although substantial increases in type 3 potency could have an adverse effect on the global supply of TOPV due to fixed production capacity, newer manufacturing methods which use continuous (Vero) cell lines in microcarrier systems would likely be sufficient to meet future demand. In view of these considerations, the World Health Organization is presently

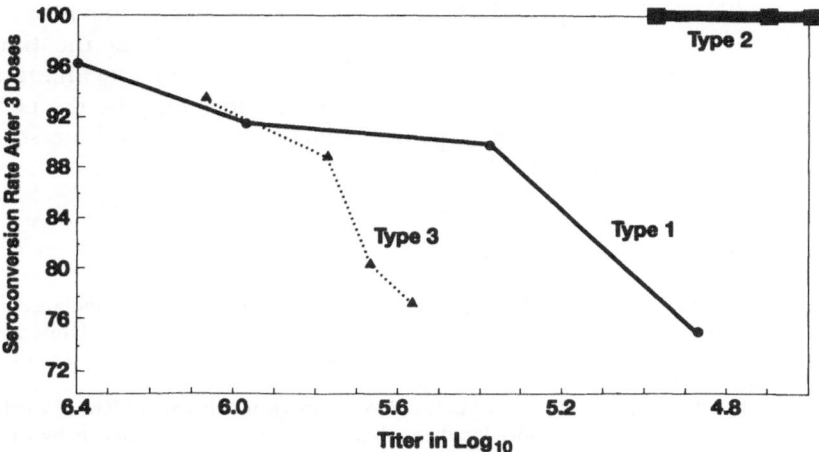

Fig. 1. Serologic response to trivalent OPV according to the dosage of each Sabin strain, Guam, 1962 (adapted from Ref. 8)

testing the higher-potency, U.S. formulation in a large-scale trial in Oman, the results of which should be available in 1993.

Although much of the work with TOPV formulation has been directed against type 3, additional studies to improve response rates to type 1 are urgently needed in view of the epidemiologic importance of type 1 polio-myelitis throughout the world [2–4]. Whether further alterations in formula-tion aimed at type 1 would lead to improved antibody responses is uncertain, but the results of the 1962 study in Guam also suggest that increasing the potency of the type 1 component beyond 10^6 $TCID_{50}$ could raise seroconver-sion rates to type 1 even further (Fig. 1). Increases in the relative dosage of type 1 may also be important because of its higher rate of inactivation (when compared with types 2 and 3) following exposure to moderate or high temperatures [37]. Because problems still remain in maintaining an adequate cold chain in some areas of the tropics, increases in the absolute dosage of type 1 would thus provide greater assurances that its higher relative potency would be maintained over a wide range of temperatures. Large-scale trials to evaluate higher-potency type 1 vaccines have recently been carried out in Brazil and The Gambia, with preliminary results indicating that seroconver-sion rates to type 1 may be increased by as much as 15% [34].

In addition to examining these and other alternative formulations of TOPV, consideration should also be given to the use of monovalent or bivalent type 1 and 3 vaccines to eliminate the problem of type 2 interference altogether. Use of such vaccines in South Africa [38] and Czechoslovakia [39, 40] has achieved good results in achieving and maintaining control of poliomyelitis and high levels of immunity in the population. Although the use of monovalent vaccines could potentially cause confusion among vaccinators in the field, the strategy adopted in South Africa whereby all newborns receive a single dose of type 1 vaccine followed by three doses of TOPV within the first 4–6 months of life has been shown to be programmatically feasible [38]. In addition, a comparison of serologic responses to monovalent type 3 vaccine versus a supplemental dose of standard TOPV at the time of measles vaccination at 9 months of age is presently under evaluation in Oman. Another potential use for monovalent vaccines may be the control of outbreaks, a strategy which appears to have been highly successful in limited evaluations to date [41].

References

1. Strebel PM, Sutter RW, Cochi SL, Biellik RJ, Brink EW, Kew OM, Pallansch MA, Orenstein WA, Hinman AR (1992) Epidemiology of poliomyelitis in the United States one decade after the last reported case of indigenous wild virus-associated disease. Clin Infect Dis 14: 568–579
2. Wright PF, Kim-Farley RJ, de Quadros CA, Robertson SE, Scott RM, Ward NA, Henderson RH (1991) Strategies for the global eradication of poliomyelitis by the year 2000. N Engl J Med 325: 1774–1779
3. World Health Organization (Expanded Programme on Immunization) (1989) Polio-myelitis in 1986, 1987, and 1988 (Part I). Wkly Epidemiol Rec 64: 273–279

4. World Health Organization (Expanded Programme on Immunization) (1989) Polio-myelitis in 1986, 1987, and 1988 (Part II). Wkly Epidemiol Rec 37: 281–285
5. John TJ (1972) Problems with oral poliovaccine in India. Indian Pediatr 9: 252–256
6. Pangi NS, Master JM, Dave KH (1977) Efficacy of oral poliovaccine in infancy. Indian Pediatr 14: 523–528
7. Montefiore DG (1971) Problems of poliomyelitis immunization in countries with warm climates. In: Proceedings of the International Conference on the Application of Vaccines against Viral, Rickettsial, and Bacterial Diseases of Man. Pan American Health Organization, Washington, DC. Scientific publication 226: 182–185
8. Patriarca PA, Wright PF, John TJ (1991) Factors affecting the immunogenicity of oral poliovirus vaccine in developing countries. Rev Infect Dis 13: 926–939
9. Sabin AB (1959) Recent studies and field tests with a live attenuated poliovirus vaccine. In: First International Conference on Live Poliovirus Vaccines. Pan American Health Organization, Washington, DC. Scientific Publication 44: 14–33
10. Cabasso VJ, Jervis GA, Moyer AW, Roca-Garcia M, Orsi EV, Cox HR (1959) Cumulative testing experience with consecutive lots of oral poliomyelitis vaccine. In First International Conference on Live Poliovirus Vaccines. Pan American Health Organization, Washington, DC. Scientific Publication 44: 102–134
11. Koprowski H, Jervis GA, Norton TW (1952) Immune responses in human volunteers upon oral administration of a rodent-adapted strain of poliomyelitis virus. Am J Hyg 55: 108–119
12. Payne AMM (1961) Field safety and efficacy of live attenuated poliovirus vaccines. In: Papers and Discussions Presented at the 5th International Poliomyelitis Conference, Copenhagen, Denmark. JB Lippincott, Philadelphia, pp 257–261
13. Horwitz A, Martins da Silva M, Bica AN (1961) Large-scale field studies with live atenuated poliovirus vaccines in the Americas. In: Papers and Discussions Presented at the 5th International Poliomyelitis Conference, Copenhagen, Denmark. JB Lippincott, Philadelphia, pp 221–228
14. Kurnosova LM, Zhilova GP (1960) Immunologic changes in the blood of children inoculated with live poliomyelitis vaccine. In: Live Vaccine Against Poliomyelitis. Institute of Experimental Medicine of the USSR Academy of Medical Sciences, Leningrad.
15. Voroshilova MK (1961) Influence of dose and schedule of oral immunization of people with live poliovirus vaccine on antibody response. In: Papers and Discussions Presented at the 5th International Poliomyelitis Conference, Copenhagen, Denmark. JB Lippincott. Philadelphia, pp 296–303
16. Verlinde JD, Wilterdink JB (1959) Vaccination and revaccination with live attenuated poliovirus in the Netherlands. In: First International Conference on Live Poliovirus Vaccines. Pan American Health Organization, Washington, DC Scientific Publication 44: 355–366
17. Cox HR, Cabasso VJ, Markham FS, Moses MJ, Moyer AW, Roca-Garcia M, Ruegsegger JM (1959) Immunologic response to trivalent oral poliomyelitis vaccine. In: First International Conference on Live Poliovirus Vaccines. Pan American Health Organization, Washington, DC. Scientific Publication 44: 229–248
18. Paul JR, Horstmann DM, Riordan JT, Opton EM, Green RH (1960) The capacity of live attenuated polioviruses to cause human infection and to spread within families. In: Second International Conference on Live Poliovirus Vaccines. Pan American Health Organization, Washington, DC Scientific Publication 50: 174–186
19. Zhdanov VM, Chumakov MP, Smorodintsev AA (1960) Large-scale ptacical trials and use of live poliovirus vaccine in the USSR. In: Second International Conference on Live Poliovirus Vaccines. Pan American Health Organization, Washington, DC Scientific Publication 50: 576–590
20. Voroshilova MK, Zhevandrova VI, Tolskaya EA, Koroleva GA, Taranova GP (1960) Virologic and serologic investigations of children immunized with trivalent live vaccine from A.B. Sabin's strains. In: Second International Conference on Live Poliovirus

Vaccines. Pan American Health Organization, Washington, DC. Scientific Publication 50: 240–265

21. Tomlinson AJH, Davies J (1961) Trial of living attenuated poliovirus vaccine. A report to the Public Health Laboratory Service to the Poliomyelitis Vaccines Committee of the Medical Research Council. Br J Med ii: 1037–1044

22. Ramos Alvarez M, Bustammante ME, Alvarez Alba R (1960) Use of Sabin's live poliovirus vaccine in Mexico. Results of a large-scale trial. In: Second International Conference on Live Poliovirus Vaccines. Pan American Health Organization, Washington, DC. Scientific Publication 50: 386–409

23. Robertson HE, Acker MS, Dillenberg HO, Woodrow R, Wilson RJ, Ing WK, MacLeod DRE (1962) Community-wide use of a "balanced" trivalent oral poliovirus vaccine (Sabin). Can J Public Health 53: 179–191

24. Belcourt RJP (1967) Development of oral polio vaccines. Part 3: Field studies. Can J Public Health 58: 152–157

25. Lebrun A, Cerf J, Gelfand HM, Courtois G, Koprowski H (1959) Preliminary report on mass vaccination with live attenuated poliomyelitis vaccine in Leopoldville, Belgian Congo. In: First International Conference on Live Poliovirus Vaccines. Pan American Health Organization, Washington, DC. Scientific Publication 44: 410–418

26. Plotkin SA, Lebrun A, Courtios G, Koprowski H (1960) Vaccination with the CHAT strain of type 1 attenuated poliovirus vaccine in Leopoldville, Belgian Congo. In: Second International Conference on Live Poliovirus Vaccines. Pan American Health Organization, Washington, DC. Scientific Publication 50: 466–473

27. Horstmann DM (1961) Factors affecting optimum dosage levels of live poliovirus vaccines. In: Papers and Discussions Presented at the 5th International Poliomyelitis Conference, Copenhagen, Denmark. JB Lippincott, Philadelphia, pp 304–310

28. Sangkawibha N, Tuchina P, Sakuntanaga P, Sunthornsratul A, Kunasol P, Bukkavesa S, Ochasanonda P (1974) Poliomyelitis vaccination in Thailand. I. A pilot study of the administration of live poliomyelitis vaccine. SE Asian J Trop Med Pub Hlth 5: 171–178

29. John TJ, Devarajan LV, Balasubramanyan A (1974) Immunization in India with trivalent and monvalent oral poliovirus vaccines of enhanced potency. Bull WHO 54: 115–117

30. John TJ, Jayabal P (1972) Oral polio vaccination of children in the tropics. I. The poor seroconversion rates and the absence of viral interference. Am J Epidemiol 96: 263–269

31. Patriarca PA, Laender F, Palmeira G, Gouto Oliveira MJ, Lima Filho J, de Souza Dantes MC, Tenorio Cordeiro M, Risi JB, Orenstein Wa (1988) Randomised trial of alternative formulations of oral poliovaccine in Brazil. Lancet i: 429–433

32. de Quadros CA, Andrus JK, Olive JM, de Macedo CG, Henderson DA (1992) Polio eradication from the Western Hemisphere. Ann Rev Public Health 13: 239–252

33. Expanded Programme on Immunization (1990) Report of the 13th Global Advisory Group Meeting. World Health Organization, Geneva (Doc no. WHO/EPI/GEN/91.3), p 5

34. WHO Collaborative Study Group on Oral Poliovirus Vaccine (1992) A randomized trial of alternative formulations of oral poliovirus vaccine in Brazil and The Gambia. In: Program and Abstracts of the 32nd Interscience Conference on Antimicrobial Agents and Chemotherapy. American Society of Microbiology, Washington, DC. (Abstract #916) page 263

35. Esteves K (1988) Safety of oral poliomyelitis vaccine: results of a WHO enquiry. Bull WHO 66: 739–746

36. Swartz TA, Skalska P, Gerichter CG, Cockburn WC (1972) Routine administration of oral polio vaccine in a subtropical area. Factors possibly influencing seroconversion rates. J Hyg Camb 70: 719–726

37. Mauler R, Gruschkau H (1978) On stability of oral poliovirus vaccine. Devel Biol Stand 41: 267–270

38. Schoub BD, Johnson S, McAnerney J, Gilbertson L, Klaassen KIM, Reinach SG (1988) Monovalent neonatal polio immunization. A strategy for the developing world. J Infect Dis 157: 836–839

39. Kucharska Z (1985) Excretion of living attenuated polioviruses in the faeces of orally vaccinated children. Comparison of two immunization schedules. J Hyg Epidemiol Microbiol Immunol 28: 211–218
40. Kucharska Z, Zdrazilek J, Koci V, Dvorak J, Skovranek V (1986) Effect of raised type 3 poliovaccine dose on virus excretion. J Hyg Epidemiol Microbiol Immunol 30: 395–403
41. Hale JH, Lee LH, Gardner PS (1961) A study of interference among enteroviruses during a mass immunization campaign with attenuated poliovirus type 2. In: Papers and Discussions Presented at the 5th International Poliomyelitis Conference, Copenhagen, Denmark, JB Lippincott, Philadelphia, pp 257–261

22

Inactivated and live, attenuated poliovirus vaccines: mucosal immunity

R. W. Sutter and **P. A. Patriarca**

Division of Immunization, Centers for Disease Control and Prevention, Atlanta, Georgia, USA

Summary

Both oral poliovirus vaccine (OPV) and inactivated poliovirus vaccine (IPV) induce some degree of mucosal immunity. However, OPV-induced mucosal immunity appears similar to that following natural exposure to wild poliovirus, while multiple doses of IPV may be necessary to induce measurable secretory antibody responses. Differences in mucosal immunity following vaccination with OPV and IPV are most pronounced in the intestine and least pronounced in the pharynx. Data on long-term persistence of mucosal antibodies are limited. However, long-term resistance to fecal and pharyngeal excretion is associated with high humoral antibody levels. Humoral immunity to poliovirus is strongly associated with individual protection against paralytic disease, while mucosal immunity decreases or eliminates fecal and pharyngeal excretion of polioviruses. Therefore, mucosal immunity is largely associated with community protection against circulation and transmission of poliovirus. To achieve the goal of eradication of poliomyelitis by the year 2000, strategies to amplify both mucosal immunity and humoral immunity are essential.

Introduction

Vaccination against poliomyelitis aims to maximize both humoral and mucosal immunity, that is, to induce lifelong persistence of both circulating and secretory antibody that will result in individual protection against paralytic poliomyelitis and community protection against spread and circulation of wild polioviruses [1]. While these objectives have been achieved to a large extent in developed countries with routine vaccination schedules with either OPV or IPV, progress in the developing world has been more difficult, and has required additional strategies such as national mass immunization campaigns and/or door-to-door vaccination efforts [2]. In this chapter, we will briefly review antibody responses following poliovirus infection; describe the mucosal immunity induced by OPV and IPV; and contrast vaccine-induced immunity to mucosal immunity following natural exposure with wild poliovirus. Our discussion then will focus on the function of mucosal immunity; the means by which mucosal immunity may be enhanced; and the possibility of reducing circulation of polioviruses in poliomyelitis-endemic countries.

Immune response following poliovirus infection

Natural infection with poliovirus initiates a complex process that eventually results both in humoral (systemic) and mucosal (local) immunity. Following replication of wild poliovirus, a relatively specific immune response is observed in the serum for each poliovirus serotype. The kinetics of this immune response have been reviewed in detail by Ogra and Karzon [3]. Production of IgM poliovirus antibody predominates initially, can be detected as early as 1–3 days after infection, and disappears after 2–3 months (Fig. 1). IgG antibody levels increase during this same period, eventually constitute the predominant class of persisting antibody, and may last for life [4]. The properties and kinetics of the immune response after vaccination or natural infection appear similar. The humoral immune response is not completely serotype-specific [5,6], and a mild degree of cross-protection (heterotypic cross reaction) has been observed. For example, antibodies to poliovirus type 2 may modify the risk of paralytic disease after exposure to poliovirus type 1 [7].

Presentation of polioviruses to the epithelium of the intestinal mucosa initiates a process that eventually results in mucosal immunity to these antigens. These virions are probably captured by M cells [8, 9, 10] and presented to the lymphoid follicles, where initial activation and proliferation of

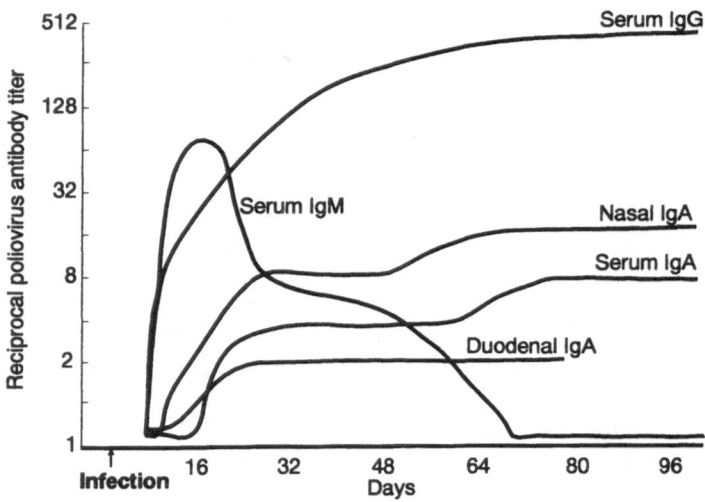

Fig. 1. Kinetics of humoral and local antibody response following poliovirus infection. Adapted from Ref. [3]

IgA precursor B cells begin [11]. Expression of a specific cellular receptor seems to restrict poliovirus replication to a few sites, including the oropharyngeal mucosa, the Peyer's patches in the ileum and motor neurons in the CNS [12]. Sensitized B cells reach the peripheral circulation through the lymphatics and the regional lymph nodes. The majority of sensitized cells return to the lamina propria in the mucosa of initial antigenic exposure, and as plasma cells, initiate synthesis of specific IgA antibodies [13]. These plasma cells synthesize IgA as a dimer with the J chain, which acquires the secretory component during passage across the mucosal epithelium [7]. Although these steps resulting in mucosal immunity are fairly uniform, evidence from other live viral vaccines (e.g., rubella [14] and varicella [15]) suggest that different vaccine strains can influence the quality and magnitude of the mucosal immune response. In addition, whether a virus is presented to the immune system following a natural route or is injected can also affect the mucosal immune response [11]. The significance of the cell-mediated immunity to poliovirus exposure remains to be shown [16], although cytotoxic T-cell responses may contribute to inflammation and cell necrosis that characterize poliovirus infections of the central nervous system.

Quantifying mucosal immunity

Mucosal immunity can be measured either *directly* by quantifying secretory antibody or *indirectly* through poliovirus challenge studies. The latter studies provide an estimate of the degree of mucosal immunity following natural infection with wild poliovirus or in response to challenge with vaccine poliovirus by measuring the proportion of study participants excreting challenge virus, as well as the duration and magnitude of such excretion from the intestine or the pharynx. These studies thus provide an indirect indication of viral replication in the intestine or the pharynx that may parallel exposure and excretion under conditions of endemic or epidemic poliovirus spread. Observational studies in populations that had received IPV or OPV may also provide indirect information on the importance of mucosal immunity in populations as a whole.

Secretory antibody

The predominant class of immunoglobulin in secretions (stool, urine, saliva, nasopharyngeal secretions, tears, colostrum and breast milk) is IgA. Secretory antibody can be assessed by measuring total secretory IgA (SIgA) or poliovirus type-specific IgA (pSIgA) by direct enzyme immunoassay (EIA) or sandwich EIA [18,19,20]. However, these methods are not widely available and are technically difficult to perform. Neutralizing antibody activity in secretions is predominantly associated with IgA in the intestine and the pharynx [21,22,23,24].

Zhaori et al [25] examined pharyngeal immunity following vaccination with OPV or enhanced-potency inactivated poliovirus vaccine (eIPV). Neutralizing antibody response in the nasopharynx appeared to be more effectively induced by OPV (70% of OPV recipients had neutralizing antibodies) than by eIPV (27% of eIPV recipients had neutralizing antibodies), with higher antibody titers in the OPV group. However, specific EIA antibody to poliovirus virion proteins, particularly VP1 and VP2, could be frequently induced in secretory sites by both vaccines. After vaccination with three doses of either eIPV or OPV, Onorato et al [26] reported that similar proportions of eIPV and OPV recipients had poliovirus type 1 secretory IgA (p1SIgA) in pharyngeal and stool specimens. However, while high levels of p1SIgA restricted excretion following monovalent type 1 OPV challenge (m1OPV) in OPV vaccinees, no such correlation was detected in eIPV vaccinees. These findings suggest that OPV-induced p1SIgA may be qualitatively superior to p1SIgA induced by eIPV or the p1SIgA in OPV recipients was a marker for another aspect of the immune response which inhibited excretion.

Faden et al [27] obtained nasopharyngeal specimens from newborns and all lacked secretory netralizing or IgA antibody. These findings appear to explain why the rate of fecal excretion in newborns following mOPV challenge (or OPV vaccination) is high. However, seroconversion rates in newborns are lower probably due to interference by maternally-derived antibody [28,29,30]) (see also Chapter No. 24, "Optimal schedule for the administration of oral poliovirus vaccine"). Interestingly, maternally-derived antibody seems to belong to the IgG immunoglobulin class, with activity directed primarily against VP1 capsid polypeptide [25,27].

The influence of vaccination with either eIPV or OPV on levels of SIgA in breast milk in women with presumed natural immunity and IPV-induced immunity is less clear. Svennerholm et al [31] and Hanson et al [32] reported depression of SIgA levels in breast milk after vaccination with OPV in presumed naturally immune mothers from Pakistan. However, whether these findings have practical implications remains to be shown.

Challenge with natural infection

Under conditions of *natural exposure* prior to the vaccine era, Gelfand et al [33] demonstrated high poliovirus infection rates (89–97%) in susceptible household contacts, with excretion of poliovirus lasting for more than three weeks. In these household settings, infection was also documented frequently in previously immune individuals* (37% of children) but fecal excretion of polioviruses in this group was low (10%). This early study suggested that natural infection with poliovirus induced effective, but not absolute mucosal immunity, which resulted in a high degree of intestinal resistance to homotypic

---

*Determined by serological assessments

poliovirus excretion, but not to homotypic reinfection. Marine et al [34] reported that IPV-vaccinated children with high circulating antibody titers excreted poliovirus for a shorter period, particularly from the oropharynx, than children without detectable levels of circulating antibody. However, while vaccination with IPV did not reduce household infection rates, a possible effect on extrafamilial spread was noted. The authors suggested that extrafamilial spread may be more easily disrupted because of less intimate contact and shorter duration of exposure than transmission within households [34].

Challenge with vaccine poliovirus

Glezen et al [35] reported that *pharyngeal excretion* after monovalent type 1 OPV (mOPV1) challenge was largely dependent on levels of pre-existing homotypic circulating antibodies. Moderate levels of humoral antibody (titer $\geq 1:128$) were associated with lack of excretion. Onorato et al [26] provided more recent information following vaccination with eIPV and OPV. After challenge with mOPV1, similar rates of pharyngeal excretion were observed (1%; [eIPV group]; and 4% [OPV group]). The only children excreting challenge virus were from the group that received the high challenge dose (560 000–600 000 $TCID_{50}$), suggesting that pharyngeal excretion may be dependent on challenge virus dosage, although the proportion of children excreting virus was small.

Ghendon and Sanakoyeva [36] suggested that the quality of *intestinal immunity* following OPV is similar to that following natural infection, based on the proportion of children resisting reinfection and the duration and magnitude of excretion (Table 1). Of 33 OPV-vaccinated children, 21 (64%) resisted reinfection after challenge with mOPV1, as did 21/32 (66%) of naturally immune children. In addition, OPV-vaccinated children excreted challenge poliovirus for a mean of 4.6 days, while naturally immune children excreted virus for a mean of 5.4 days. The mean titer of virus in feces was also similar (2.18 versus 2.03 log TCD_{50}/g for the OPV and naturally immune groups, respectively). In contrast, few IPV-vaccinated children (8/31 [26%])

Table 1. Poliovirus isolation following challenge with monovalent type 1 OPV, by vaccination group, compared with susceptible and naturally immune control groups

	Proportion with fecal excretion of Sabin type 1 (%)			
	Naturally immune	OPV-vaccinated	IPV-vaccinated	Susceptible controls
Ghendon [36]	34%	36%	74%	80%
Henry [37]	ND#	32%	65–86%*	83%
Onorato [26]	ND#	25%	63%	ND#

#Not done
*86% after 3 doses of IPV; 65% after 4 doses of IPV

Table 2. Evaluation of intestinal immunity in vaccinated (OPV or IPV), naturally immune and susceptible control children[1]

Group	Mean duration (days)	Mean titer (log TCD_{50})	Excretion index[2]	Risk ratio
Naturally immune	5.4	2.03	11.0	Reference
OPV-vaccinated	4.6	2.18	10.0	0.9
IPV-vaccinated	12.0	4.11	49.3	4.5
Susceptible controls	20.0	5.15	103.0	9.4

[1]Based on data from Ghendon and Sanakoyeva [36]
[2]Excretion index (mean duration of excretion in days multiplied by mean titer of poliovirus excretion [log TCD_{50}/g feces])

resisted infection, comparable to 6/30 (20%) of control (unvaccinated and susceptible) children who resisted infection after mOPV1 challenge. However, control children excreted virus for a mean of 20 days (with mean of 5.15 log TCD_{50} poliovirus/g of stool), while IPV-vaccinated children excreted poliovirus for a mean of 12 days only (with a mean of 4.11 log TCD_{50}/g of stool). Although these differences between the IPV-group and the control group appeared small, we made an attempt to quantify the magnitude of fecal excretion by multiplying the duration of fecal excretion (in days) with the mean titer of virus in stool (log TCD_{50}/g). The data displayed in Table 2 suggest that intestinal resistance or intestinal immunity following IPV was substantial, the epidemiological implication of which was subsequently borne out in countries which controlled wild poliovirus relying solely on IPV.

Glezen et al [34] reported that fecal excretion of poliovirus may also be affected by circulating antibody. However, in contrast to pharyngeal excretion, which could be decreased markedly by moderate levels of humoral antibody, decreases in fecal excretion may require substantially higher titers of circulating antibody; while 56% (52/93) of children with titers < 1:8 excreted poliovirus, only 9% (6/69) of children with pre-existing antibody titers 1:256 or greater excreted virus. Henry et al [37] reported marked differences in excretion of poliovirus after challenge, depending on which vaccine the study group had received and the pre-challenge levels of humoral antibody. Among IPV-vaccinated children (IPV administered at 2, 3, and 4 mos), 86% (42/49) excreted poliovirus after mOPV1 challenge, a rate similar to that observed in unvaccinated controls (83% [40/48]). Among children receiving four doses of IPV at 2, 3, 4, and 15 mos., 65% (28/43) excreted challenge virus compared with 32% (16/50) of children receiving 3 doses of OPV at 7, 8, and 9 mos. The duration and magnitude of fecal excretion was highest in control children, intermediate in the four-dose IPV group, and lowest in the OPV group (Fig. 2). Regardless of the type of vaccine received, children with higher antibody titers tended to excrete poliovirus for a shorter period. This study also reported that fecal excretion was dependent on the challenge poliovirus dosage. More recently, Onorato

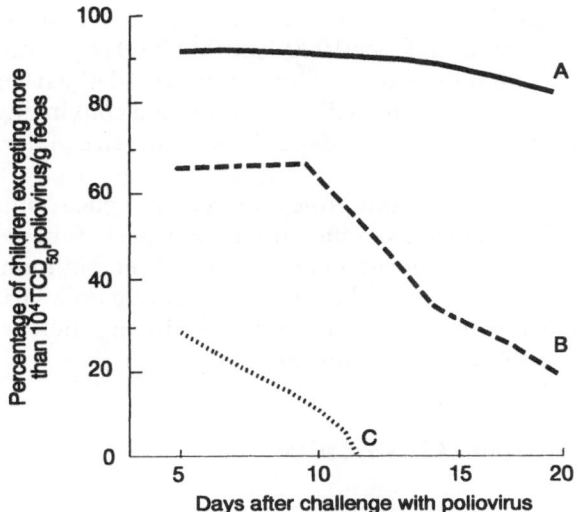

Fig. 2. Excretion of poliovirus following mOPV1 challenge, by vaccination group (*A*: control group [susceptible, unvaccinated children]; *B*: IPV group [4 doses of IPV]; *C*: OPV group [3 doses of OPV]) Adapted from Ref. [37]

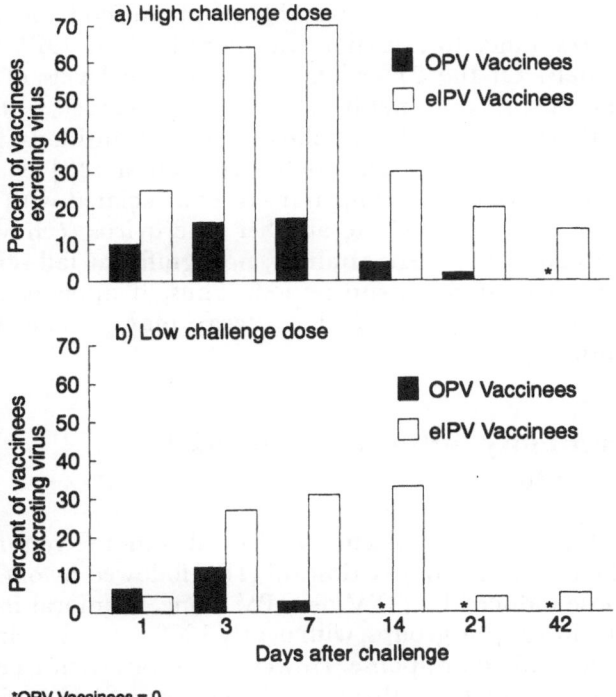

*OPV Vaccinees = 0

Fig. 3. Duration of poliovirus excretion by vaccination group (eIPV or OPV) challenged with low- and high-dose monovalent type 1 vaccine Adapted from Ref. [67]

et al [26] examined fecal isolation rates among eIPV and OPV vaccinated children according to a high (560 000–600 000 $TCID_{50}$) and low (500–800 $TCID_{50}$) mOPV1 challenge dose. This study extended earlier findings with regard to mucosal immunity to eIPV, and provided convincing evidence that intestinal resistance to excretion is dependent on the size of the challenge dose (Fig. 3).

The role of SIgA in terminating poliovirus replication is not known. Savilahti et al [38] reported significantly longer periods of excretion in IgA deficient persons than for normal controls; however, six months after OPV receipt, all IgA deficient persons had ceased excreting poliovirus. In addition, IgA deficient persons may compensate by producing increased amounts of IgM in the saliva [39], and the intestine [40].

Duration of mucosal immunity

Few studies have provided data on the persistence of mucosal antibodies. Nishio et al [20] suggested that children were completely resistant to intestinal infection 10 years after OPV vaccination, unless their serum antibody titers had dropped to 1:8 or less. Smith et al [41] studied 165 cadets vaccinated 8–16 years earlier with IPV or OPV, who were then challenged with trivalent OPV. In cadets with little or no detectable serum antibody, no evidence was obtained that resistance to excretion was more likely in OPV than in IPV vaccinated cadets. Of the 21 cadets who had low levels of antibody to poliovirus type 1 (≤ 0.1 units per ml), 20 (95%) responded with excretion of poliovirus; of 81 cadets with high poliovirus type 1 antibody (≥ 12.8 units per ml), only 12 (15%) excreted poliovirus. Ogra [42] examined local resistance in the pharynx and alimentary tract associated with the development of secretory IgA antibody to another enterovirus (echovirus type 6); after peak antibody titers were attained, no significant fall-off in titer was observed over a 4-year follow-up period. Thus, it appears that mucosal antibody may persist for years and its decay may parallel the decay of circulating antibody.

Mucosal immunity following sequential IPV-OPV schedules

Faden et al [27] compared systemic and local immune responses ensuing sequential schedules of one or two doses of eIPV followed by one or two doses of OPV to those induced by OPV or eIPV alone. Humoral immunity was induced in all four vaccine groups, with nearly 100% of all vaccinees developing a detectable antibody response. However, no consistent trend was found for pharyngeal secretory neutralizing antibodies and pSIgA antibodies. After three doses of OPV, 70%, 85% and 75% of the children developed secretory neutralizing antibodies to poliovirus types 1, 2, and 3, respectively. After three doses of eIPV, the corresponding percentages were 43%, 60%, and 66%. The percentages for the two sequential schedules were, in general, between these values depending on the number of OPV doses received. Similarly, pSIgA antibody was detected in a higher proportion of OPV vaccinees, but followed a distribution similar to the one observed for secretory neutralizing antibody. Based on these data, and information on geometric mean titers, the authors suggested that incorporation of at least one dose of eIPV at the start of the immunization schedule was able to enhance both systemic as well as local antibody production.

Enhancing intestinal immunity with trypsin-cleaved antigen

The outbreak of type 3 poliomyelitis in Finland in 1984–85 gave rise to a series of explanations why the outbreak had occurred [43,44]. Several hypotheses for the widespread circulation of poliovirus during the outbreak were generated, including the apparent lack of effective mucosal immunity [43]. Following the report of Icenogle et al [45], who showed that the intestinal protease trypsin cleaved protein of type 3 poliovirus at antigenic site 1 in vitro, Roivainen and Hovi [46] examined subjects that received eIPV exclusively; older persons exposed to natural infection; children exposed to type 3 poliovirus during 1984–85 epidemic; and children who received a single dose of OPV. With the exception of the first group (the eIPV group), all other groups neutralized the trypsin-cleaved poliovirus readily. This study may offer an alternative explanation for the relatively weak intestinal mucosal immunity obtained with eIPV, and Roivainen and Hovi [46] suggested that the addition of trypsin-treated type 3 polioviruses in the inactivated poliovirus vaccine should be studied.

Differences in pharyngeal immunity by vaccine type

Zhaori et al [47] reported that poliovirus type 3 virion-specific polypeptide response in pharyngeal secretions was different among OPV only, eIPV only, and a sequential vaccine group. While all study groups developed similar secretory IgA responses to VP1 and VP2, an antibody response to VP3 was observed in 77% of the OPV group, in approximately 60% of the eIPV-OPV group, and in 13% of the eIPV group. Significant differences were also observed when trypsin-treated poliovirus was used as the antigen for neutralization or for EIA in vitro. Both neutralizing and EIA antibody activity against cleaved poliovirus were significantly lower than against the whole virus in eIVP vaccinated subjects.

Significance of pharyngeal immunity

Pharyngeal immunity may be influenced by the availability of immunocompetent tissue in the oropharynx (i.e. nasopharyngeal tonsils and adenoids). Ogra [3,48] reported that removal of tonsils and adenoids was followed by significant decreases of secretory IgA in the nasopharynx. In addition, Ogra and Karzon [3] reported the pharyngeal mucosal response following OPV was substantially lower in children with tonsillectomy. These findings may provide an immunological explanation for the increased susceptibility of individuals with recent tonsillectomies for paralytic disease, particularly bulbar poliomyelitis [49]. Alternative hypotheses for this phenomenon include direct access of poliovirus to injured nerves or trauma, both of which were shown to increase the risk of paralytic disease independently.

Limitations of current studies

Limitations of currently available studies are numerous and include: 1) methodological issues (varying study designs); 2) laboratory issues (procedures and reagents not standardized, specimen collection, use of blocking agents, and assessment of bound and free poliovirus [50]); 3) vaccine issues (which vaccines, potency of OPV and eIPV); 4) selection of study subjects (with potential selection bias); and 5) analysis and reporting issues (adequate analysis, and uniform reporting of similar studies). Because of these limitations, substantial "gaps" in current knowledge remain in relation to: 1) long-term mucosal immunity; 2) validity of extrapolating studies conducted in industrialized countries to developing countries; 3) the epidemiological significance of IPV- or eIPV-induced mucosal immunity, particularly for developing countries; 4) intestinal resistance to challenge with other poliovirus serotypes (particularly type 3); and 5) the significance of intestinal immunity in the absence of circulating antibody.

Conclusions and future directions

Although data on mucosal immunity following poliovirus infection or IPV administration are incomplete and open to interpretation, several general conclusions can be drawn from available information. First, it appears that OPV-induced mucosal immunity closely resembles that following natural infection with wild poliovirus. Second, mucosal immunity is only partially effective as a barrier to infection, and approximately one-third of OPV vaccinees still excrete poliovirus after challenge. Third, in vaccinated individuals fecal excretion is considerably shorter in duration with a substantially decreased poliovirus load. Fourth, mucosal immunity following IPV seems most effective in decreasing pharyngeal shedding of virus (similar in magnitude to OPV), and intermediate between OPV-vaccinated and unvaccinated control children in decreasing the intestinal shedding of virus. However, one report suggests that mucosal immunity induced to one vaccine poliovirus strain may be less effective against another vaccine poliovirus strain [51].

Mucosal immunity, whether induced by OPV or IPV, may therefore have different implications for the control of poliomyelitis in developing and developed countries. It has been hypothesized that the predominant route of poliovirus transmission in developing countries is fecal–oral. In contrast, oral–oral seems to be the predominant mode of transmission in developed countries [28,52,53]. If these hypotheses are valid, then IPV or OPV induced mucosal immunity should be equally effective in controlling poliomyelitis in developed countries by minimizing pharyngeal shedding. This hypothesis appears to have been borne out in industrialized countries relying exclusively on IPV to control poliomyelitis [54,55]. However, even in developing countries with high vaccination coverage, the superior mucosal immunity induced by OPV in the intestine has not always been sufficient enough to prevent epidemic transmission of poliovirus [56,57,58]. Poliomyelitis is much more difficult to control in these settings for a number of reasons, including: 1) the less than optimal immunogenicity of OPV in developing countries [59] compared with industrialized countries [60]; 2) a predominance of fecal-oral transmission of poliovirus; and 3) a large inoculum size of poliovirus under conditions of suboptimal hygiene and sanitation. These immunological and epidemiological considerations provide a scientific rationale for the need for additional vaccination strategies to control poliomyelitis in developing countries, particularly those located in tropical areas [2].

While data on mucosal immunity following poliovirus type 1 infection are most complete, data following poliovirus types 2 and 3 infection are generally lacking. During the course of the poliomyelitis eradication program in the Americas, the appearance of type 3 epidemics in Brazil in 1986 [61], Mexico in 1990 [62], and Guatemala in 1990 [62] were of particular concern. Thus, additional research on mucosal immunity to poliovirus type 3 seems urgently needed. Given that the seroconversion rates to type 3 after administration of three doses of OPV in these settings are particularly low [59], improvements in type-specific immunity may result in accelerated control. It would be important to know whether an OPV-vaccinated child who does not develop a detectable circulating antibody may mount an adequate secretory antibody response.

Recent advances in studying differences between mucosal responses following eIPV and OPV [46,47] have provided a new hypothesis to explain variations in local responses according to anatomical location (i.e. pharynx and intestine). This hypothesis proposes that trypsin-cleaved poliovirus type 3 may escape IPV induced intestinal immunity, but is effectively neutralized by OPV induced intestinal immunity. Thus, one means to improve IPV-induced intestinal immunity may be to add trypsin-cleaved inactivated poliovirus type 3 to IPV, as has been proposed by Roivainen and Hovi [46]. Whether this approach will indeed result in improved IPV-induced intestinal immunity remains to be shown.

While the addition of trypsin-cleaved poliovirus to IPV may be possible, this approach is not likely to be developed in time for the eradication program. In the meantime, mucosal immunity induced by eIPV is clearly not sufficient to achieve eradication in developing countries. Thus, the best methods for improving mucosal immunity in these settings at the present time may include: 1) the administration of OPV to a high proportion of infants (increasing vaccination coverage); 2) the addition of one or more supplemental doses of OPV to the routine vaccination schedule to increase the likelihood of adequate antibody responses; 3) the administration of OPV simultaneously to a high proportion of children in mass vaccination campaigns (i.e. national immunization days); and 4) house-to-house provision of OPV to children determined at high risk for poliomyelitis (i.e. "mop-up" operations). These and other approaches pioneered by PAHO led to the virtual elimination of poliomyelitis from the WHO region of the Americas [2]. These achievements indicate that eIPV is not required to ensure the success of the poliomyelitis eradication program. Adoption of the strategies developed to control poliomyelitis in the Americas offer the best means currently available to eliminate and finally eradicate poliomyelitis from the other areas of the globe, and all poliomyelitis-endemic areas should plan to implement this approach [63]. Full implementation of the PAHO approach may take some time. In the meantime research should be supported to determine if supplementary methods such as increasing doses of OPV routinely or sequential or combined schedules of OPV and eIPV can help leading to more rapid eradication.

The possible role of eIPV in increasing humoral immunity, with subsequent enhanced individual protection against poliomyelitis, is currently being reviewed by the WHO for the eradication program [64]. Does eIPV have a separate role in increasing mucosal immunity in developing countries? It appears that one dose of eIPV may prime the mucosal system, and thus increase secretory antibody response following subsequent OPV administration [3,65]. One study assessing different sequential schedules of eIPV followed by OPV [27] reported no interference or immune tolerance following these regimens, but the sequential groups did not mount a better mucosal response compared with the group which received OPV alone. Thus, at this juncture, it appears that sequential vaccination schedules are not superior in terms of mucosal immunity to OPV-only schedules. Furthermore, sequential schedules of one or two doses of eIPV followed by OPV have been considered primarily for use in developed countries to reduce the risk of recipient cases of vaccine-associated paralysis, and are currently not considered programmatically feasible within the context of the Expanded Programme on Immunization. However, another approach which would not cause confusion among field workers would involve the simultaneous administration of both vaccines at the time of DTP administration, a regimen currently being evaluated in a multi-center trial in The Gambia, Oman, and Thailand (Personal communication, WHO, 1991). Such a combined schedule may enhance both humoral and secretory immunity by taking advantage of the known benefits derived from each vaccine alone. In addition, a quadrivalent preparation consisting of DTP and eIPV could be used instead of DTP vaccine alone, which would eliminate the need for a separate injection for eIPV. A final approach would involve the administration of one or more doses of eIPV after completion of the primary OPV series. Limited evaluation of such an approach by Sabin [66] showed that infants resisted reinfection with vaccine virus upon subsequent challenge with OPV, an observation compatible with the induction and/or enhancement of immunologic memory.

While it is clear that the use of eIPV alone should never be considered a viable programmatic option to control poliomyelitis in developing countries, complementary use of eIPV with OPV in a combined or sequential vaccination schedule may enhance humoral and mucosal immunity. However, rigorously controlled challenge studies to evaluate this approach in developing countries are urgently needed to provide more definitive data, which may permit the re-evaluation of the role of eIPV for developing countries. The data reviewed in this chapter would suggest, however, that, in terms of induction of mucosal immunity, the role of eIPV in the poliomyelitis eradication program may be only supplementary to that of OPV.

References

1. Institute of Medicine (1988) An evaluation of poliomyelitis vaccine policy options. Washington, DC: National Academy of Sciences (Publication No. IOM 88–04)
2. de Quadros CA, Andrus JK, Olive JM, de Macedo CG, Henderson DA (1992) Polio eradication from the Western Hemisphere. Ann Rev Publ Health 13: 239–252
3. Ogra PL, Karzon DT (1971) Formation and function of poliovirus antibody in different tissues. Prog Med Virol 13: 156–193
4. Paul JR, Riordan JT, Melnick JL (1951) Antibodies to three different antigenic types of poliomyelitis virus in sera from North Alaskan Eskimos. Am J Hygiene 54: 275–285
5. Sabin AB (1952) Transitory appearance of type 2 neutralizing antibody in patients infected with type 1 poliomyelitis virus. J Experiment Med 96: 99–106
6. Salk J (1956) Requirements for persistent immunity to poliomyelitis. Trans Assn Am Phys 69: 105–114
7. Hammon WM, Ludwig EH (1957) Possible protective effect of previous type 2 infection against paralytic poliomyelitis due to type 1 virus. Am J Hygiene 66: 274–280
8. Owen RL, Jones AL (1974) Epithelial cell specialization within human Peyer's patches: an ultrastructural study of intestinal lymphoid follicles. Gastroenterology 66: 189–203
9. Sicinski P, Rowinski J, Warchol JB, Jarzabek Z, Gut W, Szczyiel B, Bielecki K, Koch G (1990) Poliovirus type 1 enters the human host through intestinal M cells. Gastroenterology 98: 56–58
10. Owen RL (1977) Sequential uptake of horseradish peroxidase by lymphoid follicle epithelium of Peyer's patches in the normal unobstructed mouse intestine: an ultrastructural study. Gastroenterology 72: 440–451
11. Dhar R, Ogra PL (1985) Local immune responses. Brit Med Bull 41: 28–33
12. Mendelsohn CL, Wimmer E, Racaniello VR (1989) Cellular receptor for poliovirus: Molecular cloning, nucleotide sequence, and expression of a new member of the immunoglobulin superfamily. Cell 56: 855–865
13. Mestecky J, McGhee JR, Michalek SM, Arnold RR, Crago SS, Babb JL (1978) Concept of local and common mucosal immune response. In: McGhee JR, Mestecky J, Babb JL

(eds) Secretory immunity and infection. Proceedings of the International Symposium on the Secretory Immune System and Caries Immunity. Plenum Press, New York, pp 185–192

14. Mestecky J, McGhee JR, Crago SS, Jackson S, Kilian M, Kiyono H, Babb JL, Michalek SM (1980) Molecular-cellular interactions in the secretory IgA response. J Reticuloendothel Soc 28: 45–60

15. Losonsky GA, Fishaut JM, Stussenberg J, Ogra PL (1982) Effect of immunization against rubella in lactation products. I. Development and characterization of specific immunologic reactivity in breast milk. J Inf Dis 145: 654–660

16. Bogger-Goren S, Baba K, Hurley P, Yabuuchi H, Takahashi M, Ogra PL (1982) Antibody response to varicella-zoster virus after natural or vaccine-induced infection. J Inf Dis 146; 260–265

17. Wang K, Sun L, Jubelt B, Waltenbaugh C (1989) Cell-mediated immune responses to poliovirus. I. Conditions for induction, characterization of effector cells, and cross-reactivity between serotypes for delayed hypersensitivity and T cell proliferative responses. Cell Immunol 119: 252–262

18. Inouye S, Matsuno S, Yamaguchi H (1984) Efficient coating of the solid phase with rotavirus antigens for enzyme-linked immunosorbent assay of immunoglobulin A antibody in feces. J Clin Microbiol 19: 259–263

19. Losonsky GA, Rennels MB, Lim Y, Krall G, Kapikian AZ, Levine MM (1988) Systemic and mucosal immune response to rhesus rotavirus vaccine MMU 18006. Pediatr Infect Dis J 7: 388–393

20. Nishio O, Ishihara Y, Sakae K, Nonmura Y, Kuno A, Yasukawa S, Inoue H, Miyamura K, Kono R (1984) The trend of acquired immunity with live poliovirus vaccine and the effect of revaccination: Follow-up of vaccinees for 10 years. J Biol Stand 12: 1–10

21. Ogra PL, Karzon DT, Righthand F, MacGillivray M (1968) Immunoglobulin response in serum and secretions after immunization with live and inactivated poliovaccine and natural infection. N Engl J Med 279: 893–900

22. Keller R, Dwyer J (1968) Neutralization of poliovirus by IgA coproantibodies. J Immunol 101: 192–202

23. Keller R, Dwyer J, Oh W, D'Amodio M (1969) Intestinal IgA neutralizing antibodies in newborn infants following poliovirus immunization. Pediatrics 43: 330–338

24. Berger R, Ainbender E, Hodes HL, Zepp HD, Hevizy MM (1967) Demonstration of IgA polio antibody in saliva, duodenal fluid, and urine. Nature (London) 214: 420–422

25. Zhaori G, Sun M, Ogra PL (1988) Characteristics of the immune response to poliovirus virion polypeptides after immunization with live or inactivated polio vaccines. J Infect Dis 158: 160–165

26. Onorato IM, Modlin JF, McBean AM, Thoms ML, Losonsky GA, Bernier RH (1991) Mucosal immunity induced by enhanced-potency inactivated and oral polio vaccines. J Infect Dis 163: 1–6

27. Faden H, Modlin JF, Thoms ML, McBean AM, Ferdon MB, Ogra PL (1990) Comparative evaluation of immunization with live attenuated and enhanced-potency inactivated trivalent poliovirus vaccines in childhood: Systemic and local immune responses. J Infect Dis 162: 1291–1297

28. Lepow ML, Warren RJ, Gray N, Ingram VG, Robbins (1961) Effect of Sabin type 1 poliomyelitis vaccine administered by mouth to newborn infants. N Engl J Med 264: 1071–1078

29. Pagano JS, Plotkin SA, Koprowski H (1960) Variation of response in early life to vaccination with living attenuated poliovirus and lack of immunologic tolerance. Lancet i: 1224–1226

30. John TJ (1984) Immune response of neonates to oral poliomyelitis vaccine. Brit Med J 289: 881

31. Svennerholm AM, Hanson LA, Holmgren J, Jalil F, Lindblad BS, Khan SR, Nilsson A, Svennerholm B (1981) Antibody response to live and killed poliovirus vaccines in the milk of Pakistani and Swedish women. J Infect Dis 143: 707–711

32. Hanson LA, Carlsson B, Jalil F, Lindblad BS, Khan SR, van Wezel AL (1984) Different secretory IgA antibody responses after vaccination with inactivated and live poliovirus vaccines. Rev Infect Dis 6 [Suppl]: 356–360
33. Gelfand HM, LeBlanc DR, Fox JP, Conwell DP (1957) Studies on the development of natural immunity to poliomyelitis in Louisiana. II. Description and analysis of episodes of infection observed in study group households. Am J Hygiene 65: 367–385
34. Marine WM, Chin TDY, Gravelle CR (1962) Limitation of fecal and pharyngeal excretion in Salk-vaccinated children. A family study during a type 1 poliomyelitis epidemic. Am J Hygiene 76: 173–195
35. Glezen WP, Lamb GA, Belden EA, Chin TDY (1966) Quantitative relationship of preexisting homotypic antibodies to the excretion of attenuated poliovirus type 1. Am J Epidemiol 53: 224–237
36. Ghendon YUZ, Sanakoyeva II (1961) Comparison of the resistance of the intestinal tract to poliomyelitis vaccine (Sabin's strains) in persons after naturally and experimentally acquired immunity. Acta Virol (Praha) 5: 265–273
37. Henry JL, Jaikaran ES, Davies JR, Thomlinson AJH, Mason PJ, Barnes JM, Beale AJ (1966) A study of poliovaccination in infancy: excretion following challenge with live virus by children given killed or living poliovaccine. J Hygiene (Cambridge) 64: 105–120
38. Savilahti E, Klemola T, Carlsson B, Mellander L, Stenvik M, Hovi T (1988) Inadequacy of mucosal IgM antibodies in selective IgA deficiency: Excretion of attenuated polio viruses is prolonged. J Clin Immunol 8: 89–94
39. Mellander L, Björkander J, Carlsson B, Hansson LA (1985) Serum and secretory antibodies in IgA deficient and immunosuppressed individuals. J Clin Immunol 4: 284–291
40. Savilahti E (1973) IgA deficiency in children. Immunoglobulin-containing cells in the intestinal mucosa, immunoglobulins in secretions and IgA levels. Clin Exp Immunol 13: 395–406
41. Smith JWG, Lee JA, Fletcher WB, Morris CA, Parker DA, Yetts R, Magrath DI, Perkins FT (1976) The response to oral poliovaccine in persons aged 16–18 years. J Hygiene (Cambridge) 76: 235–247
42. Ogra PL (1970) Distribution of echovirus antibody in serum, nasopharynx, rectum and spinal fluid after natural infection with echovirus type 6. Infection Immunol 2: 150–155
43. Hovi T, Huovilainen A, Kuronen T, Poyry T, Salama N, Cantell K, Kinnunen E, Lapinleimu K, Roivainen M, Stenvik M (1986) Outbreak of paralytic poliomyelitis in Finland: Widespread circulation of antigenically altered poliovirus type 3 in a vaccinated population. Lancet i: 1427–1432
44. Centers for Disease Control (1986) Update: Poliomyelitis outbreak – Finland, 1984–1985. MMWR 35: 82–86
45. Icenogle JP, Minor PD, Ferguson M, Hogle JM (1986) Modulation of humoral response to a 12-amino acid site on the poliovirus. J Virology 60: 297–301
46. Roivainen M, Hovi T (1987) Intestinal trypsin can significantly modify antigenic properties of polioviruses: implication for the use of inactivated poliovirus vaccine. J Virology 61: 3749–3753
47. Zhaori G, Sun M, Faden HS, Ogra PL (1989) Nasopharyngeal secretory antibody response to poliovirus type 3 virion proteins exhibit different specificities after immunization with live or inactivated poliovirus vaccines. J Infect Dis 159: 1018–1024
48. Ogra P (1970) The role of nasopharyngeal tonsils and adenoids in the mechanism of mucosal immunity to poliovirus. American Pediatric Society, 80th Annual Meeting, Program and Abstracts
49. Aycock WL (1942) Tonsillectomy and poliomyelitis. I. Epidemiologic considerations. Medicine 21: 65–94
50. Melnick JL, Proctor RO, Ocampo AR, Diwan AR, Ben-Porath E (1966) Free and bound virus in serum after administration of oral poliovirus vaccine. Am J Epidemiol 84: 329–342
51. Janda Z, Adam E, Vonka V (1967) Properties of a new type 3 attenuated poliovirus. VI. Alimentary tract resistance in children fed previously with type 3 Sabin vaccine to rein-

fection with homologous and heterologous type 3 attenuated poliovirus, Arch Virusforschung 20: 87–98

52. Isacson P, Melnick JL, Walton M (1957) Environmental studies of endemic enteric viral infections. II. Poliovirus infections in household units. Am J Hygiene 65: 29–42
53. Salk JE (1959) Preconceptions about vaccination against paralytic poliomyelitis. Ann Intern Med 50: 843–861
54. Lapinleimu K (1984) Elimination of poliomyelitis in Finland. Rev Infect Dis 6 [Suppl]: 457–460
55. Böttiger M (1984) Long-term immunity following vaccination with killed poliovirus vaccine in Sweden, a country with no circulating poliovirus. Rev Infect Dis 6 [Suppl]: 548–551
56. Sutter RW, Patriarca PA, Brogan S, Malankar PG, Pallansch PA, Kew OM, Bass AG, Cochi SL, Alexander JP, Hall DB, Suleiman AJM, Al-Ghassani AAK, El-Bualy MS (1991) Outbreak of paralytic poliomyelitis in Oman: evidence for widespread transmission among fully vaccinated children. Lancet 338: 715–720
57. Otten MW, Deming MS, Jaiteh KO, Flagg EW, Forgie I, Sanyang Y, Sillah B, Brogan D, Gowers P (1992) Epidemic poliomyelitis in The Gambia following the control of poliomyelitis as an endemic disease. I. Descriptive findings. Am J Epidemiol 135: 381–392
58. Deming MS, Jaiteh KO, Otten MW, Flagg EW, Jaallow M, Cham M, Brogan D, N'jie H (1992) Epidemic poliomyelitis in The Gambia following the control of poliomyelitis as an endemic disease. II. Clinical efficacy of trivalent oral polio vaccine. Am J Epidemiol 135: 393–408
59. Patriarca PA, Wright PF, John TJ (1991) Factors affecting the immunogenicity of oral poliovirus vaccine in developing countries: review. Rev Infect Dis 13: 926–939
60. McBean AM, Thoms ML, Albrecht P, Cuthie JC, Bernier R, The Field Staff and Coordinating Committee (1988) Serologic response to oral polio vaccine and enhanced-potency inactivated polio vaccines. Am J Epidemiol 128: 615–628
61. Patriarca PA, Laender F, Palmeira G, Couto Oliveira MJ, Lima Filho J, de Souza Dantes MC, Tenorio Cordeiro M, Risi JB, Orenstein WA (1988) Randomized trial of alternative formulations of oral poliovaccine in Brazil. Lancet i: 429–432
62. Expanded Programme on Immunization, Pan American Health Organization (1990) Surveillance of wild poliovirus in the Americas. EPI Newsletter 12: 1–3
63. Expanded Programme on Immunization (1992) Progress towards poliomyelitis eradication. Philippines. Weekly Epidemiol Rec 67: 205–208
64. World Health Organization (1988) Global eradication of poliomyelitis by the year 2000. Weekly Epidemiol Rec 63: 161–162
65. Carlson B, Zaman S, Mellander L, Jalil F, Hanson LA (1985) Secretory and serum immunoglobulin class-specific antibodies to poliovirus after vaccination. J Infect Dis 152: 1238–1244
66. Sabin AB, Michaels RH, Ziring P, Krugman S, Warren J (1963) Effect of oral poliovirus vaccine in newborn children. II. Intestinal resistance and antibody response at 6 months in children fed type 1 vaccine at birth. Pediatrics 31: 641–650
67. Ogra PL, Karzon DT, Righthand F, MacGillivray M (1968) Immunoglobulin response in serum and secretions after immunization with live and inactivated poliovaccine and natural infection. N Engl J Med 279: 893–900

23

Immunogenicity of poliovirus vaccines with special reference to the choice between oral and inactivated vaccines in developing countries

B. D. Schoub

National Institute for Virology and Department of Virology, University of the Witwatersrand, Johannesburg, South Africa

Summary

Both inactivated and oral poliovirus vaccines have successfully eradicated poliomyelitis in the developed world. Inactivated vaccine may appear to have some advantages for the developing world, particularly with regard to interference and greater stability, but fails to address the major reservoir of faecal-orally spread virus, the gastrointestinal tract. Inactivated vaccine used to supplement oral vaccine schedules increases cost and could cause confusion in administration. Interference in oral vaccine schedules could be addressed by a supplementary neonatal dose. In most of Africa, type 1 poliovirus is the dominant endemic strain and also the cause of outbreaks, and neonatal supplementation with type 1 would have particular advantages.

Introduction

Poliomyelitis and measles are both diseases which, at first glance, one has some difficulty in understanding why they have not yet been eradicated. Both are preventable by highly effective, extremely safe and easily administerable vaccines, both cause acute self-limiting infections and both are found only in man. Both have, indeed, been almost eliminated in much of the developed world but, unfortunately, both remain as major infectious diseases of the developing world. While the satisfactory provision of adequately potent vaccines to a sufficiently high proportion of the population remains the major reason for the substantial continuance of both infections in mankind, compromises in the effectiveness of the vaccines in the developing world also contribute to a significant extent.

Immunization has been somewhat more successful with poliomyelitis than with measles in controlling and eliminating infection. Both inactivated poliovirus vaccine (IPV) and live attenuated oral poliovirus vaccine (OPV) have successfully eliminated wild-type virus from a number of developed countries. In the developing world, OPV has also achieved considerable success in dramatically reducing the incidence of wild-type poliovirus in much of the Third World. However, the final push to total eradication of the remnants of poliomyelitis necessitates a re-evaluation of vaccine strategies for the developing world. In particular, two facets require to be critically examined – the type of vaccine used and the dosage scheduling.

With the advent of enhanced potency inactivated poliovirus vaccine (e-IPV) there have been calls to consider the replacement of OPV with e-IPV in developing countries. At first glance, there are a number of features of e-IPV which appear to be attractive for use in developing countries, either, as some have proposed, to replace OPV or, as is more widely suggested, to supplement OPV, and indeed a number of successful trials of e-IPV as sole immunization have been carried out in African countries such as Senegal and more recently in Kenya [1].

What I would like to do in this paper is to review the last decade of studies of poliomyelitis in South Africa looking at the national epidemiological history of polio as it moved from endemic to controlled polio with two intervening epidemics. Much has been learnt regarding the relationship between the immunogenicity of vaccine on the one hand, and the epidemiological requirements for control and elimination of infection on the other.

Poliomyelitis in South Africa in the eighties

Prior to the mid 1970s the pattern of notified poliomyelitis in South Africa followed the three year cyclical upswings characteristically seen in the pre-vaccine era. With advancing immunization the expected upswing in 1978 did not materialize and the incidence of polio declined progressively until the major polio epidemic in Gazankulu from May to September 1982 [2]. (Gazankulu is a territory of rural black communities with a population of some 700 000 in the northern part of South Africa.) This epidemic of type 1 polio, which was responsible for 262 paralytic cases including 42 deaths, was superimposed on a relatively low level of endemic polio throughout the rest of the country. Nevertheless, surveillance carried out during the soon after the epidemic revealed substantial underimmunization, as low as 43% coverage in some rural areas. Similarly, seroprevalence studies on cluster samples of sera revealed sizable immunity gaps, especially to type 1. Vaccine recalled from 13 primary health care clinics in various rural districts in the Transvaal showed that in seven of them the titre of vaccine was below the WHO minimal requirement. Finally and perhaps most revealing of all, was the fact that in 74% of patients in the epidemic, no antibodies to types 2 and 3 were detectable while there were high titre antibodies (over 1:160) to the causative type 1 poliovirus. Clearly the Gazankulu epidemic was the result of underimmunization due mainly to insufficient vaccine coverage and, to some extent, inadequately potent vaccine.

A nationwide investigation of immunity throughout the country was carried out a year later examining central vaccine distribution, peripheral vaccine usage on health records and patient cards as well as sero-surveillance and vaccine recall [3]. Some improvements were noted in all parameters following intensified polio immunization campaigns in response to the Gazankulu epidemic. Nevertheless, pockets of unsatisfactory vaccine coverage, worrying immunity gaps and evidence of cold-chain breakdowns were still causes of considerable concern. In addition, vaccine failure evidenced by a sero-immunity gap of 22% in recipients of all three doses was observed. (Unfortunately, how much was due to inadequately potent vaccine and how much to interference could not be established.) No subsequent structured surveillance studies were carried out although data published from the Department of Health claimed increasing vaccine distribution, delivery and coverage.

From December 1987 until May 1988 a massive type 1 polio epidemic took place in KwaZulu/Natal on the eastern seaboard of South Africa involving 412 paralytic cases and 34 deaths [4]. The features of this epidemic contrasted markedly with the earlier Gazankulu epidemic in a number of important respects (Table 1). In the KwaZulu/Natal epidemic 41% of patients had not received any polio immunization while in the Gazankulu epidemic 69% definitely received no vaccine and a further 15% were unsure. In the KwaZulu/Natal epidemic only 12% of patients lacked antibodies to types 2 and 3, while 74% of patients in the Gazankulu epidemic had the serological pattern of lack of immunization. Similarly, a serosurvey of healthy children in the epidemic area demonstrated that 76% had immunity to all three types of polio in contrast to a mean seroprevalence of 58% in areas surrounding Gazankulu during the time of that epidemic.

Table 1. Comparison between the Gazankulu and KwaZulu/Natal poliomyelitis epidemics in South Africa

	Gazankulu 1982	KwaZulu/Natal 1987/88	p Values
Poliovirus types	1	1	
Case fatality	42/260 (16%)	34/412 (8%)	0.0025
Absence of types 2 and 3 antibodies in patients	288/389 (74%)	16/129 (12%)	<0.0001
History of no immunization	179/260 (69%)	70/170 (41%)	<0.0001
Triple serological gaps in population survey	317/733 (43%)	46/192 (24%)	<0.0001

Lessons from the South African poliomyelitis epidemics

Thus studies over the 1980s have revealed, firstly, that low level poliomyelitis endemicity existed in the rural communities of the country with annual notification somewhat above the threshhold of 10 cases per year and 50% immunization coverage, while straddles groups B and C of the WHO schema. Secondly, that there was a vulnerability to epidemics, dramatically demonstrated by two major outbreaks of type 1 polio. These two outbreaks, however, differed qualitatively from each other. The Gazankulu epidemic showed features of epidemics similar to those in deprived populations with marked underimmunization where outbreaks result from the introduction of wild-type virus into a community which is largely susceptible to the virus. I have called this the "low immunity" type of epidemic. The KwaZulu/Natal epidemic, however, was similar to epidemics reported in Taiwan in 1982 [5], the Gambia in 1986 [6] and Oman in 1988 [7], where population immunity was high yet the wild-type 1 poliovirus was able to "ferret out" sufficient numbers of susceptibles to produce substantial epidemics. I have called these the "high immunity" type of epidemics. In the Taiwan epidemic, contamination of water supplies may have resulted in a sufficiently large wild-type virus burden to move rapidly through the remnant susceptibles in the population. In the Oman epidemic, inadequate sanitation may well also have played a causative role in that epidemic, although the introduction of a novel genotype of type 1 poliovirus was also demonstrated. In the KwaZulu/Natal epidemic devastating floods the previous year with widespread disruption of water and sewage disposal facilities may have been responsible for a massive viral burden of wild-type virus, resulting in widespread epidemic activity even in relatively well immunized communities [8]. There was no evidence that a new genotype had been introduced and no evidence of "escape mutation" as was reported to be a factor in the very much smaller outbreaks of type 3 polio in Finland in 1984 [9] and type 1 polio in Israel in 1988 [10].

Strategies for poliomyelitis immunization in developing countries

How then does this affect strategy development for the final stages of the eradication campaign in developing countries? Widespread outbreaks in populations with relatively high immunity indicate that immunization campaigns will need to address two issues:

(a) The provision of high levels of population immunity (probably exceeding 90%) measurable by vaccine coverage and preferably also by serological immunity.
(b) Control and elimination of the reservoir of infection, i.e. the human gastrointestinal tract.

Inactivated poliovirus vaccine

Bearing this in mind, e-IPV immunization alone, has a number of disadvantages for poliomyelitis control in developing countries. In particular, a relatively poor gut immunity [11] would do little to reduce the reservoir of infection in communities where there is substantial faecal-oral transmission of organisms. There is no evidence from data collected in South Africa that coverage with a combination of DPT and IPV would increase coverage for polio immunization. If anything, DPT coverage is somewhat lower than polio vaccine coverage in nearly all South African groups [12].

Oral poliovirus vaccine

On the other side, immunization programmes based on TOPV alone are faced with the serious problem of interference which, for example, may well have contributed substantially to a sero-immunity gap of 22% in triply immunized subjects in South Africa, as mentioned above. Supplementation of routine OPV programmes with a single IPV dose has been implemented in a number of routine schedules in developing countries with demonstrable success [13]. This approach does, however, have some drawbacks:

(i) The cost of routine immunization is increased significantly by IPV supplementation.
(ii) Addition to DPT may cause some degree of confusion and occasionally erroneous administration of a tetravalent vaccine when not required, i.e. after the first dose. This confusion could be compounded where hepatitis B vaccine is also added to DPT in the same vaccine. Separate administration of IPV, on the other hand, would significantly increase immunization costs, cause resistance in mothers by subjective infants to an additional injection and also increase the burden on primary health care staff.

Neonatal supplementation

Neonatal TOPV supplementation has been recommended by WHO for developing countries since 1985 both because of the relative ease of accessibility of newborns but, in particular, to address the issue of interference [14]. We have demonstrated that neonatal monovalent type 1 supplementation produces a significantly higher immune response to type 1 antibodies than trivalent supplementation, probably because intertypic interference is obviated [15]. Monovalent type 1 supplementation also has additional advantages in that vaccine-associated paralysis, which has been a concern with regard to neonatal polio immunization has almost never been observed with type 1 vaccine strains. Thus seroresponse to type 1 occurs in almost all vaccine recipients, which would not only maximise individual immunity but also

Table 2. Advantages of monovalent type 1 supplementation

Advantages of neonatal TOPV
- Reduction of interference
- Better accessibility of newborns

Advantages of neonatal MOPV-1
- Endemicity and epidemicity of type 1
- Little or no vaccine-associated paralysis
- Obviates intertypic interference
- Reduction of type 1 reservoir in community

Disadvantage of neonatal MOPV-1
- Potential to select for wild types 2 and 3

---

substantially reduce the gastrointestinal tract reservoir for type 1 virus, by far the major cause of paralytic poliomyelitis in Africa. Clearly monovalent type 1 neonatal supplementation should only be considered in countries where type 1 polio has stably been the dominant strain. This strategy does, however, run the risk of selecting for types 2 and 3 poliovirus if inadequate follow-up with the routine TOPV schedule is not ensured.

Conclusion

The final push to eradication of poliomyelitis in developing countries requires a purposeful and critical evaluation of immunization schedules. The design of immunization strategies must address two issues – firstly maximization of community immunity and secondly, reduction of the reservoir of virus, i.e. the human gastrointestinal tract. Inactivated poliovirus vaccine on its own is not suitable for eradication of poliomyelitis in developing countries where it would not address the reservoir problem due to high faecal-oral transmission of organisms. Administration schedules involving a dose of IPV followed by OPV dose have advantages, but would increase costs and perhaps cause confusion. Neonatal supplementation, preferably with monovalent type 1, poliovirus, in countries with stable dominant type 1 endemic poliovirus, has been found to be the most satisfactory protocol to address the two issues of maximising population immunity and reducing the virus reservoir.

References

1. Kok PW, Leeuwenburg J, Tukei P, van Wezel AL, Kapsenberg JG, van Steenis G, Galazka A, Robertson SE, Robinson D (1992) Serological and virological assessment of oral and inactivated poliovirus vaccines in a rural population in Kenya. Bull WHO 70: 93–103
2. Johnson S, Schoub BD, McAnerney JM, Gear JSS, Moodie JM, Garrity SL, Klaassen KIM, Küstner HGV (1984) Poliomyelitis outbreak in South Africa, 1982. II. Laboratory and vaccine aspects. Trans R Soc Trop Med Hyg 78: 26–31

3. Schoub BD, Johnson S, McAnerney JM, Küstner HGV, van der Merwe CA (1986) A comprehensive investigation of immunity to poliomyelitis in a developing country. Am J Epidemiol 123: 316–324
4. Schoub BD, Johnson S, McAnerney JM, van Middelkoop A, Küstner HGV, Windsor I, Vinsen C, McDonald K (1992) Poliomyelitis outbreak in Natal/KwaZulu, South Africa, 1987–1988. 2. Immunity aspects. Trans R Soc Trop Med Hyg 86: 83–85
5. Kim-Farley RJ, Rutherford G, Lichfield P, Hsu S–T, Orenstein WA, Schonberger LB, Bart KJ, Lui K–J, Lin C-C (1984) Outbreak of paralytic poliomyelitis, Taiwan. Lancet ii: 1322–1324
6. World Health Organizaiton (1989) Expanded Programme on Immunization (EPI). Poliomyelitis in 1986, 1987 and 1988, Part I. Wkly Epidem Rec 64: 273–279
7. Sutter RW, Patriarca PA, Brogan S, Malankar PG, Pallansch MA, Kew OM, Bass AG, Cochi SL, Alexander JP, Hall DB, Suleiman AJM, Al-Ghassany AAK, El-Bualy MS (1991) Outbreak of paralytic poliomyelitis in Oman: evidence for widespread transmission among fully vaccinated children. Lancet 338: 715–720
8. van Middelkoop A, van Wyk JE, Küstner HGV, Windsor I, Vinsen C, Schoub BD, Johnson S, McAnerney JM (1992) Poliomyelitis outbreak in Natal/KwaZulu, South Africa, 1987–1988. I. Epidemiology. Trans R Soc Trop Med Hyg 86: 80–82
9. Hovi T (1989) The outbreak of poliomyelitis in Finland in 1984–1985. Significance of antigenic variation of type 3 polioviruses and site-specificity of antibody responses in antipolio immunization. Adv Virus Res 37: 243–275
10. World Health Organization (1984) Poliomyelitis outbreak in Israel. Wkly Epidem Rec 63: 325–326
11. Onorato IM, Modlin JF, McBean AM, Thoms ML, Losonsky GA, Bernier RH (1991) Mucosal immunity induced by enhanced-potency inactivated and oral polio vaccines. J Infect Dis 163: 1–6
12. Department of National Health, South Africa (1990) The six vaccine-preventable diseases. Epidem Comments 17: 3–19
13. Tulchinsky T, Abed Y, Shaheen S, Toubassi N, Sever Y, Schoenbaum M, Handsher R (1989) A ten-year experience in control of poliomyelitis through a combination of live and killed vaccines in two developing areas. Am J Public Health 79: 1648–1652
14. World Health Organization Expanded Programme on Immunization (1985) Global advisory group. Wkly Epidem Rec 60: 13–16
15. Schoub BD, Johnson S, McAnerney J, Gilbertson L, Klaassen KIM, Reinach SG (1988) Monovalent neonatal polio immunization – a strategy for the developing world. J Infect Dis 157: 836–839

24

Optimal schedule for the administration of oral poliovirus vaccine

P. A. Patriarca, R. W. Linkins, R. W. Sutter, and **W. A. Orenstein**

Division of Immunization, Centers for Disease Control and Prevention, Atlanta, Georgia, USA

Summary

The optimal schedule for the administration of oral poliovirus vaccine (OPV) can be based on a simple paradigm in which the period of maximum risk from natural infection is balanced by the influence of factors which may affect the immune response to vaccination. Surveillance of paralytic poliomyelitis and seroprevalence data indicate that the maximum risk of wild poliovirus infection in most developing countries occurs between 6 and 24 months of age, suggesting that the primary vaccination series for OPV should be completed as early in life as possible. Although scientific evidence and programmatic considerations provide strong support for the currently recommended schedule of OPV at birth, 6, 10, and 14 weeks of age, as many as 30–40% of recipients may still remain susceptible to poliovirus types 1 and 3 after the fourth dose. Because age-specific gaps in immunity have often been associated with epidemic disease, additional strategies to increase population levels of immunity should be considered as progress continues towards the goal of global eradication of poliomyelitis by the year 2000. Administration of OPV in mass vaccination campaigns has been shown to be highly successful in the virtual elimination of wild poliovirus infections in the Americas, and has been adopted as the strategy of choice for the global initiative. Further studies of other supplemental approaches that may hasten eradication should also be pursued, such as expansion of the routine schedule for OPV to five or more doses, and the combined use of both oral and inactivated poliovirus vaccines.

Introduction

The optimal schedule for the administration of most vaccines, including oral poliovirus vaccine (OPV), can be based on a simple paradigm in which the "window of risk" from natural infection is balanced with a number of factors which may influence the immune response to vaccination. Such factors may include the minimum age at which antibody responses occur; the number and interval between doses; and interference in inducing antibody responses from a variety of host and/or environmental factors. Logistical and programmatic factors related to co-administration of other vaccines in the Expanded Programme on Immunization (EPI) must also be taken into account. In this chapter, we will review each of these factors as they relate to poliomyelitis and the current schedule for OPV administration. Our discussion will focus primarily on the current schedule recommended by the EPI for use in developing countries [1,2], and will suggest additional ways in which levels of immunity in vaccine recipients may be further enhanced.

"Window of risk" from natural infection

The development of practical serologic tests for detecting and quantifying neutralizing antibody against poliovirus during the late 1940s paved the way for a series of serologic surveys in developing countries to define the age-specific risk of infection. These surveys consistently indicated that the maximum risk of infection with wild polioviruses was during the first two years of life [3,4], a finding that was later confirmed by lameness surveys conducted in many of the same countries [5]. As shown in the composite diagram in Fig. 1, almost all infants are born with detectable levels of antibody that is passively acquired from the mother prior to birth, which appears to confer protection against infection and paralytic disease during the first few months of life [3]. The antibody has been shown to decay with a relatively constant half-life of 28 days [6]. Coincident with this decay, the risk of natural infection begins to increase at about 6 months of age and accelerates rapidly thereafter, with nearly 80% of infants becoming infected by two years of age (Fig. 1); those who have managed to escape infection are almost inevitably

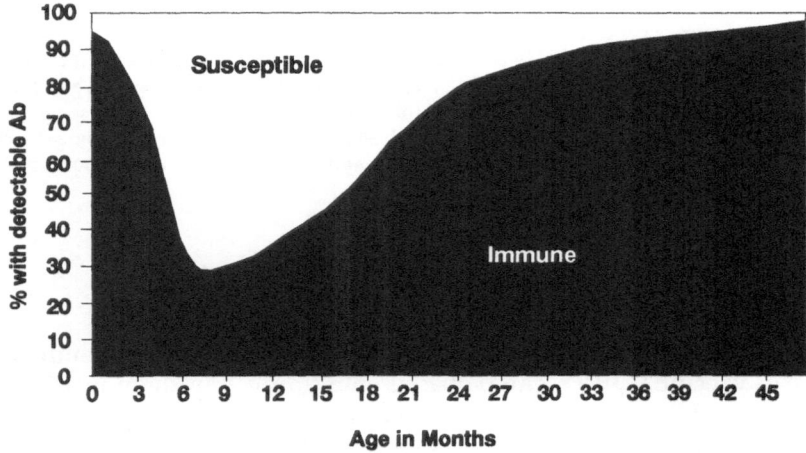

Fig. 1. Acquisition of neutralizing antibody against poliovirus type 1 in developing countries, by month of age, in the absence of an immunization program (pre–vaccination era). Data are derived from composite information reported in Ref. 3 and 4

affected over the next 1–3 year period. Although the incidence of poliomyelitis in developing countries has declined dramatically since 1974 following the advent and rapid acceleration of the EPI [2], data obtained from routine surveillance programs and outbreak investigations suggest that there has been very little change in the age distribution of poliomyelitis cases during the intervening time period [7,8]. The rapid acquisition of antibody and high incidence of disease during the first two years of life serves to illustrate the ease and rapidity with which wild poliovirus can be transmitted among young children, and underscores the need for vaccination as early in life as possible to minimize the number of susceptible infants who enter the "window" of maximum risk between 6 and 24 months of age (Fig. 1).

Factors affecting the immune response to OPV

Age at initial vaccination

Given the desirability of protecting infants from poliovirus infection as early in life as possible, the most important consideration in optimizing the vaccination schedule for OPV is determining the youngest age at which adequate antibody responses may occur. For some vaccines (e.g., measles), the presence of maternal antibody is known to impair replication of vaccine virus and the subsequent development of neutralizing antibody, so that vaccination must be deferred until a later age [9]. However, the initial site of replication of OPV is most often within the surface epithelial cells of the small intestine following attachment to specific poliovirus receptors [10], where the virus is generally shielded from binding with circulating immune globulin (IgG) derived from the mother. Indeed, a number of challenge studies in newborn infants carried out since 1955 have conclusively demonstrated that replication of vaccine virus and rates of seroconversion ($\geq$ 4-fold increases in neutralizing antibody) are only slightly lower than rates observed in older infants (Table 1). Vaccination during the newborn period has also been viewed as advantageous because of the low risk of concurrent infection with non-polio enteroviruses and other enteric pathogens which may interfere with infection with OPV [11]. Although immune tolerance has been described with some vaccines such as measles [9] and pertussis [12] – in which vaccine administration in the presence of maternal antibody may impair responses to additional doses administered later in life – no such effect has been observed with OPV [13]. In one study [14], beginning the primary vaccination series with OPV during the newborn period was associated with even higher cumulative rates of seroconversion to all three poliovirus subtypes after three doses (Table 2). Moreover, because an increasing percentage of infants are born in hospitals worldwide, the newborn period also provides easy access for vaccination. These and other observations [15] therefore provide substantial evidence in support of the current recommendation of the World Health Organization that the vaccination series should begin during the newborn period, with particular emphasis in polio-endemic countries [1,2].

Table 1. Rates of virus excretion and seroconversion to poliovirus types 1, 2, and 3 during the newborn period and at 2–3 months of age following the administration of monovalent or trivalent oral poliovirus vaccine*

Year	Country	Vaccine type	No. infants	% with antibody response to type:			% with excretion of poliovirus type:		
				1	2	3	1	2	3
Studies in newborns:									
1960	USA	TOPV	73	70	38	85	–	–	–
1961	USA	mono type 1	109	30	–	–	70	–	–
1961	USA	mono type 1	144	>50	–	–	77	–	–
1961	Israel	TOPV	53	–	–	–	70	–	70
1962	USA	mono type 1	64	47	–	–	88	–	–
1963	USA	mono type 1	109	32	–	–	66	–	–
1969	New Zealand	TOPV	22	27	36	45	–	–	–
1984	China	TOPV	108	30	30	18	41	41	27
1989	Oman†	TOPV	53	40	70	23	–	–	–
1989	Brazil†	TOPV	691	29	56	13	–	–	–
Studies in infants age 6–12 weeks:									
1959	USA	mono type 1/3	18	77	–	60	–	–	–
1962	USA	TOPV	16	44	81	50	–	–	–
1976	India	TOPV	26	42	85	31	–	–	–

*In most studies, "seroconversion" was defined as a ≥ 4–fold rise in neutralizing antibody above the projected decline in baseline (maternal) antibody. Table adapted from reference 1, which lists all publications in which original data were reported, excluding countries denoted by (†) [unpublished data]

Table 2. Rates of seroconversion to poliovirus types 1, 2, and 3 following three doses of OPV, according to the age at which the initial doses was given*

Age at first dose (in weeks)	No. infants	Seroconversion rate (%) [95% CI[a]] after 3 doses to:		
		Type 1	Type 2	Type 3
1	23	83 [67, 98]	83 [67, 98]	78 [61, 95]
2	30	80 [65, 94]	90 [79, 100]	70 [53, 86]
3	25	64 [45, 83]	96 [88, 100]	56 [36, 75]
4	26	90 [76, 100]	95 [89, 100]	65 [47, 84]
5	19	47 [25, 70]	68 [47, 89]	42 [20, 64]
6	16	69 [46, 91]	81 [62, 100]	63 [39, 84]

*Adapted from TJ John [14]
[a]CI = confidence interval

Number of doses

A second consideration in optimizing the vaccination schedule for OPV is the number of doses that should be given to maximize both short- and long-term antibody responses to all three serotypes. In contrast to most inactivated vaccines and toxoids, multiple doses of OPV have been shown to be required not so much for a "booster" effect (i.e., secondary rises in neutralizing antibody), but rather to ensure that a high proportion of recipients will seroconvert to types 1, 2, and 3. Based on extensive experience in industrialized countries, the minimum number of doses of OPV required to achieve high seroconversion rates is three [1], although some studies in the United States have shown that as few as two doses will induce detectable levels of antibody in virtually 100% of recipients [16]. Once seroconversion occurs, antibody appears to persist for at least 15 years, and probably for life [15]. Although studies in developing countries have also demonstrated incremental increases in cumulative seroconversion rates with each successive dose of OPV (Table 3), a recent review of 44 publications from such countries [15] found that an average of 27% and 30% of OPV recipients may still remain susceptible to poliovirus types 1 and 3, respectively, after three doses. These findings, combined with the recent emergence of large-scale outbreaks of poliomyelitis in highly vaccinated populations [15,17], have raised concerns that the currently recommended primary series of three doses of OPV may be inadequate to achieve the goal of global eradication of wild poliovirus infection by the year 2000 [18]. Additional strategies to overcome "gaps" in immunity associated with the 3-dose routine schedule for OPV in developing countries are addressed in more detail later in this chapter.

Interval between doses

A third consideration in the OPV vaccination schedule is the interval between doses to optimize antibody responses to all three serotypes. Such a

Table 3. Cumulative rates of seroconversion to poliovirus types 1, 2, and 3 following one, two, three, and five doses of OPV*

No. doses	Seoconversion rate (in %)		
	Type 1	Type 2	Type 3
1	19	61	30
2	35	78	48
3	69	90	76
5	83	96	82

*Adapted from John TJ (1976) Antibody response of infants in tropics to five doses of oral polio vaccine. Br Med J 1:812

consideration is particularly important for OPV, given that continued excretion of one or more types of vaccine virus may interfere with vaccine-virus infection following subsequent doses [15]. Indeed, longitudinal studies of primary vaccinees indicate that while there is a rapid decline in vaccine-virus excretion rates during the two-week period following vaccination, as many as 30%, 50%, and 38% of vaccinees will continue to shed types 1, 2, and 3, respectively, for another 4–6 week period [19]. Moreover, a recent review of seroprevalence studies carried out in developing countries since the early 1960s showed mean absolute increases in antibody prevalence of up to 16% in children who had received three doses of OPV at 6–8 week intervals compared with those who had received the same number of doses at shorter intervals [15]. An important limitation in the interpretation of these data, however, was that no studies had yet been carried out which compared antibody responses to OPV administered at different intervals within the same population [15]. To obtain more definitive information regarding the influence of interval between doses on the immune response, the World Health Organization sponsored two such trials in the United States (Nashville, Tennessee) and The Gambia in 1990. Preliminary results available from each of these studies have shown no significant differences in seroconversion rates to any of the three poliovirus serotypes when three doses of OPV were administered at 4-weeks vs. $\geq$ 6-week intervals (P.F. Wright, personal communication, 1992).

Host and environmental factors

A series of investigations carried out in both industrialized and developing countries during the past three decades [15] have also revealed a number of host and environmental factors which may influence antibody responses to OPV. Such factors include high levels of maternal antibody at the time of vaccination (titer $\geq$ 256); secretory antibody (IgA) present in colostrum or breast milk; nonspecific interference from poorly characterized substances present in saliva; poor nutritional status; and concurrent infection with a wide variety of enteric pathogens. Aside from interference due to high levels of maternal antibody–which can eventually be overcome by administering additional doses later in life after maternal antibody has decayed [1,15] – perhaps the most important of these factors is interference from non-polio enteroviruses.

Studies carried out by Pangi et al. in India [20] and Swartz et al. in Israel [21] have provided the most conclusive evidence that coinfections with non-polio enteroviruses adversely affect response rates to OPV, particularly to types 1 and 3. In India, 71%, 92%, and 86% of infants who received three doses of OPV during the cooler months of the year (November–January) seroconverted to poliovirus types 1, 2, and 3, respectively, compared with only 23%, 55%, and 24% of infants immunized during the warmer, rainy months (p < .05 for each comparison, chi-square test). These observations could not be attributed to any factor other than significantly higher rates of enterovirus excretion during the warmer months (60%) than in the cooler months (32%). Similar findings were reported from Israel, where only 50%, 83%, and 65% of infants who were shedding non-polio enteroviruses at the time of vaccination seroconverted to poliovirus types 1, 2, and 3, respectively, compared with 94%, 100%, and 94% of infants who were not co-infected [21]. These differences underscore the importance of OPV administration as early in life as possible, when enterovirus carriage rates are low [11], and also have important implications for the optimal scheduling of mass vaccination campaigns, a strategy which has assumed increasing importance in accelerating progress towards the eradication goal [18,22].

Logistic and programmatic considerations

Additional considerations in optimizing the vaccination schedule for OPV are logistical and programmatic factors related to (1) achieving the highest possible levels of vaccination coverage; and (2) maximizing program efficiency and clinical protection against other EPI diseases through simultaneous administration of other vaccines (e.g., DTP, BCG, and hepatitis B). A series of studies carried out in the United States [23] and a number of developing countries [24,25] have shown, in almost all instances, that there is an inverse relationship between the age of the child and the likelihood that he or she has received the recommended number of doses of each vaccine, including OPV. The current emphasis of primary health care programs in most developing countries focuses on children less than 12 months of age, so that opportunities to complete the primary immunization series later in life are often limited. Aside from poliomyelitis, available data also indicate that the "window of risk" from severe pertussis illness is highest during the first few months of life [26], and that in many countries in Asia and Africa, administration of the first dose of hepatitis B vaccine in the newborn period is considered essential to reduce the risk of carriage of hepatitis B surface antigen and the development of hepatocellular carcinoma during adulthood [27]. These and other logistical and programmatic considerations further underscore the importance of an integrated approach within the EPI and the provision of a primary series of all vaccines (aside from measles) as early in life as possible.

Future directions

Although the balance of scientific evidence and programmatic considerations thus provide a strong rationale for the currently recommended schedule for OPV administration at birth, 6, 10, and 14 weeks of age, extensive experience in developing countries over the past three decades also indicates that a relatively high proportion of four-dose recipients may still remain susceptible to one or more poliovirus types [15,28]. Variation in seroconversion rates has not only been observed among various countries throughout the world, but has also been documented within the same country, particularly among high-risk, low socioeconomic groups (Sutter RW, unpublished data, 1992). These observations suggest that a uniform strategy to deliver three or four doses in routine vaccination programs may not achieve levels of immunity that will be high enough to break the remaining chains of wild poliovirus transmission. Furthermore, as the contribution of natural infection to population immunity levels continues to decline over the next decade, additional epidemics of poliomyelitis following importation of wild virus from endemic areas will remain a constant threat [17].

Given the desirability of raising levels of immunity against poliovirus even further, what options are currently available? Among the most promising is the delivery of additional doses of OPV in mass campaigns, a strategy used by the Pan American Health Organization (PAHO) during the highly successful eradication campaign in Central and South America [22]. Since 1980 in Brazil, and 1985–86 in other polio-endemic countries in the region, annual campaigns have been conducted in which all children less than 5 years of age have been given two doses of OPV spaced 1–4 months apart (usually in a single day or over several days), regardless of the number of doses they may have received previously. This strategy was later supplemented by house-to-house vaccination campaigns ("mop-up" operations) in high-risk areas, which have appeared to rapidly eliminate any remaining foci of wild virus transmission [22]. Although the effectiveness of these approaches in the Americas is clearly apparent, the actual biological mechanisms involved remain unclear. Possible explanations include further reductions in the number of unimmunized children due to more aggressive approaches in vaccine delivery; higher rates of seroconversion following the administration of an average of 10–15 doses to each child over a five-year period; repeated boosting of secretory antibody (IgA) levels following multiple exposures to vaccine virus; higher levels of humoral and secretory immunity in the overall population (children and adults) due to more intensive spread of vaccine virus to susceptible persons in all age groups; and/or "displacement" of wild poliovirus and non-polio enteroviruses following massive and widespread use of OPV [29]. Regardless of the mechanism for success, mass vaccination campaigns have also been recommended by WHO for eliminating poliomyelitis in polio-endemic countries [18].

Full implementation of the PAHO strategy may take some time. In the meantime, further studies of supplemental approaches that may hasten eradication should also be pursued. If, for example, raising overall levels of humoral immunity in the target population is shown to be the most essential component in achieving eradication of wild poliovirus infection, expanding the primary series to five or more doses should improve the chances for success, since seroconversion rates increase incrementally with extra doses [15]. Although an expanded schedule may be associated with higher rates of attrition, drop-out rates could be minimized if the remaining contracts with the EPI system are utilized (e.g., at age 6 or 9 months at the time of measles vaccination), along with sporadic contacts for curative health services. An additional option would be to combine the known benefits of OPV with those of the new, enhanced-potency inactivated poliovirus vaccine (IPV), which has been shown to induce high levels of serum neutralizing antibody to all three types of poliovirus among infants in developing countries [30,31]. Such an approach has been used successfully in the Israeli-occupied West Bank and Gaza Strip for more than 12 years [32] and deserves further evaluation elsewhere. However, the high cost of IPV may preclude its use on wide scale.

The final option for improving rates of seroconversion involves changes in the vaccine itself. One approach that is particularly appealing would be to increase the potency of the type 1 and 3 components in the existing vaccine, which would not require any modification of the current vaccination schedule or delivery program. Such an approach has been associated with significantly higher rates of seroconversion compared with the currently recommended formulation [28,33], with little or no increase in cost. Vaccines containing higher concentrations of type 3 were introduced in most countries in Central and South America following the large-scale outbreak of type 3 poliomyelitis in Brazil in 1986, with relatively few cases of wild type 3 infection having been reported thereafter [22]. Dramatic advances in the fields of molecular biology and genetics have also raised the possibility of new generation vaccines with improved immunogenicity, including antigen chimeras [34] and recombinant strains [35] of poliovirus that can elicit neutralizing antibody to more than one subtype. However, in view of the extraordinary record of Sabin-derived vaccine in reducing the burden of poliomyelitis, combined with an almost unparalleled record of safety, it is doubtful that new generation vaccines would offer significant advantages during the final stages of the global eradication program in the coming decade.

References

1. Halsey N, Galazka A (1985) The efficacy of DTP and oral poliomyelitis immunization schedules initiated from birth to 12 weeks of age. Bull WHO 63: 1151–1169
2. Henderson RH, Keja J, Hayden G, Galazka A, Clements J, Chan C (1988) Immunizing the children of the world. Progress and prospects. Bull WHO 66: 535–543
3. Sabin AB (1948) Epidemiologic patterns of poliomyelitis in different parts of the world. In: Papers and discussions presented at the First International Poliomyelitis Conference. Lippincott, Philadelphia, pp 24–30
4. Paul JR, Melnick JL, Barnett VH, Goldblum N (1952) A survey of neutralizing antibodies to poliomyelitis virus. Am J Hygiene 55: 402–413
5. Bernier RH (1984) Some observations on poliomyelitis lameness surveys. Rev Infect Dis 6 [Suppl]: 371–375
6. Gelfand HM, Fox JP, LeBlanc DR, Elveback L (1960) Studies on the development of natural immunity to poliomyelitis in Louisiana. V. Passive transfer of polioantibody from the mother to fetus, and natural decline and disappearance of antibody in the infant. Am J Hygiene 85: 46–55
7. World Health Organization (Expanded Programme on Immunization) (1989) Poliomyelitis in 1986, 1987, and 1988 (Part I). Wkly Epidemiol Rec 64: 273–279

8. World Health Organization (Expanded Programme on Immunization) (1989) Polio-myelitis in 1986, 1987, and 1988 (Part II). Wkly Epidemiol Rec 37: 281–285

9. Wilkins J, Wehrle PF (1979) Additional evidence against measles vaccine administration to infants less than 12 months of age. Altered immune response following active/passive immunization. J Pediatr 94: 865–869

10. Mendlesohn CL, Wimmer E, Racaniello VR (1989) Cellular receptor for poliovirus. Molecular cloning, nucleotide sequence, and expression of a new member of the im-munoglobulin superfamily. Cell 56: 855–865

11. Lepow ML, Warren RJ, Gray N, Ingram VG, Robbins FC (1961) Effect of Sabin type 1 poliomyelitis vaccine administered by mouth to newborn infants. N Engl J Med 264: 1071–1078

12. Burstyn DG, Baraff LJ, Peppler MS, Leake RD, St. Geme J, Manclark CR (1983) Serologic response to filamentous hemagglutinin and lymphocytosis-promoting factors of Bordetella pertussis. Infect Immun 41: 1150–1156

13. Pagano JS, Plotkin SA, Koprowski H (1960) Variation of response in early life to vac-cination with living attenuated poliovirus and lack of immunologic tolerance. Lancet i: 1224–1226

14. John TJ (1984) Immune response of neonates to oral poliomyelitis vaccine. Br Med J 289: 881

15. Patriarca PA, Wright PF, John TJ (1991) Factors affecting the immunogenicity of oral poliovirus vaccine in developing countries. Rev Infect Dis 13: 926–939

16. McBean AM, Thoms ML, Albrecht P, Cuthrie JC, Bernier R, The Field Staff and Coordinating Committee (1988) Serologic response to oral polio vaccine and enhanced-potency inactivated polio vaccines. Am J Epidemiol 128: 615–628

17. Sutter RW, Patriarca PA, Brogan S, Malankar PG, Pallansch MA, Kew OM, Bass AG, Cochi SL, Alexander JP, Hall DB, Suleiman AJM, Al-Ghassany AAK, El-Bualy MS (1991) Outbreak of paralytic poliomyelitis in Oman. Evidence for widespread transmission among fully vaccinated children. Lancet 338: 715–720

18. Wright PF, Kim-Farley RJ, de Quadros CA, Robertson SE, Scott RM, Ward NA, Henderson RH (1991) Strategies for the global eradication of poliomyelitis by the year 2000. N Engl J Med 325: 1174–1179

19. Benyesh-Melnick M, Melnick JL, Rawls WE (1967) Studies of the immonogenicity, communicability and genetic stability of oral poliovaccine administered during the winter. Am J Epidemiol 86: 112–136

20. Pangi NS, Master JM, Dave KH (1977) Efficacy of oral poliovaccine in infancy. Indian Ped 14: 523–528

21. Swartz TA, Skalska P, Gerichter CG, Cockburn WC (1972) Routine administration of oral polio vaccine in a subtropical area. Factors possibly influencing seroconversion rates. J Hyg Camb 70: 719–726

22. de Quadros CA, Andrus JK, Olive JM, de Macedo CG, Henderson DA (1992) Polio eradication from the Western Hemisphere. Ann Rev Publ Health 13: 239–252

23. Cutts FT, Zell ER, Mason D, Bernier RH, Dini EF, Orenstein WA (1992) Monitoring progress toward U.S. preschool Immunization goals. JAMA 267: 1952–1955

24. Cutts FT, Glik DC, Gordon A (1990) Application of multiple methods to study the immunization programme in an urban area of Guinea. Bull WHO 68: 769–776

25. Cutts FT, Phillips M, Kortbeek S, Soarea A (1990) Door-to-door canvassing for immuni-zation program acceleration in Mozambique: achievements and costs. Int J Health Serv 20: 717–725

26. Wright PF (1991) Pertussis in developing countries. Definitions of the problem and prospects for control. Rev Infect Dis 13[Suppl]: 528–534

27. Hall AJ, Inskip HM, Loik F, Chotard J, Jawara M, Vall Mayans M, Greenwood BM, Whittle H, Njie ABH, Cham K, Bosch FX, Muir CS (1989) Hepatitis B vaccine in the Expanded Programme of Immunisation. The Gambian experience. Lancet i: 1057–1060

28. WHO Collaborative Study Group on Oral Poliovirus Vaccine (1992) A randomized trial of alternative formulations of oral poliovirus vaccine in Brazil and The Gambia. In:

Program and Abstracts of the 32nd Interscience Conference on Antimicrobial Agents and Chemotherapy. American Society for Microbiology, Washington, DC (Abstract #916) page 263

29. Sabin AB, Ramos-Alvarez M, Alvarez-Amezquita J, Pelon W, Michaels RH, Spigland I, Koch MA, Barnes JM, Rhim JS (1960) Live orally given poliovirus vaccine: effects of raid mass immunization on population under conditions of massive enteric infection with other viruses. JAMA 173: 1521–1526

30. Simoes EAF, Padmini B, Steinhoff MC, Jadhav M, John TJ (1985) Antibody responses of infants to two doses of inactivated poliovirus vaccine of enhanced potency. Am J Dis Child 139: 977–980

31. Robertson SE, Traverso HP, Drucker JA, Rovira EZ, Fabre-Teste B, Sow A, N'Diaye M, Sy MTA, Diouf F (1988) Clinical efficacy of a new enhanced-potency, inactivated poliovirus vaccine. Lancet i: 897–899

32. Tulchinsky T, Abed Y, Shaheen S, Toubassi N, Sever Y, Schoenbaum M, Handsher R (1989) A ten-year experience in control of poliomyelitis through a combination of live and killed vaccines in two developing areas. Am J Public Health 79: 1648–1652

33. Patriarca PA, Laender F, Palmeira G, Couto Oliveira MJ, Lima Filho J. de Souza Dantes MC, Tenorio Cordeiro M, Risi JB, Orenstein WA (1988) Randomised trial of alternative formulations of oral poliovaccine in Brazil. Lancet i: 429–433

34. Burke KL, Dunn G, Ferguson M, Minor PD, Almond JW (1988) Antigen chimaeras of poliovirus as potential new vaccines. Nature 322: 81–82

35. Kohara M, Abe S, Komatsu T, Tago K, Arita M, Nomoto A (1988) A recombinant virus between Sabin 1 and Sabin 3 vaccine strains of poliovirus as a possible candidate for a new type 3 poliovirus live vaccine strain. J Virol 62: 2828–2835

25

Poliovirus combined immunization with inactivated and live attenuated vaccines

H. Faden

Division of Infectious Diseases, Department of Pediatrics, State University of New York, School of Medicine at Buffalo and the Children's Hospital of Buffalo, Buffalo, New York, USA

Summary

Since the introduction of inactivated poliovirus vaccine in 1955 and live oral poliovirus vaccine in 1962, the number of cases of paralytic poliomyelitis has declined dramatically. In the United States, approximately five cases of poliomyelitis are reported annually and all of them are vaccine-associated. World-wide, there are approximately 200 000 cases of natural poliomyelitis. Both the inactivated and live vaccines have proven to be highly effective. The disadvantages of inactivated vaccine are relatively poor mucosal immunity and a belief that booster doses are required to maintain immunity. Live oral vaccine, on the other hand, is associated with paralytic poliomyelitis and poor seroconversion rates in underdeveloped regions of the world. A new enhanced potency inactivated vaccine is more immunogenic, induces greater mucosal immunity, and requires less frequent boosting than earlier inactivated preparations. A sequential immunization schedule consisting of two doses of enhanced potency inactivated vaccine followed by two doses of the live oral vaccine has been shown to induce excellent systemic and local immunity that is long lasting. The introduction of sequential immunization as the schedule of choice will require additional simplification of a currently complex vaccine program for polio.

Historical background

Poliomyelitis was a very common disease in the United States more than 40 years ago. Each Summer and Fall, 10 000 to 20 000 people developed paralytic disease. In 1952 alone, the number of cases exceeded 57 000. Although, paralytic illness represented less than 1% of the total poliovirus infections, millions of people became infected and participated in the spread of the virus throughout the community.

The development of the first inactivated poliovirus vaccine (IPV) by Dr. Jonas Salk and its introduction in 1955 led to an immediate decline in the annual reported incidence of paralytic cases from approximately 37 cases/100 000 population to 0.8 cases per 100 000 population in 1961 [39]. The first

orally administered live attenuated poliovirus vaccine (OPV) developed by
Dr. Albert Sabin was introduced in 1962 and became the vaccine of choice
by 1965. Oral poliovirus vaccine replaced IPV as the vaccine of choice in the
United States for a number of reasons. In the first place, it was easier to administer and cost less. Because it was a live replicating virus, it possessed
the added advantages of inducing local immunity in the gut and inadvertently
immunizing contacts through shedding of virus in the nasopharyngeal secretions and stool. Currently, more than 99% of poliovirus vaccines administered
in the United States are OPV.

Sixty-eight percent of the world's population presently live in countries
where more than 50% of the people are immunized against poliovirus and
experience ten or more cases of poliomyelitis each year [43]. Through collaboration of many programs, the Expanded Programme on Immunization
(EPI) increased the proportion of children receiving three doses of trivalent
OPV (TOPV) in the first year of life from 55% in 1986 to 85% in 1990 [43].
As a result of mass immunization, and imaginative programs such as immunization days, EPI has almost eradicated poliomyelitis from the Americas. In
fact, only nine cases of acute flaccid paralysis in the Americas were proven
to be caused by poliovirus in 1991 [2].

Despite the success of OPV in reducing the incidence of poliomyelitis in
underdeveloped regions after mass immunization programs, seroconversion
rates have been less than satisfactory. For example, seroconversion rates of
73% for poliovirus (PV) 1 and 70% for PV 3 have been noted following
complete immunization with OPV [32]. In contrast, the seroconversion rate
for PV 2 has been reported significantly higher at 90%, still less than that
achieved in developed nations [32]. More worrisome, have been outbreaks of
polio within countries with relatively high rates of vaccine coverage [21,31].
At present, an estimated 200 000 cases of paralytic disease occur each year
in the developing world [9,10].

Possible explanations for the failure of OPV include, vaccine instability,
vaccine formulation, concurrent enteric infections with or without diarrhea,
breast feeding, malnutrition, and the presence of inhibitory substances in the
intestine. The significance of each of these factors has been reviewed recently
by Patriarca et al [32]. However, two factors, vaccine formulation, and concurrent infection, deserve further comment.

Although OPV is administered as a multi-dose trivalent vaccine today,
it was originally given as three sequential monovalent doses. Seroconversion
rates were consistently high following monovalent preparations and serologic responses were similar in recipients in temperate and tropical climates
[32]. Poliovirus 2 typically replicated to the greatest degree and induced the
highest antibody response [32]. The exhuberant growth of PV 2 in the gut,
however, interfered with the replication of PV 1 and 3 in trivalent preparations. As a result, seroconversion rates fell from 94% to 55–70%, and from
90–95% to 55–75% for PV 1 and 3, respectively [32]. By administering the
trivalent vaccine on three or more occasions, the interference by PV 2 could
be overcome. Although this was true in developed countries, it may not be

true for less developed nations. Thus, changing vaccine formulation may be beneficial for the success of the vaccine in the tropics.

A second explanation given for the failure of OPV to induce adequate seroconversion rates in the third world relates to the presence of concurrent enteric infections. Domok et al [8] observed a 15–20% reduction in seroconversion rates with OPV 1 in children co-infected with enteroviruses. In contrast, other studies failed to demonstrate interference of enteric viruses with OPV [11,19]. We recently had a chance to examine the effect of a concurrent viral infection on the immune response to poliovirus immunization in 27 children. The children were healthy and well nourished, and infected asymptomatically with enteroviruses 60%, adenoviruses 21%, and other agents 19%. Co-infection did not significantly alter the systemic or local immune responses to any of the vaccine viruses except for marginal suppression of serum neutralizing antibody, in particular PV 1 and 3, following a third dose of vaccine (Table 1) [11]. More intriguing was the observation that the concurrent infection suppressed the immune response to inactivated vaccine as well as to live vaccine. In underdeveloped countries where children may suffer from a number of medical problems such as malnutrition, concurrent enteric infection may impose a greater threat to the immune system than we observed in healthy children of an industrialized nation.

Table 1. Effect of co-infection on geometric mean antibody response to poliovirus[a] immunization[b]

Antibody and Age (mo)	Immunization group	
	Co-infected	Controls
Serum Nt Ab		
4	20	16
5	316	501
12	158	251
13	2512	6310[c]
Nasopharyngeal Nt Ab		
4	3	3
5	4	4
12	3	3
13	8	10
Nasopharyngeal IgA Ab		
4	6	8
5	10	16
12	50	40
13	40	63

[a] poliovirus types 1, 2, and 3
[b] OPV or EIPV given at 5 and 12 months of age
[c] Mann Whitney Rank test used as non-parametric statistic $p < .001$
Nt neutralizing; *Ab* antibody

Vaccine-associated paralytic poliomyelitis

Although failure of OPV has not been a worry in the United States, Americans have been concerned about vaccine-associated paralytic polio-myelitis (VAPP). In fact, this problem has become the driving force behind re-evaluation of the use of OPV in the United States, a country without natural poliomyelitis since 1979. Vaccine-associated paralytic poliomyelitis cases began to appear in recipients soon after the introduction of OPV. It quickly became obvious that adults were at higher risk than children. Sub-sequently, the use of OPV was contraindicated in subjects 18 years of age and older. The reason for increased risk to adults has never been elucidated.

Since 1961, there have been more than 260 cases of VAPP in recipients, 35%, contacts, 42%, and in immunodeficient individuals, 23%, [39]. Over-all, the annual incidence of VAPP has remained steady for the past 30 years at approximately four cases per 10 million doses of vaccine distributed. The greatest risk of VAPP, 75%, is associated with the first dose of vaccine [39]. Specifically, 87% of recipient cases and 56% of contacts occurred following the first dose [39].

Recovery of the inciting virus from individuals with VAPP demonstrated a preponderence of polioviruses 2 and 3. Type 3 poliovirus accounted for 92% of the recipient cases and 62% of the contact cases, while type 2 virus caused 62% of the immunologically deficient cases [39]. In contrast, type 1 was recovered from less than 15% of all cases [39].

Attenuated poliovirus type 3 in OPV is associated with VAPP relatively often because it is genetically more unstable than the other two viruses. For example, type 3 virus differs from its neurovirulent predecessor by 10 muta-tions and only two mutations may be necessary for neurovirulence [33,42]. In contrast, poliovirus 1, the most stable vaccine component, differs by 56 mutations and up to six may be necessary for neurovirulence [5]. Minor and Dunn [25] have indicated that reversion appears to involve changes in short sequences in a single nucleotide. For type 3 poliovirus, uridine is replaced by cytosine at position 472 in the 5' noncoding region while for type 1 poliovirus, adenine replaces guanine at position 480 [25]. Although type 2 poliovirus may also revert to a neurovirulent form by replacing an adenine with guanine at position 481, it is more likely to cause disease in immune deficient subjects because it has a tendency to induce viremia more often than the other two viruses [17].

Reexamination of inactivated vaccines

In the face of OPV failing to prevent wild poliomyelitis in the tropics and the persistence of VAPP in industrialized nations, it is reasonable to reconsider a greater role of IPV in the immunization process. Newer enhanced potency IPV (EIPV) offers distinct advantages over older preparations. Enhanced potency vaccines utilize a microcarrier culture technique first described by van Wezel et al [40] to produce larger quantities of poliovirus antigen at lower costs. The newly formulated vaccines contain 40–8–32 D antigen units of poliovirus types 1, 2, and 3, respectively, compared to 20–2–4 D antigen units in the original vaccine. The increased antigenic content of EIPV has been associated with seroconversion rates of 99% following as few as two doses, in contrast to earlier preparations of IPV that required three to four doses to achieve the same seroconversion rate [12,15,24,38]. In addition, antibody titers following EIPV often exceed those following OPV [12,24]. Efficacy of EIPV was established in Senegal in 1986–87 during an outbreak of poliomyelitis when two doses of vaccine proved clinically effective in 89% [35].

In 1986 we initiated a large comparative study of immunization with OPV, EIPV, and combinations of the two vaccines [12]. One hundred and fifty-eight children were enrolled in the study. They were assigned to one of four immunization groups: 1) OPV-OPV-OPV 2)EIPV-EIPV-EIPV 3) EIPV-OPV-OPV 4) EIPV-EIPV-OPV. The children were immunized at 2, 4, and 12 months of age. Blood, nasopharyngeal, and fecal samples were collected at the time of each immunization and one to two months after each visit. Neutralizing antibodies to PV 1, 2, and 3 were measured in serum and nasopharyngeal samples. Poliovirus specific IgA antibodies were also determined in nasopharyngeal samples. Nasopharyngeal and fecal samples were evaluated for polioviruses and co-infecting viruses. One hundred and twenty-three children were evaluable. Groups were comparable with respect to age, percent with detectable antibody, and titer of antibody at entry into the study.

One dose of OPV or EIPV was not associated with any significant rise in neutralizing antibody in two months old infants while 94% or more of the children exhibited antibody following two doses of either vaccine (Table 2) [12]. After three doses of OPV or EIPV, EIPV consistently yielded the higher antibody titer but was only statistically greater for PV 3 (Table 3) [12]. Premature babies were also immunized with EIPV as part of our study population and they responded in a manner similar to full term babies (data not shown) [1].

Inactivated vaccines have traditionally required administration of booster doses due to an expected decline in seropositivity. One long-term follow-up study of standard IPV immunization in Sweden demonstrated persistent immunity to each of the three polioviruses as long as 25 years later in a population with an immunization rate of 99% [4]. Four year follow-up evaluation in our own study population demonstrated persistence of antibody in 99% of children immunized with either EIPV or OPV (Table 4).

Despite a good record, several nations utilizing IPV have experienced outbreaks of poliomyelitis in the past several decades. In 1984, Finland experienced a unique epidemic that involved nine cases of paralytic disease due to type 3 virus. Investigation of the problem demonstrated an extremely low rate of seropositivity, 60%, to PV 3 among the population [18]. This decline in positivity rate from earlier years was attributed to an unexpected

Table 2. Percentage of children with detectable levels of serum neutralizing antibody to poliovirus types 1, 2, and 3 by immunization group

Poliovirus type, age (mo)	Immunization group			
	OPV-OPV-OPV	EIPV-EIPV-EIPV	EIPV-OPV-OPV	EIPV-EIPV-OPV
1				
2	74	80	88	89
4*	77	61	77	94
5	100	96	94	100
12	77	91	94	100
13	100	96	100	100
2				
2	96	89	94	94
4	96	93	100	100
5	100	100	100	100
12	91	96	100	100
13	100	100	100	100
3				
2	83	77	77	78
4	77	79	82	82
5	100	96	94	100
12	77	93	94	94
13	100	100	100	100

OPV oral poliovirus vaccine; *EIPV* enhanced inactivated poliovirus vaccine.
*p < .05

decline in the PV 3 content of the vaccine and a slight change in the immunologic and molecular characteristics of the epidemic virus compared to the vaccine virus [18]. Another interesting feature of the outbreak was the extensive spread of the virus in the community. It was estimated that at least 100 000 people were infected. The extensive spread of the virus suggested that the IPV vaccine failed to induce sufficient mucosal immunity to halt person to person transmission.

Although IPV has been shown to generate good systemic immunity, less certainty surrounds its ability to protect mucosal surfaces. Ghendon and Sanakoyeva [13] compared fecal shedding of poliovirus type 1 in nonimmune, naturally immune, and in children immunized with two doses of IPV or OPV. Oral polio vaccine immunized children shed virus in a manner similar to naturally immune subjects; thirty-five percent of the OPV vaccinees shed virus in titers 1000 fold less and for a period one-fourth as long as the nonimmune group. The IPV group fell somewhere in between the OPV and nonimmune groups. Henry and Jaikaran [16] observed a similar difference of shedding after three doses of vaccine. Type 1 poliovirus shedding appeared to be reduced to a greater degree than type 2 poliovirus after IPV [7].

Table 3. Geometric mean titer of serum neutralizing antibody to poliovirus types 1, 2, and 3 by immunization group

Poliovirus type, age (mo)	Immunization group				Group differences*
	OPV-OPV-OPV[a]	EIPV-EIPV-EIPV[b]	EIPV-OPV-OPV[c]	EIPV-EIPV-OPV[d]	
1					
2	21	21	22	27	–
4	36	7	8	24	a > b
5	273	185	219	283	–
12	67	62	165	129	–
13	1470	1954	2175	3044	–
2					
2	50	43	105	56	–
4	492	33	40	42	a > b, c, d
5	2727	632	1774	481	a > b, d
12	404	136	1389	334	c > b
13	3378	5835	11 110	10 693	c, d > a
3					
2	17	17	12	15	–
4	18	16	17	19	–
5	352	635	70	1133	b, d > c
12	78	102	86	151	–
13	1522	5187	887	2348	b > a, c

OPV oral poliovirus vaccine; *EIPV* enhanced inactivated poliovirus vaccine
*$P < .05$

Table 4. Percentage of children with detectable levels of serum neutralizing antibody to poliovirus types 1, 2, and 3 by immunization group four years after the third dose of vaccine

Poliovirus	OPV-OPV-OPV (n = 11)	EIPV-EIPV-EIPV (n = 27)	EIPV-OPV-OPV (n = 11)	EIPV-EIPV-OPV (n = 13)
1	100	100	91	100
2	100	100	100	100
3	100	96	91	100

OPV oral poliovirus vaccine; *EIPV* enhanced inactivated poliovirus vaccine

Inactivated poliovirus vaccine was better able to induce mucosal immunity when the dose of the challenge virus was reduced [7,13,16,18].

Several studies have demonstrated an inverse relationship between serum neutralizing antibody titers and viral shedding. Glezen [14] examined shedding of PV 1 in the nasopharynx and stool of children previously immunized with three doses of IPV and subsequently challenged with OPV 1. Infection with OPV was not detected in the nasopharynx of the children with pre-existing neutralizing antibody titers of 1 : 128 or greater. Fecal shedding was reduced, but not eliminated, in children with relatively high antibody titers. Lepow et al [22] observed reduced shedding of OPV in nonimmunized newborns possessing passively acquired maternal antibody to poliovirus with titers of 1 : 128 or greater. Chimpanzees passively immunized with low levels of neutralizing antibody demonstrated reduced pharyngeal shedding of poliovirus [3]. Fecal shedding, however, was not altered in the chimps even when serum levels of antibody approached 1 : 500. These data imply that serum antibody may protect some mucosal surfaces through an ill-defined process, such as leakage, when the titer of antibody is sufficiently high. In the case of immunization with IPV, it is also possible, that the inactivated vaccine might induce local protection through a more active process.

New enhanced potency vaccines have the potential to stimulate further an even greater degree of local protection than standard IPV. We compared the nasopharyngeal antibody production in children immunized with two and three doses of EIPV or OPV [12]. Poliovirus specific IgA and neutralizing antibody were detected generally more often in children immunized with OPV (Tables 5, 6) [12]. Titers of IgA antibody were significantly greater for the OPV group while titers of neutralizing antibody were only significantly

Table 5. Percentage of children with detectable levels of nasopharyngeal IgA antibody to poliovirus types 1, 2, and 3 by immunization group

Poliovirus type, age (mo)	Immunization group			
	OPV-OPV-OPV	EIPV-EIPV-EIPV	EIPV-OPV-OPV	EIPV-EIPV-OPV
1				
4	76	48	63	53
5	73	57	82	65
12	100	89	94	94
13	100	89	94	75
2				
4	76	46	63	53
5	73	62	82	76
12	100	90	94	94
13	100	91	100	81
3				
4*	76	43	69	59
5	77	60	88	71
12	100	92	94	94
13	100	89	100	81

OPV oral poliovirus vaccine; *EIPV* enhanced inactivated poliovirus vaccine
*OPV-OPV-OPV > EIPV-EIPV-EIPV, EIPV-OPV-OPV, EIPV-EIPV-OPV (combined)
P < .05

Table 6. Percentage of children with detectable levels of nasopharyngeal neutralizing antibody to poliovirus types 1, 2, and 3 by immunization group

Poliovirus type, age (mo)	Immunization group				Group differences
	OPV-OPV-OPV[a]	EIPV-EIPV-EIPV[b]	EIPV-OPV-OPV[c]	EIPV-EIPV-OPV[d]	
1					
4	24	0	0	0	a > b, c, d[1]
5	27	28	53	35	–
12	29	4	12	13	a > b, c, d[2]
13	70	43	53	50	–
2					
4	67	0	13	0	a > b,c,d[1]
5	68	42	88	53	c > d[3]
12	67	12	47	25	a > b, c, d, c, d, b[3]
13	85	60	82	88	c,d > b[2]
3					
4	14	2	6	0	–
5	41	49	35	59	–
12	33	12	6	13	–
13	75	66	41	44	–

OPV oral poliovirus vaccine; *EIPV* enhanced inactivated poliovirus vaccine
[1]P < .001 (groups combined)
[2]P < .05
[3]P < .005

different for PV 2 after the third dose of vaccine (Table 7, 8) [12]. Onorato et al (30) similarly compared the local antibody response in the gut in children immunized with EIPV or OPV. Poliovirus type 1 specific IgA was detected in the stools of 32% of children immunized with three doses of EIPV compared to 28% of children immunized with three doses of OPV [30].

Local antibody to the major poliovirus structural proteins VP 1, VP 2, and VP 3 and PV 3 have also been compared in EIPV and OPV immunized children. Secretory antibody to VP 1 and 2 in the nasopharynx were similar for the two groups; however, VP 3 antibodies were detected in only 13% of EIPV recipients compared to 80% of OPV recipients (Fig. 1) [45]. These data suggest that the immune response to parenterally and orally administered poliovirus vaccines vary because the host may recognize different viral components in the two preparations.

Evidence to support this conclusion first came from the laboratory of Roivainen and Hovi [36,37] who showed that sera from children immunized solely with IPV neutralized trypsin-cleaved OPV 3 poorly, while sera from naturally infected individuals were able to neutralize trypsin-cleaved type 3 polioviruses better.

Table 7. Geometric mean titer of nasopharyngeal IgA antibody to poliovirus types 1, 2, and 3 by immunization group

Poliovirus type, age (mo)	Immunization group				Group differences*
	OPV-OPV-OPV[a]	EIPV-EIPV-EIPV[b]	EIPV-OPV-OPV[c]	EIPV-EIPV-OPV[d]	
1					
4	12	4	7	5	a > b
5	15	6	16	7	–
12	60	24	22	28	a > b
13	69	24	46	19	a > b, d
2					
4	11	4	7	5	–
5	13	6	14	9	–
12	60	26	23	31	–
13	97	25	67	22	a > b, d
3					
4	12	4	9	6	a > b
5	17	7	20	9	–
12	92	30	32	40	a > b
13	128	31	79	23	a > b, d

OPV oral poliovirus vaccine; *EIPV* enhanced inactivated poliovirus vaccine
*P < .05

Table 8. Geometric mean titer of nasopharyngeal neutralizing antibody to poliovirus types 1, 2, and 3 by immunization group

Poliovirus type, age (mo)	Immunization group				Group differences*
	OPV-OPV-OPV[a]	EIPV-EIPV-EIPV[b]	EIPV-OPV-OPV[c]	EIPV-EIPV-OPV[d]	
1					
4	2	0	0	0	a > b, c, d
5	2	2	4	2	–
12	2	1	1	1	a > b
13	6	3	4	3	–
2					
4	6	0	2	0	a > b, c, d
5	7	2	16	4	c > b, d; a > b
12	7	1	3	2	a > b, d; c > b
13	17	5	17	18	a, c, d > b
3					
4	1	1	1	0	–
5	3	3	2	4	–
12	3	1	1	1	a > b, c, d
13	6	5	3	4	–

OPV oral poliovirus vaccine; *EIPV* enhanced inactivated poliovirus vaccine
*P < .05

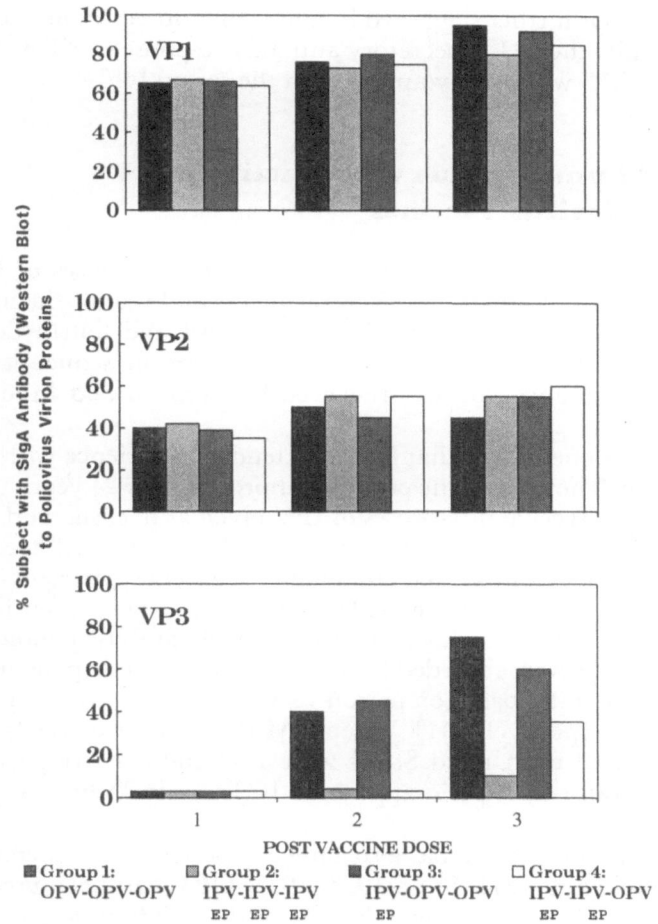

Fig. 1. Frequency of secretory IgA antibody responses to poliovirus virion proteins VP1, VP2, and VP3 in the nasopharyngeal secretions of infants after immunization with OPV, IPV-EP, or both. OPV – oral poliovirus vaccine, IPV-EP – enhanced potency inactivated poliovirus vaccine, VP – virion protein

We observed a similar pattern in the local antibody responses to PV 3 in our study population. Significant differences in poliovirus specific local antibody were observed between OPV and EPIV immunized subjects when trypsin-treated poliovirus was used as the antigen for antibody determinations. The neutralizing antibody titer against enzyme-cleaved virus was significantly higher than against whole virus in the OPV vaccinees [44]. Both neutralizing and ELISA antibody activity against cleaved virus was significantly lower than against whole virus in EIPV immunized subjects [44]. These observations suggest that the secretory antibody response was directed against distinct epitopes or antigenic sites of the poliovirus that were not available in EPIV.

This conclusion is further suggested by the earlier observation in our laboratory that despite the lack of secretory antibody response to VP 3 after EIPV, antibody to VP 3 was readily apparent in the serum [45].

Combined immunization with inactivated and live poliovirus vaccines

If an immunization schedule began with one or two doses of EIPV, this would potentially eliminate VAPP in recipients and reduce the incidence in contacts. Immunization with OPV for the third and fourth doses would insure adequate mucosal protection. The rational for this sequential approach to poliovirus immunization was reviewed by McBean and Modlin [23] in 1987.

Denmark is one nation that has an extended experience with sequential immunization. They began the program approximately 24 years ago in 1968. Danish children receive three doses of IPV given at five, six, and 15 months while OPV is administered at two, three, and four years of age. The state supplies all vaccines for free and compliance is excellent; 92% of children 15 years of age are fully immunized [41]. In the first 15 years after introducing the sequential vaccine schedule, only two cases of paralytic poliomyelitis and one imported case were recorded [41]. The risk of endemic polio in Denmark and the United States based on person years of exposure were reported to be 2.64 and 4.94, respectively [41]. When VAPP cases were considered the risks for Denmark and the United States were 1.32 and 3.20, respectively [41]. Thus, the overall risk of polio appears to be lower in Denmark than in the United States.

A second nation, Israel, has also gained experience with a combined immunization schedule [21]. From 1973–1977, OPV had been used throughout the City of Gaza. Even though 90% of the children younger than one year of age had been immunized against polio, two outbreaks of poliomyelitis occurred in 1974 and 1976 [21]. Many of the cases were due to type 1 PV. In response to the failure of OPV to protect the population, Israel initiated a vaccine program that included monovalent OPV 1 in the first month of life, TOPV and IPV at 2–3 and 3–4 months, and OPV at 5–6 and 12–14 months. The new scheduled resulted in a decline in cases of paralytic poliomyelitis from 10 cases per 100 000 population in 1977 to 2.6 cases per 100 000 population in 1982 [21].

Ogra [6,28] as early as 1984 demonstrated a serologic advantage to sequential immunization. Infants immunized with OPV at 12 months of age after a series of three doses of IPV at 2, 4, and 6 months of age developed higher levels of serum neutralizing antibody than those immunized initially with OPV [6,28]. In addition, nasopharyngeal IgA was also enhanced in the IPV primed infants [6,28]. More recently, Modlin and co-workers [26] demonstrated a similar priming effect on serum antibody in children immunized with three doses of EIPV.

As seen in Table 2, two doses of OPV, EIPV or EIPV followed by OPV, resulted in a seropositivity rate of 94% or more for each poliovirus serotype. After the third dose of vaccine, the seropositivity rate was 100% for all immunization groups except EIPV-EIPV-EIPV against PV 1 (Table 2) [12]. Antibody titers were excellent for each immunization group. The highest geometric mean antibody titers were seen in children who received one or more doses of EIPV. For example, the EIPV-EIPV-OPV produced the highest GMT for PV 1 at 1:3044, EIPV-OPV-OPV for PV 2 at 1:11, 110, and EIPV-EIPV-EIPV at 1:5187 for PV 3 (Table 3) [12]. These antibody titers were significantly greater than the OPV-OPV-OPV group for PV 2, and 3 (Table 3) [12]. A priming effect was noted after two doses of EIPV for PV 1, 2, and 3.

Sixty-two children in our original study were re-evaluated serologically at five years of age, four years after the third dose of vaccine. More than 90% of the subjects remained seropositive to the polioviruses (Table 4). Overall, the highest antibody titers were against PV 2 and the lowest against PV 1 in each of the groups (Table 9). The combined vaccine group of EPIV-EIPV-OPV possessed the highest neutralizing antibody titers to PV 1, 2, and 3 among the four groups (Table 9). Each subject was subsequently challenged with TOPV and the antibody responses to PV 1, 2, 3 determined one month later. The best responses were detected in the EIPV-EIPV-OPV and EIPV-EIPV-EIPV groups (Table 9).

Because of the importance placed on mucosal immunity when comparing inactivated and live vaccines, we examined the effects of sequential immunization on the local production of neutralizing and poliovirus specific IgA antibodies. One dose of EIPV followed by OPV resulted in excellent nasopharyngeal neutralizing and IgA antibodies to each of the three polioviruses (Tables 5–8) [12]. In fact, one dose of EIPV appeared to prime the local immune response to OPV, even though the differences were not

Table 9. Geometric mean antibody titer to polioviruses 1, 2, and 3 four years after immunization and following OPV challenge

Poliovirus type	Immunization group			
	OPV-OPV-OPV (n = 11)	EIPV-EIPV-EIPV (n = 27)	EIPV-OPV-OPV (n = 11)	EIPV-EIPV-OPV (n = 13)
Before challenge				
1	125	199	79	501
2	398	398	630	1000
3	158	251	79	630
After challenge				
1	501	1585	794	1585
2	794	3981	1995	3162
3	1000	1995	1999	2511

statistically significant. After three doses of vaccine, the two combination groups were comparable except the group receiving two doses of OPV tended to produce a better IgA antibody response to each of the three polioviruses (Table 5, 7) [12]. Although children who received OPV exclusively generally exhibited a better local immune response than children who received either EIPV exclusively or two doses of EIPV followed by one dose of OPV, their response was not statistically different from the children who received one dose of EIPV followed by two doses of OPV.

Mucosal immunity was further evaluated in our study population by determining nasopharyngeal and fecal shedding of poliovirus. Poliovirus was recovered in 40 fecal specimens and in no pharyngeal specimens. As seen in Fig. 2, 29% of children shed virus two months after the first dose of OPV in the OPV-OPV-OPV group and in 43% of the children one month after the second dose. At the same time, 62% of children in the EIPV-OPV-OPV group shed virus after the first dose of OPV. After the third dose of vaccine, 13% of children in the OPV-OPV-OPV group, 14% in the EIPV-OPV-OPV group and 20% in the EIPV-EIPV-OPV group shed virus. These differences were not statistically significant and suggest a similar degree of local production among immunization groups.

In contrast, Onorato et al [30] observed significantly more shedding of poliovirus in children originally immunized with three doses of EIPV and challenged with OPV 1 than in children immunized with three doses of OPV before challenge with OPV 1. Increased shedding of PV 1 occurred in the EIPV group despite higher serum levels of neutralizing antibody and com-

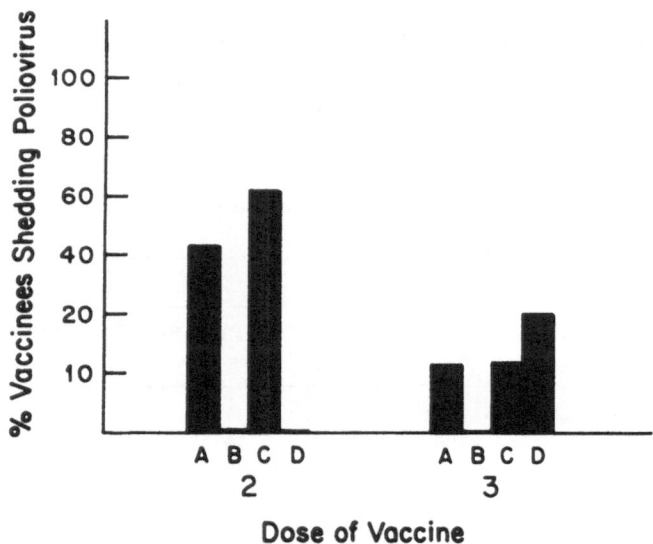

Fig. 2. Fecal shedding of polioviruses after two or three doses of vaccine in groups A (OPV-OPV-OPV), B (EIPV-EIPV-EIPV), C (EPV-OPV-OPV), and D (EIPV-EIPV-OPV). X^2 analysis and Fisher's exact test failed to show significant differences

parable fecal levels of PV 1 IgA [30]. Differences in the results of the two studies, in all likelihood, reflect differences in design. For example, we determined shedding of all polioviruses one month after challenge with TOPV while Onorato et al [30] determined shedding of PV 1 every week after challenge with OPV 1 for seven weeks. Overall, data from a number of studies that examined mucosal immunity by determining fecal shedding of virus and/or local antibody production conclude that OPV and (E) IPV both induce local immunity; however, OPV is the better stimulus. Furthermore, EIPV may prime the mucosa for challenge with OPV.

Conclusions

The cumulative data presented in this current review have reinforced the previously held belief that IPV (EIPV) and OPV are both highly effective vaccines in stimulating protective levels of immunity to the three strains of poliovirus. Both vaccines cost below $15 per dose and provide protection for prolonged periods. As the incidence of natural polio wanes, avoidance of risks associated with the live vaccine becomes more important to society. In addition, it will be necessary to optimize immunization practices in regions of high risk and where OPV has not been fully effective. The use of sequential immunization with EPIV followed by OPV may solve some of the problems.

On the other hand, it will also bring new concerns. For instance, a combined schedule will further complicate an already complex vaccine program in the United States. Record keeping will also become difficult. When a vaccination plan gets too cumbersome, compliance drops off. Already, in some urban centers in the United States, less than 80% of children under the age of two are fully immunized against polio [34].

Another potential concern, albeit minor, is the recent preliminary discovery in our laboratory that the frequency of reversions from attenuated to neurovirulent virus appears to depend on the vaccine protocol administered. For example, whereas four of nine virus isolates from children immunized with one or two doses of OPV were revertants, 11 of 12 isolates from children immunized with EIPV and OPV were revertants (Table 10) [39]. Furthermore, revertants of PV 3 were only recovered from children in the combined group (Table 10) [29]. As it turned out, local antibody responses in the nasopharynx were greatest in the specimens with revertants [29]. These early observations suggest that the development of reversion during replication of OPV in the gut may provide a more potent stimulus for the induction of mucosal immunity than the attenuated vaccine virus.

Since the use of EIPV does not protect against the development of revertants in vaccinees, another approach to immunization would be the construction of attenuated viruses that are genetically more stable. Viral hybrids of poliovirus 1 and poliovirus 2 or 3 have been created [20,27]. The recombinant viruses are fully viable, attenuated, and are antigenically similar to the parent strains [20,27]. They appear to retain the genetic stability of

Table 10. Distribution of revertant and nonrevertant vaccine poliovirus isolated in feces after different vaccination schedules

Immunization Group (n)	Nonrevertants				Revertants			
	Total	Serotype			Total	Serotype		
		1	2	3		1	2	3
OPV (9)	5	3	0	2	4*	2	2	0
EIPV + OPV (12)	1	1	0	0	11	3	4	4

*OPV vs EIPV + OPV p < .01

PV 1, and yet they can induce an immune response that is capable of neutralizing PV 1 and 3. These constructs could eliminate the interference generated by the extensive replication of PV 2 in the gut. Recombinant vaccines could then be used alone or in sequence with EIPV.

In conclusion, present conditions are ripe for an increasing role of EIPV in the campaign against poliomyelitis. Sequential use of EIPV and OPV offers several distinct advantages, but change in a vaccine schedule that has worked well will not be easy to institute. Reduction in the number of separate vaccine injections by combining the vaccines will hasten acceptance. A possible scheme for immunization might include a single injection of EIPV and HBV at two and four months, and a single injection of life DPT-HIB at two, four, and six months. In the second year of life DPT-HIB, measles-mumps-rubella-HBV, and OPV would be administered as three separate preparations. Booster doses of OPV, measles and DPT would be given upon entry into public school. Although a number of different schedules could be developed, physicians and families will not easily accept a cumbersome system that requires vaccine combinations that change year-to-year.

Acknowledgements

I would like to thank Mary Anne Borruso for her secretarial assistance.

References

1. Adenyi-Jones SC, Faden H, Ferdon MB, Kwong MS, Ogra PL (1992) Systemic and local immune responses to enhanced-potency inactivated poliovirus vaccine in premature and term infants. J Pediatr 120: 686–689
2. Andrus JK, de Quadros CA, Olive JM (1992) The Surveillance Challenge: Final Stages of Eradication of Poliomyelitis in the Americas, MMWR 41/no. SS-1: 21–26
3. Bodian D, Nathanson N (1960) Inhibitory effects of passive antibody on virulent poliovirus excretion and on immune response in chimpanzees. Bull Johns Hopkins Hosp 107: 143–162
4. Bottiger M (1987) A study of the sero-immunity that has protected the swedish population against poliomyelitis for 25 years. Scand J Infect Dis 19: 595–601

5. Christodoulose C, Colbere-Garapin G, Macadam A (1990) Mapping of mutations associated with neurovirulence in monkeys infected with Sabin 1 poliovirus revertants selected at high temperature. J Virol 64: 4922–4929
6. Dhar R, Ogra PL (1985) Local immune responses. Br Med Bull 41: 28–33
7. Dick GWA, Dane DS, McAlister J, Briggs M, Nelson R, Field CMB (1961) Vaccination against poliomyelitis with live virus vaccines. Brit Med J 266–269
8. Domok L, Balayna MS, Fayinka OA, Skrtic N, Soneji AD, Harland PSEG (xxx) Factors affecting the efficacy of live poliovirus vaccine in warm climate. Bull WHO 51: 333–347
9. Expanded Programme on Immunization. Poliomyelitis in 1986, 1987 and 1988 (Part 1). (1989) Wkly Epidemiol Record 64: 273–279
10. Expanded Programme on Immunization. Poliomyelitis in 1986, 1987 and 1988 (Part 2). (1989) Wkly Epidemiol Record 37: 281–285
11. Faden H, Duffy L. (1992) Effect of concurrent viral infection on systemic and local antibody responses to live attenuated and enhanced-potency inactivated poliovirus vaccines AJDC 146: 1320–1323
12. Faden H, Modlin JF, Thoms ML, McBean AM, Ferdon MB, Ogra PL (1990) Comparative evaluation of immunization with live attenuated and enhanced-potency inactivated trivalent poliovirus vaccines in childhood: Systemic and local immune responses. J Infect Dis 162: 1291–1297
13. Ghendon YZ, Sanakoyeva LL (1961) Comparison of the resistance of the intestinal tract to poliomyelitis virus (Sabin's strains) in persons after naturally and experimentally acquired immunity. Acta Virol 5: 265–273
14. Glezen WP, Lamb GA, Belden EA, Chin TDY (1966) Quantitative relationship of preexisting homotypic antibodies to the excretion of attenuated poliovirus type 1. Am J Epidemiol 83: 224–237
15. Grenier B, Hamza B, Xueref BC, Viarme F, Roumaintzeff M (1984) Seroimmunity following vaccination in infants by an inactivated poliovirus vaccine prepared on vero cells. Rev Infect Dis 6: S545–547
16. Henry JL, Jaikaran ES (1966) A study of poliovaccination in infancy: excretion following challenge with live virus by children given killed or living poliovaccine. J Hyg Camb 64: 105–120
17. Horstmann DM, Opton EM, Klemperen R, Llado B, Vignic AJ (1964) Viremia in infants vaccinated with oral poliovirus vaccine (Sabin). Am J Hygiene 79: 47–63
18. Hovi T, Huovilainen A, Kuronen T, Poyry T, Salama N, Cantell K, Kinnunen E, Lapinleimu K, Roivainen M, Stenvik M (1986) Outbreak of paralytic poliomyelitis in Finland: Widespread circulation of antigenically altered poliovirus type 3 in a vaccinated population. The Lancet II: 1427–1432
19. John TJ, Christopher S (1975) Oral polio vaccination of children in the tropics. III. Intercurrent enterovirus infections, vaccine virus take, and antibody response. Am J Epidemiol 102: 422–428
20. Kohara M, Abe S, Kamatsu T, Tago K, Arita M, Nomoto A (1988) A recombinant virus between the Sabin 1 and Sabin 3 vaccine strains of poliovirus as a possible candidate for a new type 3 poliovirus live vaccine strain. J Virol 62: 2828–2835
21. Lasch EE, Abed Y, Abdulla K, Tibbi AG, Marcus O, El Massri MR, Gerichter CB, Melnick JL (1984) Successful results of a program combining live and inactivated poliovirus vaccines to control poliomyelitis in Gaza. Rev Infect Dis 6: 2; 467–470
22. Lepow ML, Warren RJ, Gray N, Ingram VG, Robbins FC (1961) Effect of Sabin type 1 poliomyelitis vaccine administered by mouth to newborn infants. N Engl J Med 264: 1071–1078
23. McBean AM, Modlin JF (1987) Rationale for the sequential use of inactivated poliovirus vaccine and live attenuated poliovirus vaccine for routine poliomyelitis immunization in the United States. Pediatr Infect Dis J 6: 881–887
24. McBean AM, Thoms ML, Albrecht P, Cuthie JC, Bernier R, The Field Staff and Coordinating Committee (1988) Serologic response to oral polio vaccine and enhanced-potency inactivated polio vaccines. Amer J Epidemiol 128: 615–628

25. Minor PD, Dunn G (1988) The effect of sequences in the 5' non-coding region on the replication of polioviruses in the human gut. J Gen Virol 69: 1091–1096
26. Modlin JF, Onorato IM, McBean AM, Albrecht P, Thoms ML, Nerhood L, Bernier R (1990) The humoral immune response to type 1 oral poliovirus vaccine in children previously immunized with enhanced potency inactivated poliovirus vaccine or live oral poliovirus vaccine. AJDC 144: 480–484
27. Murray MG, Kuhn RJ, Arita M, Kawamura N, Nomoto A, Wimmer E (1988) Poliovirus type 1/type 3 antigenic hybrid virus constructed in vitro elicits type 1 and type 3 neutralizing antibodies in rabbits and monkeys. Proc Natl Acad Sci USA 85: 3203–3207
28. Ogra PL (1984) Mucosal immune response to poliovirus vaccines in childhood. Rev Infect Dis 6: S361–368
29. Ogra PL, Faden HS, Abraham R, Duffy LC, Sun M, Minor PD (1991) Effect of prior immunity on the shedding of virulent revertant virus in feces after oral immunization with live attenuated poliovirus vaccines. J Infect Dis 164: 191–194
30. Onorato IM, Modlin JF, McBean AM, Thoms ML, Losonsky GA, Bernier RH (1991) Mucosal immunity induced by enhanced-potency inactivated and oral polio vaccines. J Infect Dis 163: 1–6
31. Pan American Health Organization (1986) Polio incidence rises in Brazil's northeast. EPI News 8: 4–5
32. Patriarca PA, Wright PF, John TJ (1991) Factors affecting the immunogenicity of oral poliovirus vaccine in developing countries: Review. Rev Infect Dis 13: 926–939
33. Raconiello VR (1988) Poliovirus neurovirulence. Adv Virus Res 34: 217–246
34. Retrospective assessment of vaccination coverage among school-aged children – selected U.S. cities, 1991 MMWR 41(6): 103–107
35. Robertson SE, Drucker JA, Farbe-Teste B, D'Diaye M, Traverso HP, Rovira EZ, Sow A, Sy MTA, Diouf F. (1988) Clinical efficacy of a new, enhanced-potency, inactivated poliovirus vaccine. The Lancet 1: 897–900
36. Roivainen M, Hovi T (1987) Intestinal trypsin can significantly modify antigenic properties of polioviruses: Implications for the use of inactivated poliovirus vaccine. J Virol 61: 3749–3753
37. Roivainen M, Hovi T (1988) Cleavage of VP1 and modification of antigenic site 1 of type 2 polioviruses by intestinal trypsin. J Virol 62: 3536–3539
38. Simoes EA, Padmini B, Steinhoff MC, Jadhav M, John TJ (1985) Antibody response of infants to two doses of inactivated poliovirus vaccine of enhanced potency. AJDC 139: 977–980
39. Strebel PM, Sutter RW, Cochi SL, Biellik RJ, Brink EW, Kew OM, Pallansch MA, Orenstein WA, Hinman AR (1992) Epidemiology of poliomyelitis in the United States one decade after the last reported case of indigenous wild virus-associated disease. Clin Infect Dis 14: 568–579
40. van Wezel AL, van Stenis G, Hannik CA (1978). New approach to the production of concentrated and purified inactivated polio and rabies tissue culture vaccines. Div Biol Stand 41: 159–168
41. von Magnus H, Petersen I (1984). Vaccination with inactivated poliovirus vaccine and oral poliovirus vaccine in Denmark. Rev Infect Dis 6: 471–474
42. Westrop GD, Wareham KA, Evans DMA, Dunn G, Minor PD, Magrath DI, Taffs F, Marsden S, Skinner MA, Schild GC, Almond JW (1989) Genetic basis for attenuation of the Sabin type 3 oral poliovirus vaccines. J Virol 63: 1338–1344
43. Wright PF, Kim-Farley RJ, de Quadros CA, Robertson SE, Scott RM, Ward NA, Henderson RH (1991) Strategies for the global eradication of poliomyelitis by the year 2000. N Engl J Med 325: 1772–1779
44. Zhaori G, Sun M, Faden H, Ogra PL (1989) Nasopharyngeal secretory antibody response to poliovirus type 3 virion proteins exhibit different specificities after immunization with live or inactivated poliovirus vaccines. J Infect Dis 159: 1018–1024
45. Zhaori G, Sun M, Ogra PL (1988) Characteristics of the immune response to poliovirus virion polypeptides after immunization with live or inactivated polio vaccines. J Infect Dis 158: 160–165

26

Antigenicity and immunogenicity of inactivated poliovirus vaccine in combined vaccine trials

N. Ajjan and **P. Saliou**

Pasteur Mérieux Serums and Vaccines, Marnes-La-Coquette, France

Summary

In developed countries the use of OPV is associated with the risk of paralysis, whereas in developing countries its use is associated with relative inefficacy.

– Various studies demonstrate that the seroconversion rate and level of protective antibodies induced by the E-IPV are higher than with the OPV. After 2 or 3 doses of E-IPV the proportion of subjects with detectable neutralizing antibodies are similar. The third dose is necessary however in order to increase the geometric mean titer of antibodies and thus the duration of immunity. A schedule with 2 doses may be useful in rural areas where there are no operational health delivery services.
– The potential value of OPV in limiting wild virus circulation is contrabalanced by the higher and more liable production antibodies of E-IPV. The incorporation of at least one dose of E-IPV at the start of the immunization schedule increases humoral as well as local antibody production. The sequential or combined use of the E-IPV and OPV could reduce or even eliminate the risk of vaccine associated poliomyelitis and maintain a high degree of humoral and local immunity.

Introduction

Poliomyelitis is the cause of serious socio-economic and public health problems in developing countries, due to the disabilities that it causes.

The disease has been virtually eradicated from developed countries by extensive use of either inactivated or live attenuated vaccines or of both. In contrast, in developing countries, where an estimated 160 000 cases of paralytic poliomyelitis still to occur each year, the disease remains a major cause of mortality and permanent disability.

Two types of vaccine are available: trivalent oral poliovirus (OPV) and trivalent enhanced inactivated poliovirus vaccine (E-IPV) given by injection.

The choice between the two vaccines is a source of frequent ongoing and intense debates.

A look at the efforts to control poliomyelitis around the world shows that the disease has been eliminated in the United States using OPV, in Northern European countries using IPV and in France using both OPV and IPV. However, since 1985 the only vaccine recommended by the department of Health, is E-IPV combined with DPT. The oral poliovirus vaccine is reserved for adults as booster. This proves that the large-scale application of either or both oral and injectable vaccines can eliminate poliomyelitis in developed countries.

The World Health Organization, in the framework of EPI [1], recommends the use of the oral poliovirus vaccine (OPV) at birth and then at six, ten and fourteen weeks in association with the DPT. This vaccine has however recently been the subject of controversy mainly due to:

(a) Postvaccinal neurological accidents observed in developed countries: 260 cases of vaccine-associated paralytic poliomyelitis (VAPP) were reported in the United States [36] between 1961 and 1989. The risk of VAPP ranged from 3.5 cases per 10 million doses of OPV distributed. The risk was 9 times higher after the first dose than after subsequent doses and highest in immune – deficient infants. Ninety-three percent of recipient cases and 76% of all VAPP cases were associated with administration of the first or second dose of OPV whereas VAPP is not known to exist with IPV [28,29].

(b) Relative inefficacy of the OPV evaluated by the appearence of specific antibodies after vaccination. The seroconversion rates are often low.

What is the protective effect of each of the 3 doses?

Although OPV is highly successful in temperate countries where it produces seroconversion in at least 95% of vaccinees after 2 or 3 doses, in tropical countries it fails to induce serum antibody formation; the results recorded during well-conducted surveys show major variations in rates of seroconversion. The level of seroconversion after 3 oral doses can be as low as 42%

Several factors have been held responsible for the failure of oral poliomyelitis vaccination [29].

– *The presence of passive maternal antibodies* during the first few months of life. According to the surveys older children respond better to vaccination;
– *Breastfeeding:* it is in fact only the colostrum which contains a large amount of poliomyelitis secretory antibody and the effect is only damaging when vaccination is given at birth or during the first few days of life;
– *The four-week interval* between doses recommended by the EPI may play some role in poor seroconversion;

- *Intercurrent intestinal infections* caused by enteroviruses;
- *Vaccine stability:* certain failures in tropical countries resulting from the incorrect storage of oral vaccine and its inactivation by heat. It should be pointed out that the thermostability of the vaccine has been greatly improved over the past few years.

Epidemiological failure

The best measure of a poliovirus vaccine quality is its ability to prevent clinical disease.

In addition to the epidemiological failures referred to above, the rare epidemiological surveillance surveys carried out have shown failure in oral vaccination. In Taiwan [14] and in the Gambia [12,13,27], after 6 years of low incidence of paralytic poliomyelitis, 1031 and 305 people were respectively affected. In 1989 a widespread outbreak occurred in Oman, despite 87% coverage with 3 doses of oral poliovirus vaccine.

This experience suggests that a several-year period of excellent control of endemic poliomyelitis by a vaccination programme may still be followed by a major epidemic and that a mass vaccination campaign may be only partially successful in ending the epidemic.

Despite the decline in the incidence of poliomyelitis, a large percentage of children with the disease received two, three or more doses of OPV.

The occurrence of neurological sequelae in industrialized countries and the relative inefficacy of the oral vaccine have called this vaccine into question and stimulating research on inactivated vaccines.

In recent years there has been renewed interest in the Salk inactivated poliovirus vaccine. This is partly due to the excellent results obtained with the IPV in those countries where it is used exclusively, and due to the concern caused by the failures of the OPV in tropical countries, as well as by the paralytic accidents induced by the OPV.

Over the last ten years, Pasteur-Mérieux has developed a new technique for the industrial production of a new concentrated IPV [23], perfectly tolerated and efficient with only two injections. However, three injections are recommended in order to increase the level of antibodies and then the duration of immunity.

The cells used for the production of this more potent vaccine are Vero cells which is a cell line derived from cell culture taken from kidneys of African green monkeys.

The efficacy of this new E-IPV has been evaluated on the vaccine injected alone or combined with the DPT after two or three injections for primary vaccination, with or without booster, and has sometimes been compared with OPV.

A large number of immunization schedules with 2 or 3 doses have been carried out in order to evaluate the E-IPV alone or combined with the DPT.

Serological evaluation with varying numbers of doses has shown that after two injections, 94 to 100% of infants are protected for polio 1, 88 to 100% for polio 2 and 90 to 100% for polio 3 [6]. After the first dose, 44 to 94% of the infants are seropositive, depending on their serotypes [6]. In their study Grenier et al [10] reported results in 36 infants immunized with E-IPV. All vaccinees had significant responses after two doses of IPV given one month apart with 100% seropositivity for types 1, 2 and 3. Evaluation of immune response in relation to age evidences similar results in infants vaccinated as early as 6 to 10 weeks of age [35]. The seroconversion rates were higher in infants who were immunized with an eight-week interval between doses.

McBean et al [21] compared the results obtained with the conventional IPV and with OPV, both administered in two doses at 2 and 4 months of age. Antibody against the three types of poliovirus was detectable in 98.7% after the second dose.

In Brazil, Schatzmayr et al [33] compared the E-IPV and OPV both given 3 times at 2, 4, and 6 months of age with a serological evaluation before and after immunization; all children had antibody against the 3 types after only two doses.

A clinical evaluation of the DPT-E-EPV was conducted in Mali [7] among 320 infants 2 to 24 months of age. The infants were randomly assigned to 3 groups: one group received 2 doses of DPT-E-IPV, a second group 2 doses of DPT and E-IPV given in two separate injection sites, the third group 2 doses of DPT and oral poliovirus vaccine (OPV).

A seroconversion rate of 100% was observed in the group which received E-IPV. After two doses, all children had antibodies against types 1, 2 and 3. The antibody response was significantly better than that observed with OPV. The injection of a third dose (booster) of DPT-E-IPV determined a booster effect proved by a great increase in antibody titres.

In the study carried out by Swartz et al [37], a booster dose given 10 months after the end of primary immunization determined a booster effect proved by at least a ten fold increase in antibody titers.

The same results were observed in the studies in the USA [21,22], and India [13]. In all the studies, IPV was administered alone or combined with DPT vaccine in a quadruple vaccine DPT-Polio. The results showed no significant difference between the immunogenic response to E-IPV or to combined DPT-E-IPV.

Is there intestinal immunity after IPV immunization?

The potential value of immunization with OPV is its ability to induce intestinal immunity and thus limit the spread of wild poliovirus in the community [5].

This potential value of OPV to limit wild virus circulation is counterbalanced however by the fact that antibody levels are lower than E-IPV and that OPV may cause VAPP [26].

Some studies of feacal poliovirus excretion by recipients of earlier inactivated poliovirus vaccines have shown a decrease in the amount of virus shed and in the duration of excretion after OPV challenge compared with unvaccinated controls, but to a lesser degree than in OPV recipients [9,11,26]. The presence, amount, and duration of poliovirus shedding in stools has been shown to be inversely related to the titer of homologous neutralizing antibody [11]. Lepow et al [17] have demonstrated reduced shedding of poliovirus in non-immunized newborns possessing maternal antibody to poliovirus.

As early as 1962, Dick [5] and Marine [19] observed a reduction in the quantity and duration of faecal poliovirus excretion in individuals previously immunized with the conventional IPV and challenged by OPV1.

In previous studies, Ogra [24,25] and Dhar [4] suggested that the secretory IgA (S-IgA) response to poliovirus in the nasopharynx with IPV was enhanced by subsequent immunization with OPV.

Recently, E-IPV has been shown to induce neutralizing secretory antibodies in nasopharyngeal secretions that may limit pharyngeal spread of poliovirus. In developing countries little information concerning the E-IPV intestinal immunity is available.

It seems, however, that IPV conferred gastrointestinal mucosal immunity, when compared with a control group of unvaccinated children, but to a lesser degree than OPV, with reduction in the circulation of poliovirus in the community and the protection of unvaccinated persons.

Faden [8] studied serum neutralizing and S-IgA antibodies in infants immunized with one of four schedules involving live oral vaccine (OPV) inactivated vaccine, or combinations of the two trivalent poliovirus vaccines OPVx3, IPVx3, IPVx1, OPVx2, IPVx2, OPVx1. The highest geometric mean titer of antibody to the 3 types of poliovirus occurred in the groups receiving IPV alone or combined IPV + OPV.

OPV was more likely than IPV to induce a local antibody response but the incorporation of at least one dose of IPV at the start of the immunization schedule tended to increase systemic as well as local antibody production.

Comparable results were observed in the Zhaori et al [40] study. After three doses of vaccine, infants in all vaccine groups (OPV, E-IPV alone or combined) developed similar secretory IgA. Combined immunization with E-IPV and OPV resulted in a higher secretory antibody response than observed after E-IPV only.

Combined or sequential schedule

What policy for poliomyelitis vaccination in the future?

(1) Immunization with OPV: this means the acceptance of vaccine-associated poliomyelitis and the failure of immunization in developing countries.

(2) Change from OPV to IPV. This requires a high rate of immunization but the association of the IPV with DT or DTP increases the uptake of IPV [24].

(3) One policy option under consideration is the sequential or combined use of E-IPV and OPV, as an alternative to the exclusive use of either vaccine [1].

The combined IPV–OPV policy could reduce or even eliminate the risk of vaccine-associated poliomyelitis in recipients and their contacts and the high rate of failure after OPV and could maintain a high degree of intestinal immunity in the population in order to prevent the spread of wild viruses, which remains an important factor in the epidemiology of the disease.

For these reasons, IPV should be given before or with the first OPV dose, since the greatest danger of vaccine-associated disease comes from the first dose of OPV in individuals previously unvaccinated against poliomyelitis [8,22].

Little is known yet about immunogenicity following sequential administration of IPV and OPV vaccines [8,16,20,22].

Denmark [18], is the only country which has, since 1968, extensive experience with a combined schedule of poliovirus immunization. Only 2 cases of endemic paralytic poliomyelitis were reported over a 16-year period. More than 95% of fully-vaccinated school children have antibodies against the three types of poliovirus.

In a study conducted by Ogra et al [25], a single dose of OPV given at 12 months of age to infants who previously received 3 doses of conventional IPV at 2, 3 and 4 months, produced higher levels of antibodies than when infants had previously received 3 doses of OPV, and nasopharyngeal anti PV_1 IgA levels were also higher after the OPV booster than in infants primed with 3 doses of OPV.

Modlin et al [22] report the results of humoral immune response to type1 oral poliovirus vaccine in children who have previously received either three doses of enhanced inactivated poliovirus vaccine or 3 doses of live oral poliovirus vaccine at 2, 4 and 18 months and were challenged with a single dose of monovalent type 1 OPV between 19 and 52 months.

This trial showed the OPV administration effectively increases the level of serum antibody in children previously immunized with 3 doses of E-IPV, especially in children with lower levels of pre-existing antibody. Children previously vaccinated with E-IPV experienced considerably higher titers of PV_1 antibody than children vaccinated with OPV.

During the 1970s, in two developing areas, the West Bank and Gaza, the Government of Israel established in both areas an expanded childhood immunization program. This induced four feedings of oral poliovirus vaccine (OPV) in the first year of life. By 1975, 85 percent of infants were receiving polio vaccination and by 1977 coverage had exceed 90 percent. Despite thus many cases occurred in children who had received at least one and up to four doses of OPV. In Gaza in 1976, half of the 77 poliomyelitis cases and eight of 13 cases in 1977 had been fully immunized with OPV.

The continued occurrence of endemic poliomyelitis during the late 1970s led to the institution, since 1978, of a new immunization program including a combination of inactivated poliovirus vaccine IPV and OPV for both the West Bank and Gaza. Immunization has reached 95% of the infant population in both areas and paralytic poliomyelitis has been controlled, despite exposure to wild poliovirus from neighbouring countries including an outbreak in Israel in 1988. The Israeli experience suggests that a vaccination schedule using the combination of IPV and OPV is effective and that may make eradication of poliomyelitis possible, even in developing countries [39].

Continuation of immunization with OPV in the second half of the schedule will ensure prolonged protection and perhaps reduce the spread of wild virus during potential outbreaks of disease [8]. Moreover, replacement of the first two doses of OPV by IPV should avoid the incidence of vaccine-associated paralytic poliomyelitis which occurs in 93% of recipient cases and 76% of all VAPP cases after the administration of the first or second dose of OPV.

Duration of immunity

One of the main questions concerning all vaccine-induced immunity is: how long does it last? [26].

Vaccination with IPV in a number of European countries (Finland [15], Sweden [2]) has been practised continuously since the beginning of immunization in 1957. These countries have used vaccines from different manufacturers and applied several vaccination schedules. No-one who has received 3 doses of IPV has contracted paralytic poliomyelitis during the last few decades.

The immunologic defense conferred by vaccination of the general population with inactivated poliovirus vaccine has offered individual protection as well as herd immunity in the above mentioned countries where the coverage rate is close to 100% [15]. This indicates that it is possible to evaluate circulation of poliovirus by use of inactivated vaccine.

In Sweden, circulation of wild poliovirus virtually ceased within 6 years of the implementation of the immunization program, in 1957. This situation persists after 25 years and no natural boosting could therefore be expected.

Evaluation of the duration of immunity 18 to 25 years after routine immunization with IPV indicates that all children still had demonstrable antibodies. A slight fall in antibody titers occurred within 2–5 years after primary immunization, after which the immune levels remained stable or declined very slowly.

Salk et al [32] affirm however that the persistence of circulating antibodies is not essential for long-term protection against paralytic poliomyelitis. Durable immunity to paralysis is associated with immunological memory, which may be induced without the production of detectable circulating antibody.

There is no reason why the new enhanced inactivated poliovirus vaccine (E-IPV) should not show, in the long run, the same or better performance than the conventional IPV.

References

1. Beale AJ (1990) Polio vaccines: time for a change in immunization policy. Lancet 335:839–842
2. Böttiger M (1984) long term immunity following vaccination with killed polio vaccine in Sweden, a country with no circulating poliovirus. Rev Infect Dis 6[Suppl]: 548–551
3. Deming MS, Jaiteh OK, Otten MW et al (1992) Epidemic poliomyelitis in the Gambia following the control of poliomyelitis as an endemic disease. II clinical efficacy of trivalent oral polio vaccine. Am J Epidemiol 135: 393–408
4. Dhar R, Ogra PL (1985) Local immune responses. Br Med Bull 41: 28–33
5. Dick GWA, Dane OS, McAlister J, et al (1961) Vaccination against poliomyelitis with live virus vaccine. 7. Effect of previous Salk vaccination on virus excretion. Br Med J 2: 266
6. Drucker (1987) Utilisation en milieu tropical d'un vaccin injectable concentré et inactivé contre la poliomyélite. Résultats sérologiques- 3éme Séminaire International sur la Vaccination en Afrique. Ed Foundation Marcel Mérieux, 163–172
7. Drucker J, Soula G, Diallo, et al (1986) Evaluation of a new combined inactivated DPT Polio vaccine. Dev Biol Stand 65: 145–158
8. Faden H, Modlin JF, Thoms ML, et al (1990) Comparative evaluation of immunization with live attenuated and enhanced potency in activated trivalent poliovirus vaccines in childhood: systemic and local immune response. Infect Dis 162: 1291–1297
9. Glezen WP, Lamb GA, Belden EA, et al (1966) Quantitative relationship of pre-existing homotypic antibodies to the excretion of attenuated poliovirus type 1. Am J Epidemiol 83: 224–227
10. Grenier B, Hamza B, Biron G, et al (1984) Sero-immunity following vaccination in infants by an inactivated poliovirus vaccine prepared on Vero cells. Rev Infect Dis 6[Suppl]: 545–547
11. Henry JL, Jaikaran ES, Davies JR Tomlinson AJH, et al (1966) A study of polio vaccination in infancy: excretion following challenge with live virus by children given killed or living polio vaccine. J Hyg 64: 105
12. Jaitech KO, Otten, Deming M (1986) Epidemic poliomyelitis in the Gambia. In progress report of Epidemiologic Findings
13. John J, Sevvakumar R, Balraj V et al (1987) Field studies using killed poliovirus vaccine. 3rd InternSeminar on immunization in Africa Collection – Foundation Marcel Mérieux 173–181
14. Kim Farley RJ, Rutherford G, Litchfiel D, et al (1984) Outbreak of paralytic poliomyelitis in Taîwan. Lancet 2: 1322–1324
15. Lapinleimu K, Stenvik M (1984) Experiences with polio vaccination and herd immunity in Finland. Dev Biol Stand 47: 241
16. Lasch E, Abed Y, Gerichter CB, et al (1983) Results of a program sucessfully combining live and killed polio vaccines. Israel J Med Sc 19: 1021–1023
17. Lepow ML, Warren RJ, Grayn et al (1961) Effect of Sabin type 1 poliomyelitis vaccine administered by month for newborn infants. N Engl J Med 284: 1071–1078
18. Mugnus HV, Petersen I (1984) Vaccination with inactivated poliovirus vaccine and oral poliovirus vaccine in Denmark. Rev Infect Dis 6[Suppl]: 471–474
19. Marine WM, Chin TDY, Gravelle CR (1962) Limitation of fecal and pharyngeal poliovirus excretion in Salk vaccinated children. A family study during a type 1 poliomyelitis epidemic. Am J Hyg 173–195
20. McBean AM, Modlin JF (1987) Rationale for the sequential use of inactivated poliovirus vaccine and live attenuated poliovirus vaccine for routine poliomyelitis immunization in the United States. Pediatr Infect Dis J 6: 881–887

21. McBean AM, Thoms ML, Albrecht P, et al (1988) Serologic response to oral polio vaccine and enhanced potency inactivated polio vaccines. Am J Epidemiol 128: 615–628

22. Modlin JF, Onorato IM, McBean M, et al (1990) The humoral immune response to type 1 oral poliovirus vaccine in children previously immunized with enhanced potency inactivated poliovirus vaccine or live oral polio. A J D C 144: 480–484

23. Montagnon B, Fanget B, Vincent Falquet JC (1984) Industrial scale production of inactivated poliovirus vaccine prepared by culture of Vero cells on microcarrier. Rev Infect Dis 6[Suppl 2]: 341–344

24. Ogra PL (1984) Mucosal immune response to poliovirus vaccines in childhood. Rev Infect Dis 6[Suppl 2]: 361–368

25. Ogra PL, Faden HS, Abraham R (1991) Effect of prior immunity on the shedding of virulent revertant virus in faeces after oral immunization with live attenuated poliovirus vaccines. J Infect Dis 164: 191–194

26. Onorato I, Modlin J, McBean M, et al (1991) Mucosal immunity induced by enhanced potency, inactivated and oral polio vaccines. J Infect Dis 163: 1–6

27. Otten MW, Deming MS, Jaitech KO, et al (1992) Epidemic poliomyelitis in the Gambia following the control of poliomyelitis as an endemic disease. I Descriptive Findings. Am J Epidemiol 135: 381–392

28. Patriarca PA, Laender F, Plameira G, et al (1988) Randomized trial of alternative formulation of oral polio vaccines in Brazil. Lancet 8583: 429–433

29. Patriarca PA, Wright PF, John TJ, et al (1991) Factors affecting the immunogenicity of oral poliovirus vaccine in developing countries. Rev Infect Dis 13: 926–939

30. Plotkin SA, Katz M, Brown RE, et al (1966) Oral poliovirus vaccination in newborn African infants: the inhibitory effect of breast feeding. Am J Dis Child 111: 27–30

31. Roberston SE, Traverso HP, Drucker JA, et al (1988) Clinical efficacy of a new enhanced potency, inactivated poliovirus vaccine. Lancet 1: 897–899

32. Salk D, Van Wezel AL, Salk J (1984) Induction of long term immunity to paralytic poliomyelitis by use of non infectious vaccine. Lancet 2: 1317–1321

33. Schatzmayr HG, Maurice Y, Fujita M, et al (1986) Serologic evaluation of poliomyelitis oral and inactivated vaccines in an urban low income population at Rio de Janeiro Brazil. Vaccine 4: 111–113

34. Selvakumar R, Jacob JT (1987) Intestinal immunity induced by inactivated poliovirus vaccine. Vaccine 5: 141–144

35. Simoes EAF, Padminibi, Steinaoff MC, et al (1985) Antibody response of infants to two doses of inactivated poliovirus vaccine of enhanced potency. Am J Dis Child 139: 977–980

36. Strebel PM, Sutter RW, Cochi SL, et al (1992) Epidemiology of poliomyelitis in United States one decade after the last reported case of indigenne wild virus associated disease. Vlin Infect Dis 14: 568–579

37. Swartz TA, Handher R, Stoeckel P, et al (1989) Immunologic memory induced at birth by immunization with inactivated polio vaccine in a reduced schedule. Eur J Epidemiol 5: 143–145

38. Sutter RW, Patriarca PA, Brogan S, et al (1991) Outbreak of paralytic poliomyelitis in Oman: evidence for widespread transmission among fully vaccinated children. Lancet 338: 715–720

39. Tulchinsky T, Yehia Abed MPH, Shaheen S (1989) A ten year experience in control of poliomyelitis through a combination of live and killed vaccines in two developing areas. A J P H 79: 1648–1651

40. Zhaori G, Sun M, Faden HS, et al (1989) Nasopharyngeal secretory antibody response to poliovirus type 3 virion proteins exhibit different specification after immunization with live or inactivated poliovirus vaccines. J Infect Dis 159: 1018–1021

27

Studies aiming at improvement of inactivated poliovirus vaccine preparations

T. Hovi, M. Roivainen, and **L. Piirainen**

Enterovirus Laboratory, National Public Health Institute, Helsinki, Finland

Summary

It has been known for some time that inactivated poliovirus vaccine (IPV), including the recently introduced enhanced-potency preparations, in spite of inducing very high levels of neutralizing and probably protecting serum antibodies are inferior to oral poliovirus vaccine (OPV) in preventing intestinal poliovirus infection in the vaccinees. We have proposed that this deficiency, at least as regards type 3 poliovirus, is based on different antigenic sites involved in the induction of antibodies by the two different vaccines because of proteolytic cleavage of a major antigenic site during infection. An approach to circumvent this problem comprises modification of the type 3 component of IPV with trypsin before immunization. A pilot vaccine prepared according to this principle has been obtained and its clinical evaluation will be started soon.

Introduction

Inactivated poliovirus vaccine (IPV) was developed in 1950ies by Dr. Jonas Salk and co-workers and rapidly accepted for regular immunization of children. As a consequence, the incidence of paralytic poliomyelitis was significantly reduced to the extent that the disease was practically eliminated in some countries. However, in some other countries where high coverage of immunization could not be reached poliomyelitis continued to occur at a moderate level of incidence. Oral immunization with the attenuated strains of poliovirus (OPV) was later on adopted in the latter countries and has since then spread all over the world. However, a few countries in Europe and parts of Canada have successfully persisted in the exclusive use of IPV in childhood immunizations.

Immunization schedules that combine the use of IPV and OPV have been used in Denmark since late 1960ies and recently also started in Israel.

In these schedules immunizations are commenced with 2 or 3 doses of IPV and supplemented with later boosters of OPV [1,2]. Theoretically, the use of IPV guarantees relatively high levels of circulating antibodies to all three serotypes of poliovirus while OPV boosters are thought to provide the vaccinees with good mucosal immunity. Furthermore, prior immunization with IPV is likely to reduce further the as such very low risk of vaccine-associated paralytic disease due to OPV [3]. This is an important point to consider in countries where poliomyelitis caused by wild polioviruses has been eliminated, because under these conditions the public attitude to unwanted harmful effects of vaccines tends to get highly critical. Because of these aspects, a potential switch to the combined schedule is presently under evaluation in several countries.

IPV preparations used today contain more viral antigen than the original ones [4,5]; hence the designation "enhanced-potency IPV" (eIPV). However, it still seems that at least a short term resistance to intestinal poliovirus infection is more easily obtained with OPV than with eIPV, in spite of the much stronger serum antibody response after the latter type of vaccine [5]. The traditional reasoning explained the relatively weak mucosal immunity, that is reached with IPV, with a view that parenteral immunization can not stimulate the necessary local immunity in the gut and, especially, will not result in mucosal secretion of virus-specific IgA class antibodies [6]. This view was questioned already a decade ago by demonstrating secretory IgA responses after repeated IPV administration [7]. Since then, evidence has been accumulating supporting an alternative explanation: OPV and IPV induce different panels of virus-specific antibodies and it is the target distribution, rather than the class of antibodies, that makes the important difference between the two types of poliovirus vaccine.

Experimental data relevant to the latter hypothesis and their implications to potential improvement of the present eIPV preparations are discussed in the following.

Antigenic site-specific immune responses to poliovirus

Antigenic sites in polioviruses

Polioviruses are members of the *Enterovirus* genus in the *Picornaviridae* family. They are non-enveloped icosahedral RNA viruses with 60 identical structural subunits in the capsid. Three major proteins (VP1, VP2 and VP3) make up the surface mosaic of the capsid as documented by X-ray crystallographic analysis of the 3-dimensional structure of the virion [8]. Several of the surface loops of the three major capsid proteins are known to be involved in the induction of neutralizing antibodies. Antigenic sites have been identified by growing the viruses in the presence of neutralizing monoclonal antibodies (MAb), harvesting virus variants that are resistant to the MAb in question, and analysing these escape mutants for amino acid substitutions in capsid proteins. Neutralization antigenic sites identified in this manner are clustered in four separate regions, all on the virion surface [8–10], with designated names antigenic site 1 through 4 (NAg 1–4).

Interestingly, the three serotypes of poliovirus appear to differ from each other in the sense which of these four sites is immunodominant: while most MAbs neutralizing type 2 or 3 polioviruses are targeted to Nag 1 located close to the 5-fold axis of the virion other sites are clearly immunodominant in the case of type 1 poliovirus [9,10]. Antigenic sites show variation in amino acid sequence between and within the three serotypes [9,11,12]. Again, there are differences in this sense between the different antigenic sites.

It is known that the relative immunodominance of different parts of a protein antigen can depend on the species immunized [13]. In the case of poliovirus it would be naturally most pertinent to know the importance of different antigenic sites as recognised by the human immune system. Generation of the required amount of data with human monoclonal antibodies would be an enormous task. We have therefore taken another approach and used the peptide scanning technique to reveal regions of poliovirus capsid proteins recognized by human immune sera [14]. Peptides derived from the NAgs identified by murine MAbs usually also bound some human antibodies. By far the strongest reactions were, however, obtained with peptides corresponding to some other parts of the capsid proteins including regions that are hidden in the 3-dimensional crystal structure [14]. Whether these novel antigenic regions have a role in inducing a protective immune response in man is not yet known.

Modification of poliovirus surface structures by host proteases

Studies published during last few years have revealed a significant difference in the surface structure between cell culture grown poliovirus and that existing in vivo. Fricks and coworkers first showed that trypsin treatment affected the structure of the Sabin strain of type 1 poliovirus in a highly selective manner: capsid protein VP1 was cleaved at the carboxy side of amino acid 99 but the virus remained fully infectious [15]. The cleavage site is located in a surface exposed loop of VP1, often referred to as the BC-loop according to the 3-dimensional structure, that forms the central part of the designated antigenic site 1 [8,9]. Indeed, it was subsequently shown that in the case of type 3 poliovirus, with this site being immunodominant in mice, murine monoclonal antibodies targeted to this loop and capable of neutralizing untreated virus could not neutralize trypsin-treated virus [16]. In contrast, monoclonal antibodies targeted to other antigenic sites in type 3 poliovirus were able to neutralize trypsin-treated virus to the same or a greater extent than the untreated one [16,17].

The above observations indicate that selective proteolysis of poliovirus surface structures results in significant antigenic changes. A question then arises: do they occur in vivo? Trypsin is a normal constituent of the pancreatic secretions, and we have shown that incubation of polioviruses in samples of human intestinal fluid brings about similar molecular and antigenic changes as effected by trypsin [18,19]. Trypsin-sensitive strains of poliovirus are also excreted into faeces in the cleaved form [20]. Furthermore, we have shown that these changes in poliovirus surface properties are not restricted to the intestinal tract since plasmin, another serine protease, cleaves the VP1 protein of polioviruses in a similar manner [21]. The precursor of plasmin, plasminogen, is a common plasma protein and is converted to active plasmin during several physiological and pathologic phenomena. Hence, it is fair to conclude that host proteases can and do modify antigenic and other surface properties of type 2 and type 3 polioviruses in vivo. Evidence supporting this view is further substantiated in the following chapter. For reasons not fully understood today, practically all type 1 poliovirus strains are resistant to trypsin under similar conditions [20, Roivainen & Hovi, unpublished].

Target sites of human poliovirus-neutralizing antibodies induced by various modes of immunization

Several approaches could be used to analyze systematically the distribution of antigenic site specificities of human poliovirus antibodies. One way would be to produce a large number of human monoclonal neutralizing antibodies and to characterize the escape mutants that are resistant to individual MAbs, as described above for the murine antibodies. Another approach would be to use competition assays with different neutralizing murine MAbs as labelled indicators. Virus-binding and virus-neutralization are, however, not identical phenomena and, therefore, this approach has some caveats. The same is true for the third approach, peptide scanning that we have used [14], which is also an assay measuring only binding of antibodies. Regions of capsid proteins found antigenic in the peptide scan should be subjected to further analysis, for instance, characterization of the human antibodies captured to and eluted from the active peptides, to really assess their putative role in the protective immune response. All these approaches are rather tedious and this may be the main reason that no systematic analysis on this matter has been published.

The most pertinent part of the question is, of course, are the antibodies induced by an immunization scheme capable of neutralizing the virus in vivo. Fortunately, it is possible to measure accurately the antibodies that are capable of neutralizing the protease-modified polioviruses similar to those naturally occurring in the body. A straightforward use of the standard several-day-incubation microneutralization assays may result in falsely high apparent levels because of the presence of antibodies to the uncleaved BC-loop, and the fact that the progeny of the cleaved virus in the culture will have an intact BC-loop. To avoid this we initially used the plaque reduction assay where the unbound antibody is removed after the adsorption period of the neutralized virus, and cannot confuse the later formation of the plaques [18]. Later on we developed an alternative assay, the radiometric cytolysis inhibition assay (RACINA), assessing the amount of the residual infectious virus during the first replication cycle [22]. In essence, these methods can be used to measure two subfractions of neutralizing antibodies in a serum specimen, those targeted outside the BC-loop of VP1 (neutralizing cleaved virus) and those specific for the BC-loop (titer against the untreated virus minus that against the cleaved one).

Children immunized with four doses of IPV only had moderate or high levels of antibodies to the untreated, intact type 3 poliovirus but very low if any antibodies to the corresponding trypsin-cleaved virus. Children who on top of this had experienced a natural type 3 poliovirus infection had equal levels of neutralizing antibodies to both virus preparations. One dose of OPV significantly increased the antibody levels to the cleaved poliovirus in IPV-only group [18]. This indicated that in recipients of IPV the BC-loop is immunodominant to an extent that antibodies targeted to other sites, and necessary to neutralize the natural trypsin-cleaved virus, are not properly induced. In contrast, these other sites appear to be better exposed to the host immune system during poliovirus infection, both natural and OPV-derived. Consequently, antibodies are induced that are capable of neutralizing protease-cleaved poliovirus as well as the intact virus [18]. Type 2 poliovirus antibodies specific for the BC-loop were less prominent and practically all IPV-only-immunized children had a significant level of neutralizing antibodies to the cleaved virus [19]. Zhaori and coworkers have observed a similar difference between eIPV and OPV vaccinees in poliovirus type 3 specific antibodies in nasopharyngeal secreta (NPS) [23]. They also found that eIPV recipients had less antibodies to the capsid protein VP3 in their NPS specimens than OPV vaccinees. Neutralization antigenic sites located in the VP3 protein [9] remain functional in trypsin-treated type 3 poliovirus [18]. Onorato and coworkers have analyzed fecal secretory IgA antibodies in groups of eIPV and OPV vaccinees. Several of the IPV recipients had these antibodies but when challenged with a dose of type 1 poliovirus (Sabin strain) no correlation was observed between the level of SIgA and resistance to intestinal infection [5].

In conclusion, these studies suggest that the resistance of IPV vaccinees to intestinal type 3 poliovirus infection is relatively weak because most of the antibodies induced by the immunization are targeted to the trypsin-sensitive site and hence, useless in the intestines and several other locations in the body. This hypothesis does not explain why the intestinal immunity of IPV vaccinees to type 1 poliovirus infection is also poor [5], as most type 1 poliovirus strains are resistant to trypsin anyhow.

Trypsin-tailored type 3 poliovirus as immunogen

Most, if not all of the antigenic sites involved in the neutralization of trypsin-cleaved type 3 poliovirus are also present in the intact poliovirus [17]. The observed difference in the antigenic site specificities of poliovirus neutralizing antibodies between IPV and OPVvaccinees could be a result of several factors, acting independently or in combination. Firstly, there could be simply a quantitative difference: replication of the virus could provide the host immune system with much more antigen than what is given with the parenteral standard dose immunization. Antibodies neutralizing the cleaved virus can be induced by IPV, at least transiently, if the immunization dose of viral antigen is large enough [18]. Secondly, the route of immunization might influence antigenic processing of poliovirus in such a way that the non-BC-loop sites would be more favourably presented in the intestinal mucosa. Thirdly, a qualitative difference between the immunogen could be important: cleavage of the BC-loop might improve the immunogenicity of the non-BC-loop antigenic sites. This view is supported by the observation that monoclonal antibodies with specificities outside the BC-loop of VP1 were obtained more readily if mice were immunized with trypsin-cleaved virus preparations rather than with the regular intact virus [16].

We have started to evaluate if the overall antibody response to poliovirus could be modified by using trypsin-tailored type 3 poliovirus as immunogen. Groups of Balb/c mice were immunized with small amounts of intact or trypsin-cleaved type 3 poliovirus, strain Saukett (kindly provided by Institut Merieux, Lyon, France, and by Dr. A. C. Beuvery, National Institute of Public Health and Environmental Protection [RIVM], Bilthoven, The Netherlands). Antibodies induced by the intact virus readily neutralized the intact virus but not the cleaved virus, while those induced by the cleaved virus equally well neutralized both virus preparations [17]. Similar results have been observed elsewhere in rats (A. C. Beuvery, personal communication).

Table 1. Type 3 poliovirus-specific antibodies in Balb/c mice months after third dose of untreated or trypsin-treated type 3 poliovirus immunogen

Test	Immunogen	N	Antibodies tested against			
			Untreated virus		Trypsin-treated virus	
			Percent of positives	Median titre	Percent of positives	Median titre
Neutralization assay	Untreated virus	12	100	250	33	<4
	Trypsin-treated virus	12	100	250	92	50
Enzyme immunoassay (IgG)	Untreated virus	16	100	430	19	<50
	Trypsin-treated virus	18	100	500	89	320
Enzyme immunoassay (IgM)	Untreated virus	16	43	<50	0	<50
	Trypsin-treated virus	18	89	450	83	450

In later studies we have analyzed the effect of additional doses of immunogen on the persistence of the induced antibody levels and on immunoglobulin class distribution. Three remarkable observations were made in these studies. First, there was an antigenic site-specific difference in the persistence of the response with the BC-loop-specific antibodies declining more slowly than those capable of neutralizing the cleaved virus. Secondly, while it was possible to induce a long-lasting neutralizing antibody response to the cleaved virus with the cleaved immunogen, most of the antibodies appeared to belong to the IgM class still several months after the third dose of the immunogen [24]. Thirdly and most importantly, levels of neutralizing antibodies towards the protease-cleaved type 3 poliovirus were persistently clearly higher in the group of mice immunized with the cleaved virus than in the recipients of the intact immunogen (Table 1). The reason for the persistence of the IgM response is not known but it cannot be due to a plausible chronic infection as the Saukett strain of type 3 poliovirus does not replicate in mice and because results obtained by immunizing mice with formalin-killed virus were practically identical with those described above for the live-virus immunized animals.

Conclusions and future prospects

It seems rather well established that there are differences in the antigenic site specificities of antibodies induced by IPV and OPV in man. Host proteases, trypsin in the intestinal tract and plasmin – potentially anywhere in the body – cleave a major loop in the designated antigenic site 1 of type 3 poliovirus which is immunodominant in IPV vaccinees. As a result, a large part of IPV-induced type 3 poliovirus-neutralizing antibodies can be useless in vivo. On the other hand, antibodies prompted by a poliovirus infection, both natural and OPV-derived, are mostly targeted to other sites on the surface of the virion and remain effective against the protease-cleaved form of the virus as well. On the other hand, our experiments in mice support the view that at least some of these differences can be overcome by fine-tailoring the immunogen virus with trypsin before inactivation. Whether this is the key question in the human immunity to poliovirus infection is, of course, another matter. Anyhow, these results justify similar immunogenicity studies in man and if equally promising, a protection assay to compare efficacy of the trypsin-tailored eIPV with that of the regular eIPV. A pilot IPV containing trypsin-treated Saukett strain as the type 3 component has been prepared by RIVM and we plan to evaluate it in groups of children in Finland. Protection against intestinal poliovirus infection after 2 and 3 doses of the IPV preparations will be assessed by measuring the take rates and excretion times of a challenge type 3/Sabin virus. This study design also provides us with an opportunity to evaluate possible correlations between different pre-challenge serology parameters and resistance to the challenge dose of an attenuated poliovirus.

References

1. von Magnus H, Pedersen I (1984) Vaccination with inactivated poliovirus vaccine and oral poliovirus vaccine in Denmark. Rev Infec Dis 6(2): 471–474
2. Slater PE, Orenstein WA, Morag A, Avni A, Handsher R, Green MS, Costin C, Yarrow A, Rishpon S, Havkin O, Ben-Zvi T, Kew OM, Rey M, Epstein I, Swartz TA, Melnick JL (1990) Poliomyelitis outbreak in Israel in 1988: a report with two commentaries. Lancet 335: 1192–1198
3. Nkowane BM, Wassilak SGF, Orenstein WA, Bart KJ, Schonberger LB, Hinman AR, Kew OM (1987) Vaccine-associated poliomyelitis, United States: 1973–1984. JAMA 254: 1335–1340
4. Salk D, van Wezel AL, Salk J (1984) Induction of long term immunity to paralytic polio myelitis by use of noninfectious vaccine. Lancet ii: 1317–1321
5. Onorato IM, Modlin JF, McBean AM, Thoms ML, Losonsky GA, Bernier RH (1991) Mucosal immunity induced by enhanced-potency inactivated and oral polio vaccines. J Infect Dis 163: 1–6
6. Ogra PL, Karlzon DT, Righthand F, MacGillivray M (1968) Immunoglobulin response in serum and secretions after immunization with live and inactivated poliovaccine and natural infection. New Engl J Med 279: 893–900
7. Svennerholm A–M, Hanson L–Å, Holmgren J, Jalil F, Lindblad BS, Khan SR, Nilsson A, Svennerholm B (1981) Antibody responses to live and killed poliovirus vaccine in the milk of Pakistani and Swedish women. J Infect Dis 143: 707–711
8. Hogle JM, Chow M, Filman J (1985) Three dimensional structure of poliovirus at 2.9 Å resolution. Science 229: 1358–1365
9. Minor PD, Ferguson M, Evans DMA, Almond JW, Icenogle JP (1986) Antigenic structure of poliovirus serotypes 1, 2 and 3. J Gen Virol 67: 1283–1291
10. Page GS, Mosser AG, Hogle JM, Filman DI, Rueckert RR, Chow M (1988) Three-dimensional structure of poliovirus serotype 1 neutralizing determinants. J Virol 62: 1781–1794
11. Huovilainen A, Kinnunen L, Ferguson M, Hovi T (1988) Antigenic variation among 173 strains of type 3 poliovirus ioslated in Finland during the 1984 to 1985 outbreak. J Gen Virol 69: 1941–1948
12. Pöyry T, Kinnunen L, Kapsenberg J, Kew O, Hovi T (1990) The type 3 poliovirus strain responsible for the outbreak in Finland in 1984–1985 is genetically related to common Mediterranean strains. J Gen Virol 71: 2535–2541
13. Getzoff ED, Tanner JA, Lerner RA, Geysen HM (1988) The chemistry and mechanism of antibody binding to protein antigens. Adv Immunol 43: 1–99
14. Roivainen M, Närvänen A, Korkolainen M, Huhtala M–L, Hovi T (1991) Antigenic regions of poliovirus type 3/Sabin capsid proteins recognized by human sera in the peptide scanning technique. Virology 180: 99–107
15. Fricks CE, Icenogle JP, Hogle JM (1985) Trypsin sensitivity of the Sabin strain of type 1 poliovirus: cleavage sites in virions and related particles. J Virol 54: 856–859
16. Icenogle JP, Minor PD, Ferguson M, Hogle JM (1986) Modulation of humoral response to a 12-amino acid site on the poliovirus virion. J Virol 60: 297–301
17. Roivainen M, Montagnon B, Chalumeau H, Murray M, Wimmer E, Hovi T (1990) Improved distribution of antigenic site-specificity of poliovirus neutralizing antibodies induced by a protease-cleaved immunogen in mice. J Virol 64: 559–562
18. Roivainen M, Hovi T (1987) Intestinal trypsin significantly modifies antigenic sites of polioviruses: Implications for the use of inactivated vaccines. J Virol 61: 3749–3753
19. Roivainen M, Hovi T (1988) Cleavage of VP1 and modification of antigenic site 1 of type 2 poliovirus by intestinal trypsin. J Virol 62: 3536–3539
20. Minor PD, Ferguson M, Phillips A, Magrath DI, Huovilainen A, Hovi T (1987) Conservation in vivo of protease cleavage sites in antigenic sites of polioviruses. J Gen Virol 68: 1857–1865

21. Roivainen M, Huovilainen A, Hovi T (1990) Antigenic modification of polioviruses by host proteolytic enzymes. Arch Virol 111: 115–126
22. Hovi T, Roivainen M (1989) Radiometric cytolysis inhibition assay (RACINA): A new rapid test for neutralizing antibodies to intact and trypsin-cleaved poliovirus. J Clin Microbiol 27: 709–715
23. Zhaori G, Sun M, Faden HS, Ogra PL (1989) Nasopharyngeal secretory antibody response to poliovirus type 3 virion proteins exhibit different specificity after immunization with live or inactivated poliovaccines. J Infect Dis 159: 1021–1024
24. Roivainen M, Piirainen L, Hovi T (1993) Persistence and class-specificity of neutralizing antibody response induced by trypsin-cleaved type 3 poliovirus in mice. Vaccine (in press)

28

Development of candidates for new type 2 and type 3 oral poliovirus recombinant vaccines

M. Kohara[1] and **A. Nomoto**[1,2]

[1] Department of Microbiology, The Tokyo Metropolitan Institute of Medical Science, Honkomagome, Bunkyo-ku, Tokyo, Japan
[2] Department of Microbiology, The Institute of Medical Science, The University of Tokyo, Shirokanedai, Minato-ku, Tokyo, Japan

Summary

A number of recombinants between the virulent Mahoney and attenuated Sabin 1 strains of type 1 polioviruses were constructed. Results of biological tests including the monkey neurovirulence test on there recombinant viruses indicated that major determinants of the genome. The viral capsid proteins that contribute to viral antigenicity and immunogenicity, had a little correlation with the neurovirulence or attenuation phenotype of the virus. These results suggested that more safer vaccine strains of the Sabin type 2 and type 3 viruses can be constructed by replacing the sequence of viral capsid proteins of the Sabin 1 poliovirus by the corresponding genome sequences of the Sabin type 2 and type 3. Accordingly, we constructed recombinant viruses between the Sabin type 1 and type 2 or type 3, designated PV1/2(SS)BB and PV1/3(SS)BN, respectively, as candidates for type 2 and type 3 oral poliovirus vaccines. These recombinant viruses had antigenicity and immunogenicity of type 2 and type 3 polioviruses, respectively. The monkey neurovirulence tests and in vitro phenotypic marker tests of these recombinant virsuses indicate that PV1/2(SS)BB and PV1/3(SS)BN are possible candidates for new strain of type 2 and type 3 oral poliovirus vaccine.

Introduction

Poliovirus is a human enterovirus that belongs to the family *Picornaviridae* and is the causative agent of poliomyelitis. Poliovirus consists of a single stranded RNA of plus-strand polarity and 60 copies each of four capsid proteins, VP1, VP2, VP3, and VP4, and is classified in to three distinct serological types, type 1, type 2, and type 3. Humans are only natural host of poliovirus. Poliovirus, however, can also infect to monkeys.

For control of poliomyelitis, attenuated strains (Sabin 1, Sabin 2, and Sabin 3 strains) of all three poliovirus serotypes have been isolated and effectively used as oral live vaccines [1]. However, the vaccine strains have the inherent problem of reversion from attenuated to neurovirulent phenotype upon repeated passages. Indeed, a very small number of cases of paralytic poliomyelitis occur in countries with extensive oral poliovirus vaccine programs [2]. Experimental evidence strongly suggests that the most vaccine-associated cases are caused by the vaccines themselves, especially type 2 (Sabin 2) and type 3 (Sabin 3) vaccines [3,4]. Vaccine-associated cases clearly caused by the Sabin 1 vaccine strain has not been reported until now. Accordingly, it is desirable to develop safer vaccine strains of type 2 and type 3 polioviruses.

The whole nucleotide sequences of the genomes of both the virulent and attenuated strains of all three poliovirus serotypes have been elucidated. An important progression in the molecular biology of poliovirus was the construction of infectious cDNA clones [5]. It is possible to identify genome loci that influence the attenuation phenotype by constructing and performing monkey neurovirulence tests on recombinant viruses of the virulent and attenuated strains. This kind of attempts were performed on recombinant viruses between the virulent Mahoney and the attenuated Sabin 1 strains of poliovirus type 1. The results indicated that the genome loci of viral capsid proteins encode only weakly influence the attenuation phenotype and that strong neurovirulence determinant(s) reside in the 5' noncoding sequence (particularly, nucleotide position at 480) [7,8,10].

Stability of attenuation phenotype

Nucleotide position 480 of type 1 poliovirus was suggested to be a strong determinant important for the expression of attenuation phenotype in the 5' noncoding region [10]. Similarly, nucleotide positions 481 and 472 have been suggested to be important nucleotides for attenuation phenotype of type 2 and type 3 polioviruses, respectively [11,12]. In order to investigate the genetic stability of these nucleotides, poliovirus isolates excreted from humans who had been vaccinated with the Sabin 1, Sabin 2, or Sabin 3 strains were prepared and their 5' noncoding nucleotide sequences including these positions were determined (Table 1). 480 G in Sabin 1 was more stable than 481 A in Sabin 2 and 472 U in Sabin 3 during the propagation in the alimentary tracts. Although monkey neurovirulence tests for these isolates have not been performed, these observations may explain why the Sabin 1 virus is the safest

Table 1. Ratio of nucleotide change in a noncoding region

Strain and position	Sabin nucleotide	Detected nucleotide	
Sabin 1 (n480)	G	G 41/48	A 6/48
Sabin 2 (n481)	A	A 1/12	G 11/12
Sabin 3 (n472)	U	U 0/12	C 12/12

vaccine among the Sabin vaccines. Although the attenuation phenotype of Sabin 1 strain is determined by multiple points and scattered over the whole genome, the viral capsid proteins had little correlation with attenuation phenotype [7,8,10]. These results suggested that a possible strategy for constructing new safer type 2 and type 3 vaccine strains is the replacement of the Sabin 1 genome sequence encoding the whole-coat protein region by the corresponding sequence of the Sabin 2 and 3.

Construction and characterization of recombinant type 2 and type 3 polioviruses

Intertypic recombinant polioviruses were constructed according to the strategy described above. The recombinant viruses thus constructed were designated as PV1/2(SS)BB and PV1/3(SS)BN and the genome structures of both the recombinant viruses are shown in Fig. 1.

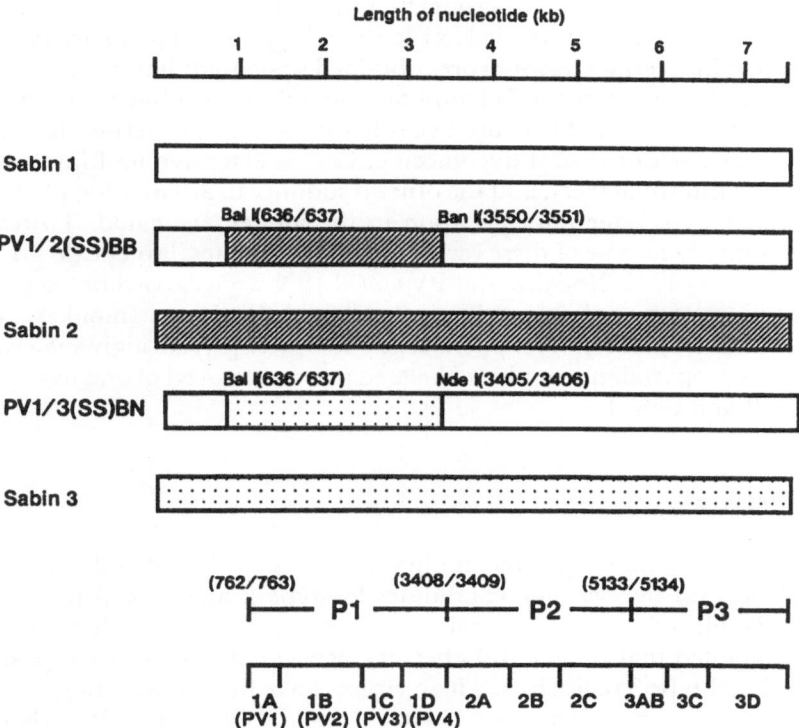

Fig. 1. Genome structures of intertypic recombinant viruses. Genome structures of intertypic recombinant polioviruses are shown. The length of nucleotides from the 5′ end of the genome is shown at the top of the figure in kilobases (kb). Open, hatched and stippled bars represent the sequences of the Sabin 1, Sabin 2 and Sabin 3 genomes

Table 2. Serological identification and monkey neurovirulence tests

Virus strain	Specifically neutralizing			Average lesion score
	anti-PV1	anti-PV2	anti-PV3	
F-209	−	+	−	0.48
PV1/2(SS)BB	−	+	−	0.42
F-310	−	−	+	0.67
PV1/3(SS)BN	−	−	+	0.26

Viruses PV1/2(SS)BB and PV1/3(SS)BN intertypic viruses were neutralized by specific antibodies against type 2 and type 3 polioviruses (Table 2). Monkey neurovirulence tests (intraspinally) were performed on these viruses. Each of the seronegative cynomolgus monkeys was inoculated intraspinally with 0.1 ml of the virus suspension (10^7 $TCID_{50}$/ml). The intensity of histological lesions was scored by established procedure [9]. The Sabin 2 and Sabin 3 reference strains used in this study were F-209 and F-310. These viruses had been prepared by three and two passages of the corresponding Sabin original strains in primary cell cultures from African green monkey kidney. They are now being used as type 2 and type 3 oral poliovirus vaccines in Japan. The average region scores obtained suggested that those two intertypic recombinants were sufficiently attenuated for oral live vaccine candidates (Table 2) [8,13]. There are two other important properties that should be associated with the oral live vaccine. One is extensive multiplication in the human intestinal tract, and the other is stability in attenuation phenotype of excreted virus after multiplication in the alimentary canal. To test the attenuation phenotype of these recombinant candidates, large scale preparations of viruses PV1/2(SS)BB and PV1/3(SS)BN were carried out according to procedures for vaccine production. All biological tests (monkey neurovirulence tests, stability in rct phenotype and others [9]) strongly suggest that these virus preparations are able to be used as a master seed of oral live vaccine for type 2 and type 3.

Conclusion

We constructed intertypic recombinant viruses PV1/2(SS)BB and PV1/3 (SS)BN as oral live vaccine candidates for type 2 and type 3 polioviruses [8,13]. Results of monkey neurovirulence tests performed on these recombinants suggested that these viruses become new condidates for vaccine strains of type 2 and type 3 poliovirus. These viruses have the 5' noncoding sequence and the region for replication proteins of the Sabin 1 virus. It is therefore possible that the interfering phenomena, which are usually recognized among the Sabin vaccine strains because of the different efficiency of multiplication between them, does not occur when new trivalent vaccine consists of the Sabin 1, PV1/2(SS)BB, and PV1/3(SS)BN.

The rate of spontaneous mutations in single-stranded RNA genome replication is especially high as compared with these of double-stranded DNA replication [15]. The high frequency of mutations that occurs during poliovirus genome replication makes it difficult to maintain the attenuation phenotype of the polio live vaccine. Genetic information of these vaccine candidates was stored in cDNAs as inserts of bacterial plasmids [8,14]. Genetic information of attenuation phenotype on plasmids can be amplified with much lower risk of reversion than those in RNA genomes in mammalian cells. Accordingly, the infectious cDNA clone can be used as stable repository of the genetic information for poliovirus vaccine strains.

References

1. Sabin AB, Boulger LR (1973) History of Sabin attenuated poliovirus oral live vaccine strains. J Biol Stand 1: 115–118
2. Melnick JL (1984) Live attenuated oral poliovirus vaccine. Rev Infect Dis 6 [Suppl]: 323–327
3. Kew OM, Nottay BK, Hatch MH, Nakano JH, Obijeski JF (1981) Multiple genetic changes can occur in the oral poliovaccines upon replication in humans. J Gen Virol 56: 337–347
4. Minor PD (1982) Characterization of strains of type 3 poliovirus by oligonucleotide mapping. J Gen Virol 59: 307–317
5. Racaniello VR, Baltimore D (1981) Cloned poliovirus complementary DNA is infectious in mammalian cells. Science 214: 916–919
6. Nomoto A, Omata T, Toyoda H, Kuge S, Horie H, Kataoka Y, Genba Y, Nakano Y, Imura N (1982) Complete nucleotide sequence of the attenuated poliovirus Sabin 1 strain genome. Proc Natl Acad Sci USA 79: 5793–5797
7. Omata T, Kohara M, Kuge S, Komatsu T, Abe S, Semler BL, Kameda A, Itoh H, Arita M, Wimmer E, Nomoto A (1986). Genetic analysis of the attenuation phenotype of poliovirus type 1. J Virol 58: 348–358
8. Kohara M, Abe S, Komatsu T, Tago K, Arita M, Nomoto A (1988) A recombinant virus between the Sabin 1 and 3 vaccine strains of poliovirus as a possible candidate for a new type 3 poliovirus live vaccine strain. J Virol 62: 2828–2835
9. World Health Organization (1983) Requirements for poliomyelitis vaccine (oral). WHO Expert Committee on Biological Standardization. Annex 4. Tech Rep Ser 687: 107–165
10. Kawamura N, Kohara M, Abe S, Komatsu T, Tago K, Arita M, Nomoto A (1989) Determinants in the 5′ noncoding region of poliovirus Sabin 1 RNA that influence the attenuation phenotype. J Virol 63: 1302–1309
11. Evans DMA, Dunn G, Minor PD, Schild GC, Cann AJ, Stanway G, Almond JW, Currey K, Maizel, J.V.Jr. (1985). Increased neurovirulence associated with a single nucleotide change in a noncoding region of the Sabin type 3 poliovaccine genome. Nature 314: 548–550
12. Ren R, Moss EG, Racaniello VR (1991) Identification of two determinants that attenuate vaccine-related type 2 poliovirus. J Virol 65: 1377–1382
13. Nomoto A, Iizuka N, Kohara M, and Arita M (1988) Strategy for construction of live picornavirus vaccines. Vaccine 6: 134–137.
14. Kohara M, Abe S, Kuge S, Semler BL, Komatsu T, Arita M, Itoh H, Nomoto A (1986) An infectious cDNA clone of the poliovirus Sabin strain could be used as a stable repository and inoculum for the oral polio live vaccine. Virology 151: 21–30
15. Holland J, Spindler K, Horodyski F, Grabau E, Nichol S, VandePol S (1982) Rapid evolution of RNA genomes. Science 215: 1577–1585

29

Intratypic differentiation of polioviruses

A. M. van Loon, A. Ras, P. Poelstra, M. Mulders,
and **H. van der Avoort**

Laboratory of Virology, National Institute of Public Health and Environmental Protection, Bilthoven, The Netherlands

Summary

The eradication of wild poliovirus from man and environment is a prerequisite for global eradication of poliomyelitis. Large scale vaccination with the live, attenuated poliovirus vaccine is a cornerstone of the eradication programme, but also leads to the ubiquitous presence of vaccine-derived virus. Therefore, characterization of polioviruses isolated from man and environment as wild or vaccine-derived is of crucial importance.

Past characterization methods were mainly based on biological or antigenic differences as determined by virus replication at supra-optimal temperatures or in the presence of partially strain-specific antisera. Because of their insufficient specificity these methods have been gradually replaced by immunological methods which use highly specific, cross-absorbed antisera or monoclonal antibodies or by molecular methods, like dot hybridization, the polymerase chain reaction or restriction fragment length polymorphism analysis.

At present the immunological methods may have some advantages because of their well-known and widely available methodology. Molecular methods however can be expected to become more important because of their ability to obtain information on relatedness between wild virus strains and therefore on possible virus reservoirs and transmission patterns.

Introduction

After the introduction of the live attenuated, oral poliovirus vaccine (OPV), vaccine viruses have found widespread dissemination in the community. Although the OPV has an excellent safety record, the vaccine is very infrequently the cause of paralytic illness. Characterization of poliovirus isolates as wild-type or vaccine-derived strains is of great importance for the analysis of paralytic cases temporally associated with the use of OPV as well as for epidemiological surveillance of wild poliovirus circulation.

In May 1988 the 41st World Health Assembly accepted a resolution for the global eradication of poliomyelitis by the year 2000. The objectives are (i) no new cases of poliomyelitis associated with wild poliovirus, and (ii) no

wild poliovirus identified worldwide through sampling of man and environment. Clearly, these objectives further stress the crucial importance of intratypic differentiation of poliovirus isolates.

Past characterization methods

A large number of in vitro methods has been used for intratypic differentiation
of poliovirus isolates [13]. In the early days of the polio vaccination era these
methods focussed on biological, physico-chemical or antigenic differences
between wild and vaccine-derived viruses. These included differences in reproductive capacity at supra-optimal temperatures (RCT marker test, [28])
or at suboptimal pH conditions [20], in elution patterns on DEAE cellulose or
aluminiumhydroxide gelchromatography [19], or in reactivity with partially
strain-specific antisera (intratypic serodifferentiation test), as originally described by McBride [9] and Wecker [21]. The RCT marker and the antigenic
marker test were the most commonly used assays.

RCT marker test

In the RCT marker test the viral titre in cell cultures incubated at increased
temperature (39.5–39.9°C) is compared with that at the standard temperature
for isolation of poliovirus (35°C). If titres at the two temperatures differ less
than 10^2 fold, a strain is RCT-positive and if the difference is at least 10^4 fold
the strain is RCT-negative. In other cases the strain is called intermediate.
Naturally occurring wild viruses are usually RCT-positive, live attenuated
viruses are RCT-negative.

Antigenic marker tests

In the early antigenic marker tests, unabsorbed antisera that were raised
against the prototype wild or vaccine viruses, were used. These sera were only
partially specific for wild or vaccine-derived poliovirus strains. Nevertheless,
they have been used quite extensively for intratypic differentiation using the
techniques described by McBride [9] or Wecker [21].

In the McBride assay viral strains are incubated for 5–15 minutes with
the immune sera, followed by determination of the infectivity in a plaque
assay. The neutralization rates of the unknown viral strains are compared
with those of reference strains.

Reduction in plaque size or numbers when the virus multiplies in the
presence of limiting concentrations of immune sera is the basis of the Wecker
[21] or modified Wecker [12] technique. Both the McBride and Wecker assays
require careful standardization and are technically quite demanding.

In the late seventies the World Health Organization started a collaborative study on markers of poliovirus strains isolated from cases temporally
associated with the use of OPV [24]. The study was organized to gain in-

formation on the comparability of results obtained in different laboratories using the same techniques for testing RCT and antigenic markers as well as to select the most reliable system for intratypic differentiation. In addition to the above mentioned methods, a serum neutralization assay using cross-absorbed antisera [6,16] was included. These sera had been prepared according to the method described by van Wezel and Hazendonk [22].

The study showed that antigenic marker tests with cross-absorbed immune sera were the most reliable for differentiation of vaccine-derived strains and naturally occurring viruses [11,24]. Clear-cut, unequivocal results were obtained in all of the different laboratories. The McBride or (modified) Wecker techniques proved to be of limited value. Unequivocal results were obtained in the participating laboratories in less than three-quarter of the strains examined. $Al(OH)_3$ gel-chromatography or RCT marker tests were found to be useless for intratypic differentiation because of the lack of specificity – causing a need for supplemental antigenic marker tests to confirm the results – and the great variation between laboratories. Particularly the RCT marker test appeared to be of little or no value since many naturally occurring poliovirus strains isolated before introduction of OPV were found to be RCT-negative. In addition, vaccine strains, particularly of type 3, were known to frequently revert to RCT-positivity upon passage in man.

Present characterization methods

The intratypic serodifferentiation was further extended by the introduction of the ELISA methodology [15] and of monoclonal antibodies [3,5,15]. Furthermore, in addition to methods focussing on the antigenic phenotype, molecular methods were introduced. These methods not only allowed intratypic characterization but also were able to provide information on genetic relatedness of poliovirus strains from different geographic regions [7].

Antigenic differentiation

(a) Cross-absorbed antisera

The original test described by van Wezel and Hazendonk [22] was a neutralization assay that used sera of rabbits immunized by two intravenous injections with a highly purified virus preparation. Such sera are type-specific and are made strain-specific by cross-adsorption with the heterologous strain, for instance anti-vaccine virus type 1-specific by adsorption with Mahoney virus. Strain specificity is determined in a neutralization assay.

These antisera have been used successfully for many years in a neutralization assay that is carried out as a standardized titration of the absorbed sera with a virus dose of 100 $TCID_{50}$ in microtiter plates. After a sufficiently long incubation period, e.g. 4 hrs at 37°C, cells are added. Neutralization is read at day 3 and 5. Results are clear-cut with most strains, showing neutralization by only one of the cross-absorbed sera. More recently, these sera have been used in an easy, rapid, indirect sandwich ELISA system [15] that allows a more economical use of the cross-absorbed antisera. In this ELISA, wells of a microtiter plate are coated with IgG antibodies of a bovine poliovirus-monotypic antiserum. Next, the virus strain to be typed is added. This may be simple culture-grown virus with a titre of at least 10^7 TCID$_{50}$. Subsequently, the cross-absorbed rabbit antisera, horseradish peroxidase-labelled anti-rabbit IgG antibodies and substrate are added. The optical density is determined at 450 nm. As a control the virus strain is also examined with a type-specific antiserum. Prototype wild and vaccine viruses are included in each test series. A typical result is shown in Table 1.

The ELISA uses a well-known methodology, is economical, quite easy to perform and yields results within 24 hours. Clear-cut results are obtained in more than 90% of isolates. However, the production of the antisera is not without problems and adsorption has to be carried out very carefully.

In the period 1988–1990, 321 viral isolates were submitted for intratypic differentiation to the Laboratory of Virology, RIVM, Bilthoven, The Netherlands. After reisolation and serotyping, poliovirus strains were intratypically characterized by the ELISA. In case of inconclusive results – no reactivity with either of the antisera or with both – a neutralization assay was carried out. The 321 isolates yielded a single poliovirus in 253 (78.8%) cases, two or more polioviruses in 45 (14.0%), one type of poliovirus as well as a non-polio enterovirus in 15 (4.7%) and a single enterovirus in 8 (2.5%) cases.

Clear-cut results in the intratypic differentiation with the ELISA were obtained in 92.6% of the strains. The neutralization test had to be carried

Table 1. Optical densities of prototype and unknown strains in the ELISA that uses cross-absorbed rabbit antisera for intratypic differentiation of polioviruses

| | Antiserum specific for poliovirus type | | | | | |
| | 1 | | 2 | | 3 | |
	NSL	SL	NSL	SL	NSL	SL
Mahoney	1.025	0.065	–	–	–	–
LSc, 2ab	0.135	1.118	–	–	–	–
MEF-1	–	–	1.478	0.327	–	–
P712 Ch, 2ab	–	–	0.115	1.504	–	–
Saukett	–	–	–	–	1.302	0.216
Leon, 12a$_1$, b	–	–	–	–	0.215	0.775
80–30 107	0.963	0.015	–	–	–	–
88–10 078	–	–	0.251	1.551	–	–
80–30 123	–	–	–	–	1.342	0.194

NSL: non Sabin-like virus; SL: Sabin-like virus

out in only 27 strains (7.4%). Of these, 15 could be characterized as either wild or vaccine-like virus.

(b) Monoclonal antibodies

Because of their epitope-specificity, monoclonal antibodies would be exquisitely suitable for intratypic differentiation of polioviruses. Indeed, monoclonal antibodies highly specific for the prototype wild and vaccine viruses have been developed [5]. However, the natural variation of wild polioviruses [3] as well as the readily changes of vaccine viruses upon passage in man [14,24] prohibit the use of single pairs of monoclonal antibodies to discriminate wild and vaccine-derived viruses within a type. Therefore, monoclonal antibody panels have been developed for each of the three serotypes [25]. For each type the panel includes six antibodies with different degrees of strain specificity (Table 2). These allow strains to be identified as a) identical to Sabin vaccine strains, b) related to, but drifted from Sabin vaccine strains,

Table 2. Percentage of wild and vaccine-related polioviruses reacting with monoclonal antibody panels [23]*

Antibody	Poliovirus wild type	Poliovirus live vaccine-related
Type 1		
11	0%	88%
12	0%	88%
13	49%	100%
14	51%	100%
15	0%	65%
16	97%	100%
Type 2		
21	95%	100%
22	81%	75%
23	93%	100%
24	0%	80%
25	67%	100%
26	100%	100%
Type 3		
31	92%	100%
32	24%	100%
33	8%	80%
34	8%	100%
35	71%	100%
36	25%	70%

*Results of studies at the National Institute of Biological Standards and Control, UK, on over fifty viruses for each serotype

c) typical wild strains, or d) antigenically unusual wild strains. At present, it is recommended that these panels be used in neutralization assays based on neutralization index measurements in which varying amounts of virus are exposed to a fixed amount of antibody. The assay uses a well-known methodology and is recommended in the WHO Manual for the Virological Investigation of Poliomyelitis [25]. However, difficulties have been experienced with regard to stability, reconstitution and specificity of the monoclonal antibodies [26]. In addition, the method is quite laborious and experience on a more or less routine basis is limited. A revised protocol using a smaller number of monoclonal antibodies for each type as well as a simplified procedure (fixed virus concentration, antibody dilutions) is currently being investigated (D. Wood, personal communication).

The recent application of an immunofluorescent assay for each, rapid detection and typing of polioviruses in patients' specimens would also offer new possibilities for rapid intratypic differentiation [23,26]. So far, however, results with cross-absorbed antisera and monoclonal antibodies have been disappointing because of non-specific reactivity (D. Hazlett, personal communication). It should, however, be possible to solve these problems.

Molecular differentiation

(a) Oligonucleotide mapping

T_1 oligonucleotide mapping was the first molecular method for genomic analysis of the origin of a poliovirus strain [8,10,14]. Digestion with ribonuclease T_1 cleaves the poliovirus RNA at guanosine residues and generates a large number of oligonucleotides. The size of the fragments varies with the position of the bases in the RNA. Upon resolution by two-dimensional electrophoresis, the larger, structurally unique, radiolabeled oligonucleotides yield a characteristic pattern of between 50 and 60 spots: a fingerprint.

Oligonucleotide mapping is a sensitive method to determine similarities or differences between strains. Even after considerable mutational drift, epidemiological relationships can still be determined unambiguously [14]. When carried out in specialized laboratories the method is well reproducible. However it is technically demanding and difficult to standardize for use in many laboratories world-wide.

(b) Probe hybridization

Nucleic acid hybridization with specific probes is a rapid, reliable and sensitive method for detection of specific viral sequences. For detection and characterization of polioviruses synthetic oligonucleotide probes [4] as well as subgenomic riboprobes [17] have been used. Both synthetic oligonucleotides as well as riboprobes are easily prepared in large quantities. In addition, the use of non-radioactive probes no longer decreases sensitivity. For intratypic differentiation of polioviruses type 1 and 3 oligonucleotide probes, 21–23 nucleotides in length, have been prepared that react specifically with Sabin strains or with different wild poliovirus genotypes endemic to Brazil [4]. The probes are complementary to sequences near the 5' terminus of the VP1 gene. These sequences are conserved among related strains only. The design of such probes, however, requires prior knowledge of the sequences of the virus under investigation [26]. This is not a problem for vaccine-derived viruses because substantial genomic evolution is not expected. Wild polioviruses, however, are genetically quite heterogenous resulting in the worldwide existence of numerous wild poliovirus genotypes. Therefore, a large collection of wild poliovirus-specific probes would be required to allow positive identification of all wild polioviruses. For surveillance purposes in areas close to control of poliomyelitis and in which the indigenous wild polioviruses have been well characterized the method has clear advantages. Wild imported viruses may, however, be missed [4], although they probably will be recognized as wild viruses by their inability to hybridize with vaccine virus-specific probes.

Intratypic differentiation by hybridization, however, still requires prior isolation and typing of the virus in cell cultures to resolve mixtures of viruses and to increase the concentration of viral RNA.

(c) Polymerase chain reaction (PCR)

The polymerase chain reaction (PCR) is an elegant method for rapid in vitro amplification of viral DNA sequences, allowing their detection in a highly sensitive and specific manner. So far, Yang et al. [27] have been the only authors to describe a PCR method that permits intratypic differentiation of poliovirus strains. They designed primer pairs for amplification of sequences specific for Sabin virus vaccine-related strains. Primers are complementary to sequences from the 5' end of the VP1 gene and permit Sabinvirus type-specific identification by the electrophoretic mobility of the amplified products (Sabin 1:97 bp, Sabin 2:77 bp, Sabin 3:53 bp). The detection limit is approximately 250 genome copies upon visualization by ethidium bromide after gelelectrophoresis and ≤ 2.5 genomes after hybridization with ^{32}P-labeled internal oligonucleotide probes.

The primer pairs were highly specific for Sabin strains, did not amplify sequences of recent wild polioviruses and even allowed the detection of Sabin-specific sequences in the presence of a 10^6-fold excess of wild virus (O. Kew, personal communication). However, sequences of prototype wild viruses and wild viruses historically related to the Sabin viruses are also amplified ([27], van Loon et al. unpublished results).

The problem of using PCR for intratypic differentiation is the positive identification of wild viruses. The failure to react with Sabinvirus specific primers is an indication that the virus has a wild genotype. As with probe hybridization, however, positive identification of wild poliovirus requires that primer pairs and probes be developed and validated for each wild-type genotype.

(d) Restriction fragment length polymorphism (RFLP)

Analysis of viral genomes by digestion with restriction enzymes followed by electrophoretic separation of the DNA fragments is frequently used to examine relatedness of viral genomes. Balanant et al. [1] have applied the method to study the natural genomic variability of polioviruses. After extraction of polio-virus RNA from the supernatant of cell cultures a 480 nucleotide sequence was reverse-transcribed and amplified by PCR. The region coding for the N-terminal half of the capsid protein VP1 was selected since it includes sequences involved in the constitution of antigenic site 1. Primers were chosen that would amplify sequences from most or all polioviruses. The endonucleases Hae III, Dde I and Hpa II were used as restriction enzymes.

Experience so far is limited but has shown amplified genomic fragments of a unique, constant size, reproducible RFLP patterns among related strains and a generally good agreement with the monoclonal antibody analysis of strains [1]. Sabin vaccine strains and related viruses can be recognized by their Sabin-specific RFLP profiles (one for each serotype). A wide variety of profiles was found for wild type viruses.

RFLP analysis of poliovirus strains is an elegant method to obtain genetic information on strains. It allows not only intratypic differentiation but may also give epidemiologically relevant information.

(e) Genomic sequencing

Genomic sequencing of (parts of) the viral genome is the ultimate method for determining relatedness between poliovirus strains. By comparing se-quences from a 150 nucleotide segment of the VP1/2A junction region of a large number of poliovirus strains, Rico-Hesse et al. [18] have been able to construct genealogic dendrograms that very elegantly show the relatedness of poliovirus strains. The VP1/2A region was chosen because it spans two functional domains, those of the major capsid protein VP1 and those of the viral protease, 2A, and has conserved sequences needed for primer binding adjacent to highly variable segments [7]. This region may also be useful to select primers for PCR amplification of wild virus genotypes.

Extension of synthetic DNA primers with reverse transcriptase in the presence of chain-terminating, radiolabeled inhibitors is mostly used for genomic sequencing of poliovirus genomes [7]. It is technically demanding and can only be carreid out in specialized laboratories. Although the use of the PCR method in combination with automated sequence instruments has greatly facilitated genomic sequencing, it is not probable that it will become the method of choice for intratypic differentiation within the next years. It will, however, be the gold standard for evaluation of methods for intratypic differentiation.

Conclusion

Intratypic differentiation becomes more and more important as the success of the WHO-polioeradication initiative continues to increase. A large array of methods has been used for this purpose. Since many laboratories worldwide will have to carry out intratypic differentiation, the method of choice should not only be specific and rapid, but also be easy to perform, to interpret and to standardize and be cheap.

The use of RCT markers or of differences in elution patterns on gelchromatography or in neutralization by partially strain specific antisera – McBride or Wecker techniques – is no longer acceptable for intratypic differentiation [24]. These methods lack sufficient specificity. Even then, a number of possible methods remain. To compare some of these methods we composed a panel of 57 poliovirus type 1 strains with a wide historical and geographical diversity. To these a number of "problem strains" were added. The methods used for intratypic differentiation included the following:

- an ELISA using cross-absorbed antisera (PoAb-E)
- a neutralization assay with cross-absorbed antisera (PoAb-N)
- a neutralization assay using a panel of monoclonal antibodies as described in the WHO Manual for the Virological Investigation of Poliomyelitis [25]
- a Sabin virus-specific PCR reaction as described by Yang et al [27].

The results are given in Table 3.

Table 3. Comparison of some methods for intratypic differentiation of 57 poliovirus type 1 strains

Method	Result	PoAb-E			
		NSL	SL	DR	NR
PoAb-N	NSL	32	0	1	0
	SL	0	13	0	1
	IM	1	0	2	1
	NR	1	1	3	1
MoAb*	NSL	34	4	5	3
	SL	0	10	1	0
PCR	NSL	27	4	2	1
	SL	7	10	4	2

PoAb-N, -E, neutralization (PoAb-N) or ELISA (PoAb-E) with cross-absorbed antisera; *MoAb-N,* neutralization with a monoclonal antibody panel; *PCR,* polymerase chain reaction (PCR); *NSL,* non Sabin like (= wild) virus; *SL,* Sabin-like virus; *IM,* intermediate strain; *DR,* double-reactive strain; *NR,* non-reactive strain
*Excluding monoclonal antibody 15 and neutralization index ≥ 2.0 (instead of ≥ 1.0)

To obtain useful results the interpretation of the tests with the monoclonal antibody panel had to be changed: results with monoclonal antibody 15 were discarded and a neutralization index of at least 2.0 was used to denote a positive reaction with a given monoclonal antibody from the panel. Even although there is an overrepresentation of "problem strains", the agreement between the different methods is disappointing, particularly between the antibody assays and the PCR.

At a recent WHO meeting on new approaches to poliovirus diagnosis using laboratory techniques [26] it was concluded that only few laboratories have experience with more than one method for intratypic differentiation. Therefore, it was recommended to start a collaborative study involving well-experienced laboratories in order to determine the relative ability to correctly identify polioviruses by each of the following methods: (i) ELISA with cross-absorbed antisera, (ii) neutralization with a reduced panel of monoclonal antibodies, (iii) probe hybridization, (iv) PCR and (v) RFLP. This study has recently been started.

The methods focussing on the antigenic phenotype for intratypic differentiation (cross-absorbed sera, monoclonal antibodies) may, at present, have technical advantages because their methodology is well-known and available in many laboratories worldwide. This would facilitate their wide-spread routine application, particularly since most of the molecular methods still require prior virusisolation and typing in cell culture. This may, however, change rapidly when molecular methods become more and more established in virusdiagnostic laboratories worldwide. A major advantage of some of the molecular methods, however, is the ability to obtain information on relatedness between wild virus strains and therefore on local, regional or global transmission patterns of poliovirus. This information is essential for the success of the WHO polio-eradication programme.

References

1. Balanant J, Guillot S, Candrea A, Delpeyroux FR and Crainic R (1991) The natural genomic variability of poliovirus analysed by a restriction fragment length polymorphism assay. Virology 184: 645–654
2. Crainic R, Couillin P, Caban N, Bové A and Horodniceau F (1982) Determination of type 1 poliovirus subtype classes with neutralizing monoclonal antibodies. Dev Biol Stand 50: 229–234
3. Crainic R, Couillin P, Blondel B, Caban N, Bové A and Horodniceau F (1983) Natural variation of poliovirus neutralization epitopes. Infect Immun 41: 1217–1225
4. Da Silva EE, Pallansch MA, Holloway BP, Oliveira MJC, Schatmayr HG and Kew OM (1991) Oligonucleotide probes for the specific detection of the wild poliovirus type 1 and 3 endemic to Brazil. Intervirology 32: 149–159
5. Ferguson M, Macgrath·DI, Minor PD and Schild GC (1986) WHO collaborative study on the use of monoclonal antibodies for the intratypic differentiation of poliovirus strains. Bull WHO 64: 239–246
6. Fogel A and Eylan E (1963) Serodifferentiation of virulent and attenuated polioviruses by adsorbed antisera. Virology 20: 533–536

7. Kew OM, Pallansch MA, Nottay BK, Rico-Hesse R, De L, Yang, Ch-F (1990). Genetic relationships among wild polioviruses from different regions of the world . In: New Aspect of Positive-Strand RNA Virus, Eds Brinton MA, Heing FX, p. 357–365. Am Soc Microbiology, Washington DC.
8. Lee YF and Wimmer E (1976) "Fingerprinting" high molecular weight RNA by two-dimensional gel electrophoresis: application to poliovirus RNA. Nucleic Acids Res 3: 1647–1658
9. McBride WD (1959) Antigenic analyses of poliovirus by kinetic studies of serum neutralization. Virology 7: 45–58
10. Minor PD (1981) Comparative biochemical studies of type 3 poliovirus. J Virol 34: 73–84
11. Minor PD and Schild GC (1982) Identification of the origin of poliovirus isolates. Report of World Health Organization informal meetings 1979–80
12. Nakano JH and Gelfand HM (1962) The use of a modified Wecker technique for the serodifferentiation of type 1 polioviruses related and unrelated to Sabin's vaccine strains. I. Standardization and evaluation of the test. Am J Hyg 75: 363–376
13. Nakano JH, Hatch MH, Thieme ML, and Nottay B (1978) Parameters for differentiating vaccine-derived and wild poliovirus strains. Prog Med Virol 24: 178–206
14. Nottay BK, Kew OM, Hatch MH, Heyward JT and Obijeski JF (1981) Molecular variation of type 1 vaccine-related and wild polioviruses during replication in humans. Virology 108: 405–423
15. Osterhaus ADME, van Wezel AL, Hazendonk AG, Uytdehaag FCGM, van Asten JAAM and van Steenis G (1983) Monoclonal antibodies to polioviruses. Comparison of intratypic strain differentiation of poliovirus type 1 using monoclonal antibodies versus cross-absorbed antisera. Intervirology 20: 129–136
16. Pervikov VY, Chumakov MP, Voroskilova MK, Rubin SG, Savinskaya SS and Gracheva LA (1974) Improvements of methods for intratypic differentiation of polioviruses. III Neutralisation test with cross-absorbed sera for differentiation between wild and vaccine strains of polioviruses. Arch Ges Virusforsch 46: 71–77
17. Petitjean F, Quibriac M, Freymuth F, Fuchs F, Lacanche N, Aymard M and Kopecka H (1990). Specific detection of polioviruses in clinical samples by molecular hybridization using poliovirus subgenomic riboprobes. J Clin Microbiol 28: 307–311
18. Rico-Hesse R, Pallansch MA, Nottay BK and Kew OM (1987) Geographic distribution of wild poliovirus type 1 genotypes. Virology 160: 311–322
19. Thomsson R and Mayer M (1965) Different pattern of elution of poliovirus strains from DEAE cellulose and aluminium hydroxide gel. Arch Ges Virusforsch 15: 735
20. Vogt M, Dulbecco R and Wenner HA (1957) Mutants of poliomyelitis viruses with reduced efficiency of plating in acid medium and reduced neuropathogenicity. Virology 4: 141–155
21. Wecker E (1960) A simple test for serodifferentiation of poliovirus strains within the same type. Virology 10: 376–79
22. van Wezel AL and Hazendonk AG (1979) Intratypic differentiation of poliomyelitis virus strains by specific antisera. Intervirology 2: 2–8
23. Whittle H, Hazlett D, Wood D and Bell E (1992) Immunofluorescence technique for the identification of polioviruses. Lancet 339: 429–430
24. World Health Organization (1981) Markers of poliovirus strains isolated from cases temporally associated with the use of live poliovirus vaccine. Report on a WHO collaborative study. J Biol Standard 9: 163–185
25. World Health Organization (1990) Manual for the Virological Investigation of Poliomyelitis. WHO Geneva, Switzerland
26. World Health Organization (1992) New approaches to poliovirus diagnosis using laboratory techniques: memorandum from a WHO meeting. Bull WHO 70: 27–33
27. Yang Ch-F, De L, Holloway BP, Pallansch MA and Kew OM (1991) Detection and identification of vaccine-related polioviruses by the polymerase chain reaction. Virus Res 20: 159–179
28. Yoshioka I, Riordan JT, Horstmann PM (1959) Thermal sensitivity of polioviruses isolated during oral vaccine field trial: comparison with monkey neurovirulence. Proc Soc Exp Biol Med 102: 342–347

30

Natural evolution of oral vaccine poliovirus strains

R. Crainic[1], **M. Furione**[1], **D. Otelea**[1], **S. Guillot**[1],
J. Balanant[1], **A. Aubert-Combiescu**[2], **M. Combiescu**[2],
and **A. Candrea**[1]

[1] Unité de Virologie Médicale, Institut Pasteur, Paris, France
[2] Cantacuzino Institute, Bucharest, Romania

Summary

The oral poliovirus vaccine, prepared with attenuated Sabin strains, is widely used to efficiently control poliomyelitis. One problem raised by its use is the occurrence of a certain, very few, number of vaccine-associated paralytic cases. This is due to the genetic instability of the three component viruses, mainly type 2 and type 3, upon multiplication in humans. Point mutations in critical nucleotide positions are generally accepted as one mechanism of reversion towards neurovirulence of Sabin strains. In this paper we present evidence that intermolecular recombination, possibly in combination with mutation, may also play a role in the loss of attenuation of vaccine polioviruses.

1. Poliomyelitis – control and eradication

A few years before the proposed term for its eradication, paralytic poliomyelitis is still a major health problem in certain areas of the world. The estimated incidence of the disease, however, decreased from approximately 400 000 new paralytic cases annually in 1983 [47], to 200 000 in 1988 [13] and to 116 000 in 1991 [56]. The developed countries, however, are no longer confronted with this debilitating disease. That sustained vaccination campaigns could interrupt the circulation of wild poliovirus in these countries constitutes the basis for attempting global eradication of poliomyelitis. Experience in South and Central America [11], where the disease caused by the wild poliovirus has practically disappeared, has already demonstrated that eradication can be extended to developing countries.

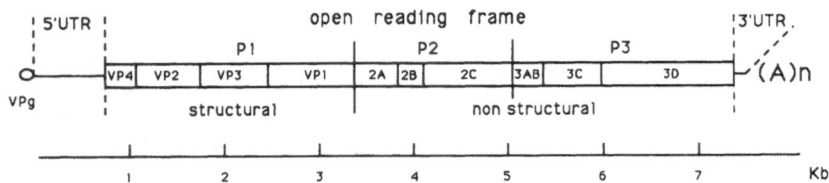

Fig. 1. Diagrammatic representation of the structure of the poliovirus genome

1.1   The virus

Poliovirus, the etiologic agent of poliomyelitis, is a member of the genus Enterovirus of the Picornavirus family. There are 3 serotypes of the virus. Based on nucleotide sequence alignment of the genomic region that encodes the capsid, the 3 poliovirus serotypes, together with coxsackievirus A21, are classified as one of the 11 subgenera of the Picornavirus family which includes members sharing more than 66% nucleotide homology [41].

The poliovirus is a non enveloped icosahedral particle of 28 nm diameter, constituted of 60 copies of each of the four viral-encoded structural polypeptides VP1, VP2, VP3 and VP4 [17,44]. The poliovirus genome (Fig. 1) is a single stranded, positively oriented RNA molecule of about 7500 nucleotides. A small polypeptide (VPg) is covalently attached at the 5' end, and the 3' end is polyadenylated [26]. About 10% of the genome at the 5' extremity and 1% at the 3' extremity are untranslated regions (UTR). The rest of the RNA is an open reading frame (ORF) translated into a single large polyprotein of 246 Kd. The polyprotein is subsequently cleaved into 3 intermediate proteins (P1, P2, and P3), and finally processed into structural (capsid) proteins (derived from P1) and non structural proteins (derived from P2 and from P3). The proteins 2A and 3C function as viral proteases. The protein 3D (derived from P3) is an RNA-dependent RNA polymerase which functions to replicate the genomic RNA via a negative-sense intermediate [45].

1.2   The disease

Humans are the only natural hosts for poliovirus multiplication. The transmission is fecal-oral. It is generally accepted [32] that, following a primary multiplication in the lymphoid organs of the naso-pharynx and in the gut, a transient viremia may occur, accompanied by proliferation of the virus in the reticuloendothelial system. The excretion in the stools may last several weeks, until finally the virus is eliminated. The period of viral multiplication is generally either clinically silent or accompanied by a minor febrile syndrome. A second viral multiplication in extraneural tissues is followed by a persistent viremia during which, in a limited number of cases (1–2%), the virus may traverse the blood-brain barrier. The virus multiplies in the central nervous system, preferentially in the motor neurons of the anterior horn of the spinal cord or in those of its equivalents in the mid-brain. The neural infection is clinically accompanied by the acute flaccid parmanent paralysis characteristic of the major illness. Exceptionally, poliovirus may multiply in other parts of the central nervous system, always with a preference for the motor neurons. Multiplication in the central nervous system is a dead-end in the natural circulation of the virus and, therefore, the neurovirulence of a strain is not required for its spread; however, it does not diminish viral transmission, because, even for neurovirulent strains, the neural multiplication remains an "accident" occurring concomitantly with multiplication in the gut.

1.3   The vaccines

The eradication of poliomyelitis relies first on the fact that man is the only virus reservoir. Second, there are two available vaccines, the efficacy of which has been proven: the inactivated poliovirus vaccine (IPV) and the live, attenuated, oral poliovirus vaccine (OPV). The immunity conferred by the vaccine is homotypic, and therefore a representative of each poliovirus sero-type has to be included in the vaccine, be it IPV or OPV.

The IPV efficiently protects against the disease and presents the advantage of not being incriminated in the etiology of vaccine-associated cases of para-lysis. Because it confers a lower degree of mucosal immunity than the OPV it does not prevent virus from multiplying in the gut. It has been shown, however, in countries using IPV exclusively, that the circulation of wild strains can be interrupted. A certain level of intestinal immunity might be acquired by its administration [19].

The OPV is the most widely used vaccine. It also has the advantage of conferring solid intestinal immunity and, thus, efficiently interrupting the transmission of wild polio virus strains. Furthermore its lower administration costs makes it more accessible to countries with endemic poliomyelitis. It was estimated that the extensive use of OPV in temperate climates for twenty years (1965–1984) prevented annually 250 000 cases of paralytic poliomyelitis [47], and prevented 440 000 cases in the whole world in 1990 [55].

The three attenuated Sabin strains globally used now as seed for the OPV were obtained by successive passages, in vivo and in vitro, of wild polioviruses [48]. They have the capacity to multiply in the gut, but they do not infect neuronal cells. As the attenuation of neurovirulence is not accompanied by a significant modification of antigenicity, these strains could successfully be used as live virus vaccine.

1.4   Variability of Sabin polioviruses

The most serious problem stemming from the use of OPV is the appearance of a certain number of paralytic cases of poliomyelitis among the vaccinees and their contacts [3], i.e. vaccine-associated poliomyelitis (VAP). The appearance of cases of VAP is due to the reversion to neurovirulence of the vaccine strains. One cause of the reversion is the intrinsic variability of polioviruses, which is, in fact, a common characteristic of all RNA viruses.

Because it has no proofreading mechanism, the RNA polymerase introduces a number of point-mutations (nucleotide substitutions) at an average frequency of 10^{-4} [18,52]. Thus, the so called strains are actually heterogeneous populations (quasispecies), even shortly after plaque purification [12].

Even if the recombination among RNA viruses seems to be a more general phenomenon than previously recognized, only a few viruses undergo homologous RNA recombination at a detectable frequency [29]. Poliovirus seems to be in a particular situation. With the OPV, three different strains of live virus are simultaneously present in the gut at rather high multiplicity. Such conditions can theoretically generate the appearance of inter or intratypic recombinant viruses, as has been shown to occur in vitro [1,16]. In fact evidence is accumulating that such recombinants, which have been occasionally isolated for some time now [6,23], represent a rather frequent phenomenon in vivo [30] (our studies); whether such an evolutionary pathway could lead to the appearance of strains with increased neurovirulence and thus capable of causing disease in vaccinees or their contacts, is a question that remains to be answered.

The epidemiological fate of vaccine-derived strain with increased neurovirulence is not yet clear. Their ability to circulate in a well-immunized population is presumably as low as that of the homotypic wild strains. Indeed, study of the nucleotide sequences of strains isolated in well-immunized populations suggested that the strains had not undergone several inter-human multiplication cycles. The situation might not be the same in populations with lower vaccination rates, where modified strains might encounter non-immunized hosts and might thus be able to undergo further modification.

2. Poliovirus genotypes

2.1 Genetic polymorphism

As for other picornaviruses, there can be considerable variation within each poliovirus serotype. Up to 15% base sequence divergence was detected, demonstrating distant evolutionary relationships among wild homotypic poliovirus isolates [43]. Camparison of selected polymorphic genomic regions by nucleotide sequencing [24] or by a restriction fragment length polymorphism (RFLP) assay [4] has revealed that strains isolated from different geographical regions and at different periods may have distinct genotypes.

The rapid evolution of poliovirus in nature does not correspond to mutations equally distributed all along the genome. Most of the 5' UTR, except the 90 residues preceding the open reading frame, is highly conserved [49]. Within the ORF, at the level of the RNA encoding the internal framework of the capsid, variability occurs mainly by nucleotide substitutions to generate synonymous codons [22], with a tendency towards conservation of the encoded aminoacid [17]. The genomic sequences coding for the surface structures forming antigenic sites are highly variable [25,34], as is the region coding for the 30 N-terminal aminoacids of VP1 [10,49,58]. Comparison of the amino acid sequences of the non structural proteins show that they are more conserved than the structural proteins. The sequence conservation at the protein level does not exclude some degree of variability at the corresponding genomic level. Silent mutations, which generally occur at the third (3' end) nucleotide position of the codons, generate a polymorphism that could profitably be used to identify poliovirus genotypes.

2.2   Sabin strains – particular poliovirus genotypes

The virulent strain P1/Mahoney-USA/41, the naturally less neurovirulent strain P2/P712 and the strain P3/Leon/37, which was isolated from a bulbo-spinal poliomyelitis, served as source for the attenuated Sabin strains: type 1 (LSc, 2ab), type 2 (P712, Ch, 2ab) and type 3 (Leon, $12a_1$ b) respectively. With the exception of P712, multiple in vivo and in vitro passages were necessary before the obtention of attenuation [48].

Apart from the difference in neurovirulence, the Sabin strains differ from the wild strains by several biological characteristics. Some of these traits were used, more or less successfully, as in vitro phenotypic markers to evaluate the safety of OPV [35] or to identify whether poliovirus isolates originated from vaccine or wild strains [53]. These characteristics include antigenic properties, studied with strain-specific antibodies (polyclonal or monoclonal), the capacity to grow at supra-optimal temperature (rct marker), the sensitivity of growth to low concentrations of bicarbonate (d marker), the plaque size, the thermostability in the presence of aluminum ions (A marker), or $Al(OH)_3$ gel elution (e marker) [35]. With the exception of antigenic characteristics, no phenotypic property was retained as a reliable marker for the distinction between strains of vaccine origin and wild strains among field poliovirus isolates [53].

Once the complete nucleotide sequence of the three Sabin strains had been determined [37,49], the molecular basis of attenuation and other biological properties could be deduced. The nucleotide sequences of the Sabin strains were compared with that of their parental viruses or with that of the neurovirulent revertants and several nucleotide sequence differences were found [2,7,31,37,42,49].

The nucleotide sequence of Sabin strains is distinct from that of their parents, which are no longer supposed to circulate, and from that of the circulating wild polioviruses [22,24,43]. Due to their particular genomic sequences, the Sabin viruses could thus be considered as distinct genotypes within each poliovirus serotype.

2.3   Molecular methods for the identification of poliovirus genotypes

Poliovirus isolates are currently characterized [53] either by their antigenic properties [9,15,40,50] or by molecular analysis of their genome. The

molecular methods include oligonucleotide fingerprinting with Tl ribonuclease (38), hybridization with oligonucleotide probes [10], partial nucleotide
sequencing of selecting geonomic segments [43], genotype-specific amplification by the polymerase chain reaction (PCR) [58] and RFLP analysis of
genomic segments amplified by PCR using generic primers [3].

3. Vaccine-associated poliomyelitis

3.1   OPV – risk over benefit

The OPV used throughout the world is prepared with attenuated Sabin
strains. Its efficacy has been demonstrated over the last three decades in many
countries. Successful control of poliomyelitis by IPV, on the other hand, was
also demonstrated in certain European countries, Iceland and parts of Canada
[19,57].
    The Sabin OPV strains have a good general safety record. However,
vaccine-associated paralytic poliomyelitis (VAP) exists. Frequencies of
1 : 500 000 to 1 : 1 200 000 of VAP after the first dose of OPV have been reported [36]; only type 2 and type 3 strains were incriminated. The risk of VAP
decreases for each individual with the number of vaccine doses received.
    The accepted rate for the frequency of mutation of the poliovirus genome
is one to two nucleotide substitutions per week, [38] and the "hot-spot"
nucleotide positions are those involved in the suppression of attenuation.
Thus, there is a risk for the viruses excreted by each vaccinated child to have
a lower degree of attenuation than that of the original Sabin strains. The risk
of contamination of a susceptible contact with a Sabin strain with increased
neurovirulence is diminished by the extent of immunization coverage of the
population.
    The diagnosis of VAP is based first on clinical observation. An acute
flaccid paralysis recorded to an OPV recipient among 6 and 30 days after
the administration of OPV should be considered as a suspected case of VAP.
Laboratory confirmation by the isolation of a vaccine-like poliovirus that is
believed to be the causes of the disease is necessary before classifying the case
as VAP (54). The number of VAP cases may be overlooked in populations
with a high incidence of wild poliovirus disease. However, during inter- or
post-epidemic periods, the incidence of VAP is easier to determine; the isolation of a unique Sabin-like virus from a polio case in the lack of circulation
of wild poliovirus strongly supports the diagnosis of VAP. The isolated virus
would, in this case, be a good candidate as the etiologic agent of the disease.

3.2   A particular high rate of VAP

Romania is one of the countries in which poliomyelitis has been successfully
controlled since 1960 by the use of OPV alone. As a general rule, until recently, primary vaccination was carried out in two one-week, nation-wide

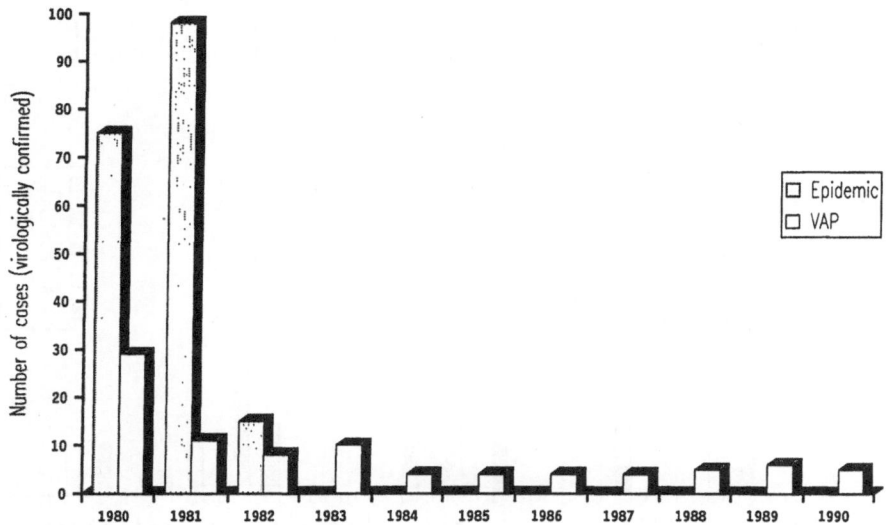

Fig. 2. Annual incidence of virologically confirmed poliomyelitis cases in Romania, in 1980–1990. Paralytic cases were classified as "epidemic" or "vaccine-associated" (VAP) according to the current clinical definition and the laboratory determined origin of the isolated virus (wild or vaccine-related)

campaigns per year (spring and fall), covering children from 6 weeks old. A "consolidation" dose six months later was followed by a booster dose at 9 years of age. From the second vaccination campaign in the fall 1978, trivalent OPV was temporarily replaced with monovalent type 1 OPV.

The incidence of virologically confirmed poliomyelitis in Romania in 1980–1990 is presented in Fig. 2. Cases were classified as either "epidemic" (wild virus) or VAP (vaccine-derived virus). While the second half of 1978 and all of 1979 no polio case was recorded, in 1980–82 two outbreaks comprising a total of about 250 clinically diagnosed cases occurred. The first outbreak, in the spring of 1980, was mostly due to a wild type 2 poliovirus. A wild type 3 poliovirus was also isolated. The second, mostly due to a wild type 1 poliovirus, began in September 1980, reached a maximum in the summer and fall of 1981 and was completely ended after September 1982. In the following years, the incidence of poliomyelitis in Romania decreased to levels of 8–16 sporadic cases (4–8 virologically confirmed). Vaccination with trivalent OPV was resumed in April 1980 during the first poliomyelitis outbreak. It is interesting that the number of VAP cases was higher in epidemic than in non-epidemic years. It is possible that some of the cases in which Sabin-like viruses were isolated were in fact epidemic cases counted as VAP, but this may not be the only explanation.

The monthly distribution of VAP cases in the years' 1980–1983 is presented in Fig. 3. It can be seen that the incidence of the VAP cases strictly correlates with the OPV vaccination campaigns, except for two cases in August 1991 in which type 1 virus was isolated.

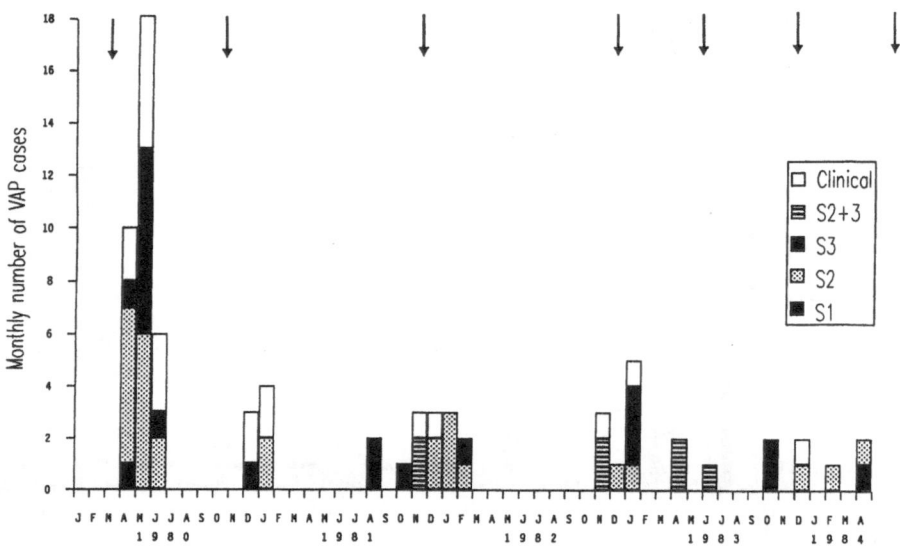

Fig. 3. Monthly incidence of vaccine-associated poliomyelitis (VAP) cases in Romania, in 1980–1983. The arrows represent the date of the beginning of the national seasonal vaccination campaigns with trivalent OPV (A seasonal campaign consisted in two one week national campaigns at six weeks interval). The serotypes of the isolated virus are indicated with different patterns, as indicated in the legend (S1 = Sabin 1, S2 = Sabin 2, S3 = Sabin 3, S2 + 3 = mixture of S2 and S3, Clinical = clinically diagnosed paralytic cases without virus isolation)

During the extra-epidemic periods (1983–1990), the annual incidence of clinically diagnosed poliomyelitis decreased to 8–16, of which more than half were virologically confirmed. All the isolated viruses were vaccine derived and no wild-type poliovirus was isolated in Romania during the 8 year's period under study. The frequency of VAP to the first dose of OPV varied from approximately 1:50000 to 1:100000, which is about 10 times higher than the generally accepted risk [36].

Whether particular local, epidemiological and social conditions or a higher neurovirulence of the OPV used was the cause of these outbreaks is not yet clear. Preliminary safety control of the vaccine based on animal and molecular tests, as well as a similar frequency of VAP after the administration of OPV from a different source seems to exclude the involvement of the locally produced vaccine.

4. Characterization of strains isolated from VAP

From the 317 cases of poliomyelitis clinically diagnosed in Romania among 1980 and 1990, 206 poliovirus strains were isolated from stools, from nasopharyngeal swabs, from blood clot, from cerebro-spinal fluid, or from the central nervous system.

4.1 Identification of vaccine origin

From the 206 poliovirus strains identified, 70 vaccine-derived strains were selected and further analyzed because they were suspected of being the etiological agent of VAP: 5 type 1 (Sabin 1), 43 type 2 (Sabin 2) and 22 type 3 (Sabin 3). Whether the origin of these strains was a wild or a vaccine strain was tested first with panels of homotypic strain-specific neutralizing monoclonal antibodies, according to a method previously described [9]. Confirmation of the strain origin or the identification of the genotype of strains giving ambiguous results with monoclonal antibodies was carried out with a recently developed RFLP test [4].

The genomic segment analyzed in RFLP-1 test was a 480-nucleotide sequence of the open reading frame of the poliovirus genome, situated between nucleotides 2401 and 2880. This segment, coding for the N-terminal half of the capsid polypeptide VP1, is highly variable and includes the coding sequence of the antigenic site 1 [5,17]. The polymorphic character of this nucleotide sequence made it possible to clearly distinguish among different poliovirus genotypes according to their restriction profiles generated by the PCR amplification product of the genomic segments (Fig. 4).

Because the vaccine Sabin strains generates specific profiles that are conserved upon intra- and inter-human passages of the virus [4], the RFLP-1 test could be used as genetic marker to identify the origin of the field isolates.

4.2 Molecular basis of suppression of attenuation

The term vaccine-associated was chosen instead of vaccine-caused poliomyelitis to express the incertitude concerning the actual etiologic agent of paralytic cases [47].

Genomic variation in the Sabin strains, which may give rise to phenotypic variation, can be caused by 2 mechanisms: mutation and recombination. Of these, mutation is more generally evoked as the cause of variation, whereas natural recombination was, until recently, considered to be an exceptional phenomenon in the evolution of poliovirus, even though natural recombinants have already been described (see 4.2.2).

4.2.1 Mutations

Because of the very high frequency of mutations (nucleotide substitutions) induced by the poliovirus RNA polymerase error, estimated to be among 5.4×10^{-3} and 7×10^{-4} [52], it can be assumed that practically every dose of OPV contains a certain level of "contaminant" genomes carrying one of the mutations critical for the stability of attenuation.

Among the three Sabin strains, the attenuated phenotype of type 3 seems to be the least stable. An explanation would be the small number of nucleotide substitutions responsible for its attenuation, increasing the frequency of selection of neurovirulent revertants. Indeed, from the 10 nu-

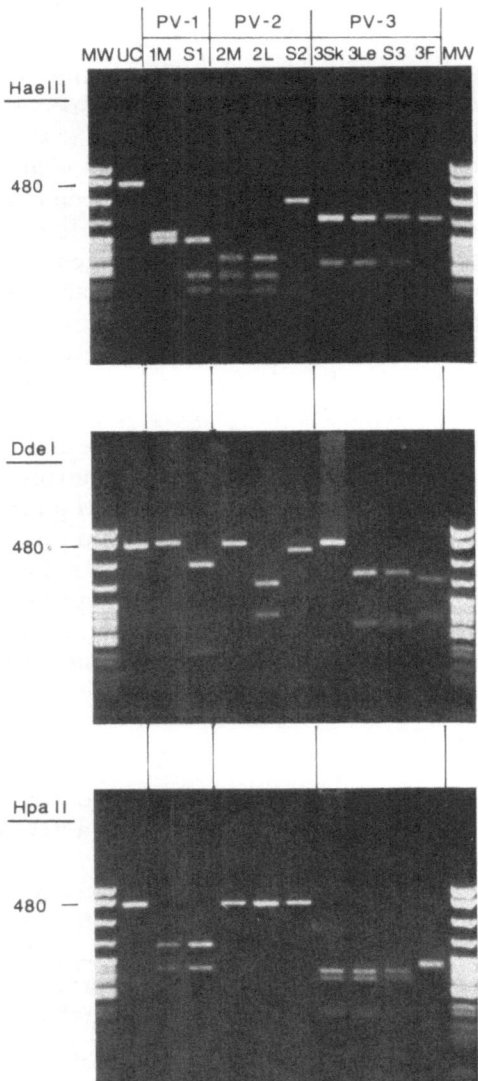

Fig. 4. RFLP-1 test. A. The analyzed genomic segment of each of the poliovirus strains tested was a 480 bases nucleotide sequence between nucleotides 2401 and 2880 coding for the N-terminal moiety of the capsid protein VP1. The genomic segment was reverse-transcribed and amplified by PCR using a pair of generic primers. The resulting DNA was digested with the indicated restriction enzymes (Balanant et al, 1991). 1M = P1/Mahoney, S1 = Sabin 1, 2L = P2/Lansing, 2M = P2/MEF1, S2 = Sabin 2, 3Sk = P3 Saukett, 3Le = P3/Leon/37, S3 = Sabin 3, 3F = P3/Finland/84

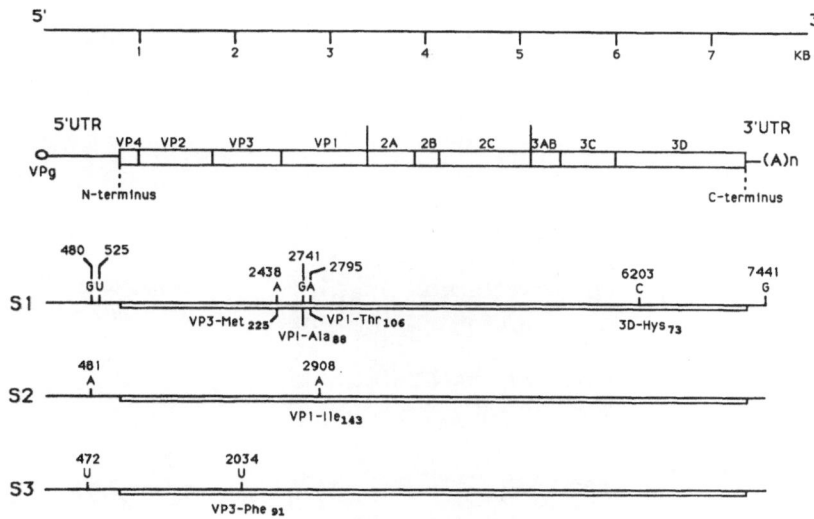

Fig. 5. Diagrammatic representation of some critical nucleotide positions involved in the attenuation of Sabin poliovirus strains. S1 = Sabin 1, S2 = Sabin 2, S3 = Sabin 3, UTR = untranslated regions [2, 7, 14, 21, 31, 42]

cleotides that differentiate the Sabin 3 strain from its neurovirulent wild P3/Leon/37 parent, only two seem to be involved in attenuation (Fig. 5): the U_{472} in the 5′ UTR [2,14] and U_{2034} in the codon of the Phe_{91} in VP3, which is involved in the inter-protomeric interaction of the capsid protein [33]. The $U_{472} > C$ reversion, accompanied by an increased neurovirulence, apparently confers a selective advantage; variants with this modification overgrow the parental population within a few days after multiplication in healthy vaccinees [25,34]. It was also demonstrated that in each dose of OPV, the Sabin 3 component contains a variable proportion of C_{472} genomes [8].

The Sabin 2 strain was derived from a field isolate with low neurovirulence [48]. Until now, only two nucleotide positions were considered to be responsible for the attenuated phenotype [31,42]. They are located in the 5′ UTR (A_{481}) and in the capsid coding region (U_{2908}) in the codon of Ile_{143} of VP3 (Fig. 5).

The Sabin 1 virus is reputed to have the most stable attenuated phenotype among the OPV strains. Very few, if any, VAP cases have been attributed to this virus [3]. The attenuating mutations among the 57 original candidates [37] appear to be spread all along the virus genome [27,39]. This in itself would decrease the probability for the selection of neurovirulent revertants. As seen in Fig. 5, sites that possibly play a role in attenuation have been located [7,21] at nucleotide positions G_{480} and U_{525} in the 5′ UTR, C_{2438} (Met_{225} in VP3), G_{2741} (Ala_{88} in VP1), A_{2795} (Thr_{106} in VP1), C_{6203} (Hys_{73} in 3D-polymerase) and G_{7441} in the 3′ UTR.

One way to determine the role of the Sabin strains in the pathogenesis of VAP is to find out whether the nucleotides that are ciritical for attenua-

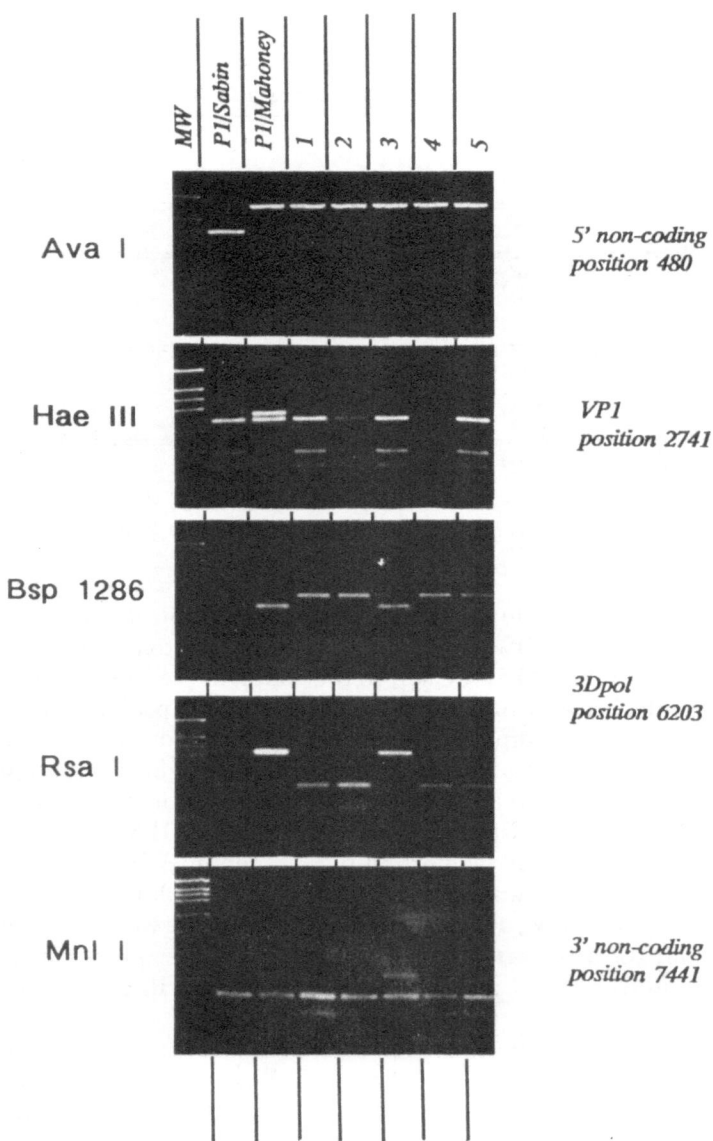

Fig. 6. Examination of some nucleotide positions critical for the attenuation of type 1 OPV strain in 5 Sabin 1-derived strains isolated from VAP cases. The genomic segments including the target nucleotide positions of each of the examined strains were amplified by PCR, after reverse transcription, using pairs of primers specific for each segment. Mutations were determined by comparing particular restriction profiles of each strain with those of the Sabin 1 strain and its neurovirulent parent Pl/Mahoney strain (modified from 40a)

tion are modified in the isolates retrieved from patients. Since a multitude of strains must be checked for several point mutations, we have proposed to use an allele-specific detection test instead of nucleotide sequencing. We could do this by taking advantage of the involvement of some of the critical nucleotides in the constitution of a restriction site in the DNA copy of the viral genome. The substitution of the targeted nucleotide by another nucleotide would suppress the original restriction site or, in some cases, create a new one. The presence or the absence of a restriction site is easily detected by analyzing the restriction profile of the reverse-transcribed, PCR-amplified genomic segment. Several critical nucleotides, i.e. those at positions 480, 2438, 2741, 6203 and 7441 in Sabin 1, fall within the sequence of a restriction site. It was thus possible to explore these positions for nucleotide substitution in the five Sabin 1-derived strains isolated from paralytic cases in Rumania from 1980 to 1984 (Fig. 6).

By accepting an average frequency of error of the RNA polymerase of 10^{-4} [18], for a genome of approximately 7500 Kb practically every virion will differ from the others at one nucleotide position. Furthermore, if a single point mutation is enough to reduce attenuation, when a child is vaccinated with one dose of OPV containing 1 million infectious units of attenuated poliovirus, he will concomitantly receive about 100 less attenuated virions.

If a mutation leaves the virion neutral with respect to the environmental conditions, the ratio of the resulting variant over the entire viral population will remain constant during infections at high multiplicity. On the contrary, if a mutation offers a selective advantage to the variant, the variant will overgrow the parental population and the mutation will become fixed. This advantage might or might not be correlated to the neurovirulence of the mutated virus and should be demonstrated for each particular nucleotide site.

In the evolution of wild poliovirus, variation may occur in the absence of strong functional selection [43] by random sampling of dilute populations [22], which in fact is a process of natural cloning. In the case of OPV strains, the conditions of virus replication are different. These viruses are administered to the host at high dose and, if the vaccine takes, they multiply and may be excreted in large quantities for several weeks. The spread of the excreted virus in the human population should theoretically be limited by the mass immunity. The close, susceptible contacts of the vaccinees, will generally be contaminated with large doses of virus. During the few inter-human passages that the OPV strain may undergo, the spread of a diluted virus is very unlikely. Thus, theoretically, in the case of OPV strains, mostly mutations that confer an advantage to the variant would be selected.

4.2.2 Recombination

Natural intertypic recombinants of Sabin poliovirus strains have been described only recently [6,23]. Recombinants constituted of both vaccine and wild poliovirus have also been identified [43]. Recombination may also occur between genetically more distant viruses. The coxsackievirus A21 virus is a putative recombinant in which the 2000 nucleotides at the 3′ extremity appear to have been replaced by a poliovirus sequence [20]. The 5′ UTR of enterovirus 70 is very similar to poliovirus, while the rest of the genome is highly divergent [46].

Recombinants were initially detected by the use of laborious techniques, such as T1 oligonucleotide fingerprinting and partial nucleotide sequencing. To screen for recombinants among a large number of strains, we developed a second RFLP test for which the 3′ extremity of the poliovirus genome is

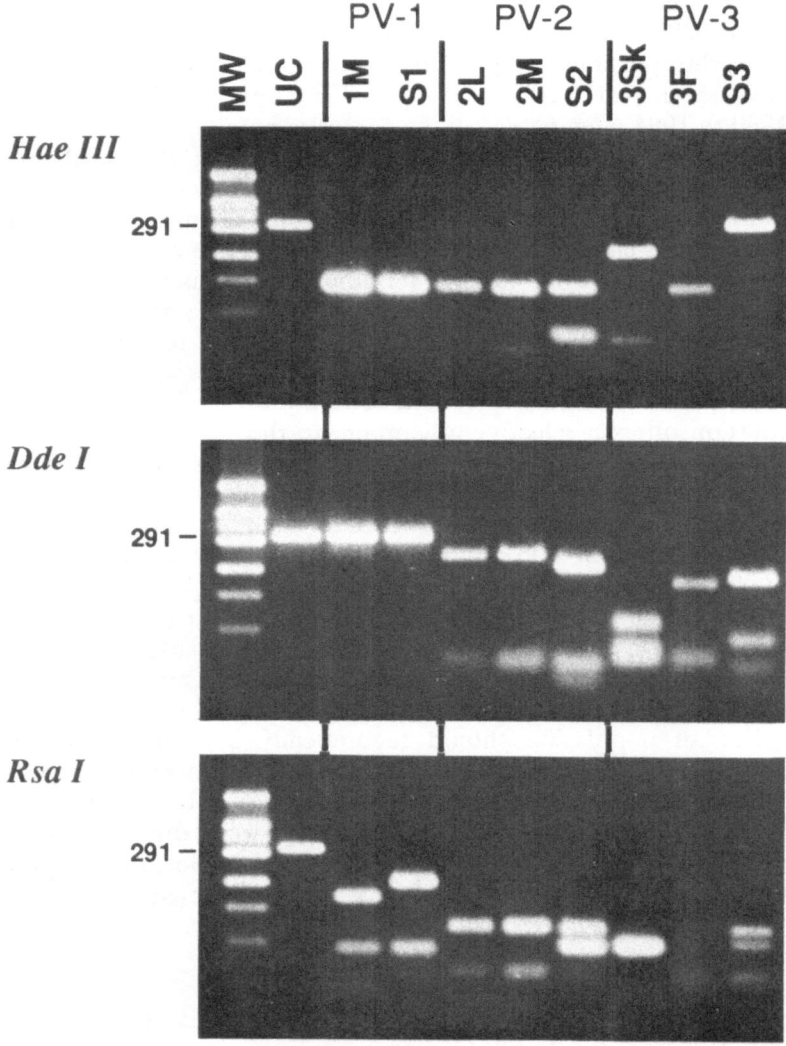

Fig. 7. RFLP-3D1 test. The analyzed genomic segment was a 291 nucleotide sequence between nucleotides 6086 and 6376, coding for a part of the non-structural protein 3D polymerase. For legends, see Fig. 4

examined (RFLP-3D1). The possibility of finding a polymorphic segment within a conserved region, such as that coding for the 3D polymerase, was suggested by the sequence heterogeneity in the 3D^{pol}-coding regions found among foot-and-mouth disease virus variants [51]. Because the conservation of 3D polymerase is not due to an intrinsically lower mutability but rather to the continuous selection of the same (or closely related) aminoacids, genomic variation may easily occur without affecting the structure of the encoded protein. A polymorphic segment of 291 bases was indeed found between nucleotides 6086 and 6376, coding for part of the 3D viral polymerase. It generates genotype-specific restriction patterns that were used to identify the origin of the examined strains (Fig. 7).

When 70 strains isolated from Romanian VAP cases were analyzed by RFLP-3D1, a very high proportion of recombinant genomes was detected (Table 1). About half of the 43 Sabin 2 strains and two-thirds of the 22 Sabin 3 strains presented a recombinant genotype.

The majority of the recombinant strains were inter-typic recombinants, but some vaccine/non-vaccine and vaccine/yet-unidentified-partner recombinants were also detected. The identification of origin of the unidentified partner, which might be a multiple mutated Sabin strain, a wild poliovirus or even a non-polio enterovirus, is in course.

We are presently attempting to determine whether the recombinant viruses isolated from VAP were the actual etiological agent of the disease. In

Table 1. Distribution of recombinant genotypes among poliovirus strains isolated from vaccine-associated poliomyelitis (VAP) cases in Romania in 1980–1990

Serotype	Genotype*	Number	Percent
Type 1	S1 × S1	5	100%
	S2 × S2	22	51%
Type 2	S2 × S1	9	
	S2 × S3	3	49%
	S2 × NS	6	
	S2 × NA	3	
	S3 × S3	7	32%
Type 3	S3 × S1	4	
	S3 × S2	9	68%
	S3 × NS	2	

*The genotype of the poliovirus strains as determined by RFLP-1 in the VP1 coding region (at the left) and by RFLP-3D1 in the 3D^{pol} coding region (at the right).
S1 = Sabin 1; S2 = Sabin 2; S3 = Sabin 3; NS = non Sabin-like profile; NA = not amplified by the pair of generic primers used

favor of this is that 20 of the 36 recombinant strains were isolated from VAP over an 8-year period (1983–1990) during which no paralytic cases caused by wild poliovirus were identified. Moreover, 5 of the 36 recombinants were isolated from the cerebro-spinal fluid or from the central nervous system of independent paralytic cases. If the neurovirulence of the isolates is greater than that of the original Sabin viruses, which remains to be determined, will also argue in favor of these strains being the agent of disease. Neurovirulence tests on transgenic mice sensitive to poliovirus (28) are in progress. An interesting observation is that 10 of the 13 recombinants having an Sabin 1 component at the 3' extremity carried a $C_{6203} > U$ mutation in the Sabin 1 moiety, a mutation that is suspected to increase the neurovirulence of the Sabin 1 virus [7]. Identifying the back or suppressor mutations known to be involved in the reversion of the neurovirulence in vaccine strains, especially in the 5' UTR, would make it possible to better evaluate the contribution of the mutations and that of recombination in the natural evolution of Sabin vaccine strains.

The high proportion of recombinant genomes found among Sabin-derived viruses isolated from VAP cases indicates not only that recombination is a frequent event in vaccine poliovirus replication, but also those recombinant viruses easily overgrow the original viral population. This strongly suggests that recombinant viruses are better adapted than the Sabin strains to grow in the human gut. The multitude of recombinant genotypes found and their large temporal and geographical spread pleads in favor of frequent recombination events rather than the diffusion of the progenies resulting from a few crosses. The detection of recombinants between Sabin strains and wild poliovirus or possibly non-polio enteroviruses indicates the compatibility in terms of recombination between genetically rather distant viruses and makes plausible the idea of reciprocal recombination.

Even if the role of recombinant viruses as etiological agent of paralytic disease is strongly suggested by these studies, more research is necessary to determine whether recombination is an actual factor of risk in the use of OPV.

5. Comments

The OPV has proved its efficacy in eliminating paralytic poliomyelitis. The remarkable experience of the Americas, which celebrated in August 1992 the first anniversary of the last paralytic case caused by wild poliovirus, is a clear demonstration that poliomyelitis can be eradicated from an entire continent.

To eradicate poliomyelitis, both wild and vaccine-associated neurological disease should be eliminated. For this reason, research efforts to obtain genetically more stable OPV strains are entirely justified. This is also one of the reasons why some countries still use IPV exclusively, even if it is somewhat more expensive and more difficult to administer than OPV [19]. Once the circulation of the wild poliovirus is stopped and its elimination is certified, the administration of the OPV will have to be suppressed so as to eradicate the poliovirus species and because the risks that it involves will no longer be worth taking. It is probable that, at that moment, the use of IPV will be reevaluate. It is not yet known whether the Sabin virus, once delivered in nature, is able to freely circulate and spread as "wild" poliovirus. The identification of multiple mutated Sabin-derived strains suggests that this is possible. Furthermore recombination, another source of "new" viruses, is a frequent phenomenon among polioviruses [30] (our results). Studies presented in this paper suggest that recombination might be a powerful mechanism of variation; a bulk of phenotypic properties might be transmitted to an innocuous virus in one round of multiplication. Under conditions in which a global polio immunization coverage would be obtained, the circulation of viruses with the antigenic properties of poliovirus will be stopped. However, it is not excluded that viruses with different antigenic properties, but expressing some of the phenotypic properties of poliovirus, such as host specificity or tissue tropism, may appear and spread in nature. Only by continuing research efforts will it be possible to determine the limits of poliovirus variation and the actual impact of this variation on its natural evolution.

With the eradication of poliovirus we assist probably at the disappearance of a virus species, first by replacing the wild with a laboratory made virus, then by getting rid of the replacing genotypes. It might be a unique opportunity to learn more about how nature accepts human intervention and, if accepted, what are the counterpart reactions. Science must not miss this occasion.

Acknowledgments

We thank Karen Pepper for the English version and Saskya Ballet for the skillful secretarial work. Part of the work presented in this article was supported by grants from World Health Organization (V26/181), from the Direction Scientifique des Applications de la Recherche of the Institut Pasteur, Paris, and from the Institut national de la Santé et de la Recherche Médicale (CRE 88.1006 and CRE 91.1301).

References

1. Agut H, Kean KM, Belloq C, Fichot O, Girard M (1987) Intratypic recombination of polioviruses: evidence for multiple crossing-over sites on the viral genome. J Virol 61:1722–1725
2. Almond J (1987) The attenuation of neurovirulence. Ann Rev Microbiol 41:153–180
3. Assaad F, Ljungars-Esteves K (1984) World overview of poliomyelitis: Regional patterns and trends. Rev Infect Dis 6 [Suppl 2]:S302–S307
4. Balanant J, Guillot S, Candrea A, Delpeyroux F, Crainic R (1991) The natural genomic variability of poliovirus analyzed by a restriction fragment length polymorphism assay. Virology 184:645–654
5. Blondel B, Akacem O, Crainic R, Couillin P, Horodniceanu F (1983) Detection by monoclonal antibodies of an antigenic determinant critical for poliovirus neutralization present on VP1 and on heat inactivated virions. Virology 126:707–710
6. Cammack N, Phillips A, Dunn G, Patel V, PD M (1989) Intertypic genomic rearrangements of poliovirus strains in vaccinees. Virology 167:507–514

7. Christodoulou C, Colbére-Garapin F, Macadam A, Taffs LF, Marsden S, Minor PD, Horaud F (1990) Mapping of mutations associated with neurovirulence in monkeys infected with Sabin 1 poliovirus revertants selected at high temperature. J Virol 64: 4922–4929

8. Chumakov KM, Powers LB, Noonan KE, IB R, Levenbook IS (1991) Correlation between amount of virus with altered nucleotide sequence and the monkey test for acceptability of oral poliovirus vaccine. Proc Natl Acad Sci USA 88: 199–203

9. Crainic R, Couillin P, Blondel B, Cabau N, Boué A, Horodniceanu F (1983) Natural variation of poliovirus neutralization epitopes. Infect Immun 41: 1217–1225

10. Da Silva EE, Pallansch MA, Holloway BP, Oliveira C, Kew OM (1991) Oligonucleotide probe for the specific detection of the wild poliovirus type 1 and 3 endemic in Brazil. Intervirology 32: 149–159

11. De Quadros CA, Andrus JK, Olive JM (1991) Eradication of poliomyelitis: progress in the Americas. Pediatr Infect Dis 10: 222–229

12. Domingo E, Martinez-Salas E, Sobrino J, de la Torre JC, Portella A, Ortin J, Lopez-Galindez C, Perez-Brena P, Villaneuva N NR, VandePol D, Steinhauer D, ND, Holland JJ (1985) The quasispecies (extremely heterogeneous) nature of viral RNA genome population: biological relevance – a review. Gene 40: 1–8

13. EPI (1989) Poliomyelitis in 1986, 1987 and 1988. Wkly Epidemiol Rec 64: 273–285

14. Evans DMA, Dunn G, Minor PD, Schild GC, Cann AJ (1985) Increased neurovirulence associated with a single nucleotide change in a noncoding region of the Sabin type 3 poliovaccine genome. Nature 314: 548–550

15. Ferguson M, Qi Y-H, Minor PD, Magrath DI, Spitz M, Schild GC (1982) Monoclonal antibodies specific for the Sabin vaccine strains of poliovirus 3. Lancet ii: 122–124

16. Hirst G (1982) Genetic recombination with Newcastle disease virus, poliovirus and influenza. Cold Spring Harbor Symp Quant Biol 27: 303–309

17. Hogle JM, Chow M, Filman DJ (1985) Three dimensional structure of poliovirus at 2.9 A resolution. Science 229: 1358–1365

18. Holland JJ, Spindler K, Horodyski F, Grabau E, Nichol S, Vandepol S (1982) Rapid evolution of RNA genomes. Science 215: 1577–1585

19. Hovi T (1991) Remaining problems before eradication of poliomyelitis can be accomplished. Prog Med Virol 38: 69–95

20. Hughes PJ, North C, Minor PD, Stanway G (1989) The nucleotide sequence of coxsackievirus A21. J Gen Virol 70: 2943–2952

21. Kawamura N, Kohara M, Abe S, Komatsu T, Tago K, Arita M, Nomoto A (1989) Determinants in the 5′ non-coding region of poliovirus Sabin 1 RNA that influence the attenuated phenotype. J Virol 63: 1302–1309

22. Kew O, De L Yang, CF, Nottay B, Pallansch M (1992) The role of virologic surveillance in the global initiative to eradicate poliomyelitis. In: Kurstak E (ed. inchief), Applied Virology Research, Vol. 2. New-York, pp. 215–246

23. Kew OM, Nottay BK (1984) Evolution of oral poliovirus vaccine strain in humans occurs by both mutation and intermolecular recombination. In: R Chanock RL (ed) Modern approach to vaccines. New-York, pp 357–362

24. Kew OM, Nottay BK, Rico-Hesse RR, Pallansch MA (1990) Molecular epidemiology of wild poliovirus transmission. Applied Virol Res 2: 199–221

25. Kinnunen L, Huovilainen A, Pöyry T, Hovi T (1990) Rapid molecular evolution of wild type 3 poliovirus during infection in individual hosts. J Gen Virol 71: 317–324

26. Kitamura N, Semler B, Rothberg PG, Larsen GR, C J Adler, A J Dorner, Emini EA, Hanecak R, Lee JJ, van der Werf S, Anderson CW, Wimmer E (1981) Primary structure, gene organization and polypeptide expression of poliovirus RNA. Nature 291: 547–553

27. Kohara M, Omata T, Kameda A, Semler BL, Itoh H, Wimmer E, Nomoto A (1985) In vitro phenotypic markers of a poliovirus recombinant constructed from infectious cDNA clones of the neurovirulent Mahoney strain and the attenuated Sabin 1 strain. J Virol 53: 786–792

28. Koike G, Taya C, Kurata T, Abe S, Ise I, Yonekawa H, Nomoto A (1991) Transgenic mice susceptible to poliovirus. Proc Natl Acad Sci USA 88: 951–955
29. Lai MMC (1992) RNA recombination in animal and plant viruses. Microbiol Rev 56: 61–79
30. Lipskaia GY, Muzychenko AR, Kutitova OK, Maslova SV, Equestre M, Drozdov SG, Perez-Bercoff R, Agol VI (1991) Frequent isolation of intertypic poliovirus recombinants with serotype 2 specificity from vaccine-associated polio cases. J Med Virol 35: 290–296
31. Macadam AJ, Pollard SR, Ferguson G, Dunn G, Skuce R, Almond JW, Minor PD (1991) The 5' noncoding region of the type 2 poliovirus vaccine strain contains determinants of attenuation and temperature sensitivity. Virology 181: 451–458
32. Melnick JL (1990) Enteroviruses: polioviruses, coxsackieviruses, echoviruses and newer enteroviruses. In: Fields BN, Knipe DM, Chanock RM, Melnick J, Roizman B, Shope RE (eds) Virology. New-York, pp 739–794
33. Minor PD, Dunn G, Evans DMA, Margrath DI, John A, Howlett J, Phillips D, Westrop G, Warcham K, Almond JW, Hogle JM (1989) The temperature sensitivity of the Sabin type 3 vaccine strain of poliovirus: molecular and structural effects of a mutation in the capsid protein VP3. J Gen Virol 70: 117–1125
34. Minor PD, John A, Ferguson M, Icenogle JP (1986) Antigenic and molecular evolution of the vaccine strain of type 3 poliovirus during the period of excretion by primary vaccine. J Gen Virol 67: 693
35. Nakano JH, Hatch MH, Thieme ML, Nottay B (1978) Parameters for differentiating vaccine-derived and wild poliovirus strains. Prog med Virol 24: 178–206
36. Nkowane BM, Wassilak SGF, Orenstein WA, Bart KJ, Schonberger LB, Hinman AR, Kew OM (1987) Vaccine-associated paralytic poliomyelitis: United States: 1973 through 1984. Jama 257: 1335–1340
37. Nomoto A, Omatoa T, Toyoda H, Kuge S, Horie H, Kataoka Y, Genba Y, Nakano Y, Imura N (1982) Complete nucleotide sequence of the attenuated poliovirus Sabin 1 strain genome. Proc Natl Acad Sci USA 79: 5793–5797
38. Nottay BK, Kew OM, Hatch MH, Heyward JT, Obijeski JF (1981) Molecular variation of type 1 vaccine-related and wild polioviruses during replication in humans. Virology 108: 405–423
39. Omata T, Kohara M, Kuge S (1986) Genetic analysis of the attenuation phenotype of poliovirus type 1. J Gen Virol 58: 348–358
40. Osterhaus ADME, Van Wezel A, Uztdehaag FGCM, Hazendonkk TG, Van Asten JAAM, Van Steenis B (1981) Monoclonal antibodies to poliovirus: production of specific monoclonal antibodies to the Sabin vaccine strains. Intervirology 16: 218–244
40a. Otelea D, Guillot S, Furione M, Aubert-Combiescu A, Balanant J, Candrea A, and Crainic R (1992) Genomic modifications in naturally occurring neurovirulent revertants of Sabin 1 polioviruses. Develop Biol Standard, 78: 33–38.
41. Palmenberg AC (1989) Sequence alignments of picornaviral capsid proteins. In: Semler BL, Ehrenfeld E (ed) Washington DC, pp 211–241
42. Racaniello VR (1988) Poliovirus neurovirulence. Adv Virus Res 34: 217–246
43. Rico-Hesse R, Pallansch MA, Nottay BK, Kew OM (1987) Geographic distribution of wild poliovirus type 1 genotypes. Virology 160: 311–322
44. Rossamann MG, Arnold E, Erickson JW, Frankenberger EA, Griffith JP, Hecht HJ, Johnson JE, Kamer G, Luo M, Mosser AG, Rueckert RR, Sherry B, Vriend G (1985) Structure of a human common cold virus and functional relationship to other picornaviruses. Nature 317: 145–153
45. Rueckert RR (1985) Picornaviruses and their replication. In: Fields BN (ed) Virology. New-York, pp 705–738
46. Ryan MD, Jenkins O, Hughes PJ, Brown A, Knowles NJ, Booth D, Minor PD, Almond JW (1990) The complete nucleotide sequence of enterovirus type 70: relationships with other Picornaviridae. J Gen Virol 71: 2291–2299
47. Sabin AB (1985) Oral poliovirus vaccine: History of its development and use and current challenge to eliminate poliomyelitis from the world. J Infect Dis 151: 420–436

48. Sabin AB, Boulger LR (1973) History of Sabin attenuated poliovirus oral live vaccine strains. J Biol Standard 1: 115–118

49. Toyoda H, Kohara M, Kataoka Y, Suganuma T, Omata T, Imura N, Nomoto A (1984) Complete nucleotide sequences for all three poliovirus serotype genomes: implication for genetic relationship, gene function and antigenic determinants. J Mol Biol 174: 561–585

50. Van Wezel AL, Hazendonk AG (1979) Intratypic serodifferentiation of poliomyelitis strains by strain-specific antisera. Intervirology 11: 2–8

51. Villaverde A, Martinez-Salas E, Domingo E (1988) 3D gene of foot-and-mouth disease virus conservation by covergence of average sequences. Gene 23: 185–194

52. Ward CD, Stoekes MA, Flanegan JB (1988) Direct measurement of the poliovirus RNA polymerase error frequency in vitro. J Virol 62: 558–562

53. WHO (1981) Markers of poliovirus strains isolated from cases temporally associated with the use of live poliovirus vaccine. J Biol Standard 9: 163–184

54. WHO (1990) Expanded programme on immunization. Progress towards the eradicating poliomyelitis from the Americas. Wkly Epidem Rec 65: 361–364

55. WHO (1992) Expanded programme on immunization. Global Advisory Group. Wkly Epidem Rec 67: 9–31

56. WHO (1992) Expanded Programme on Immunization. Poliomyelitis in 1988, 1989 and 1990. Wkly Epidem Rec 67: 113–120

57. Wright PF, Kim-Farley RJ, de Quadros GA, Robertson SE, McN Scott R, Ward NA, Henderson RH (1991) Strategies for the global eradication of poliomyelitis by the year 2000. N Engl J Med 325: 1774–1779

58. Yang CF, De L, Holloway BP, Pallansch MA, Kew OM (1991) Detection and identification of vaccine-related polioviruses by polymerase chain reaction. Virus Res. 20: 159–179

31

Stability of attenuated poliovirus strain genotypes in the human intestinal tract

P. Minor[1], **G. Dunn**[1], **J. W. Almond**[2], and **A. Macadam**[1]

[1] National Institute for Biological Standards and Control, Blanche Lane, South Mimms, Potters Bar, Herts, United Kingdom
[2] University of Reading, Whitenights, United Kingdom

Summary

The major part of the attenuation of the live poliovirus vaccine strains of Sabin is attributable to differences from virulent strains at two or three bases. The mutations revert or are suppressed by second site mutations in vaccine recipients with varying degrees of rapidity and to varying extents, employing both point mutation and recombination. Despite the high degree of genetic variability of the virus, vaccine associated paralysis is very rare.

Introduction

During the 20th century poliomyelitis occurred in epidemics in developed countries whose effect led to intensive and successful efforts to develop vaccines, firstly the formaldehyde inactivated virus preparations of Salk [20], then the live attenuated vaccine strains of Sabin [19]. The effect of the vaccines, where they have been used appropriately, has been dramatic, in both developed and developing countries. In the United Kingdom the average number of cases was about 4000 per year during the 1950s, whereas it is now of the order of 2 or 3. The continued occurrence of poliomyelitis in other parts of the world makes vaccination still necessary, but WHO has declared its goal of eliminating the disease from the world by the year 2000.

The efficacy of the vaccines has a number of implications, one of which is that the live attenuated vaccine strains of Sabin, which were developed using animal models and parental routes of administration are well attenuated for human recipients when given orally. The work to be described in this review concerns attempts to define this attenuation in molecular terms, and to examine the effect of replication in the recipient on the phenotype and genotype of the virus.

Properties of poliovirus

Polioviruses are members of the enterovirus genus of the picornaviridae. They occur in three distinct serotypes, designated type 1, type 2, and type 3 such that infection with one serotype is not believed to confer significant protection against the others. The virus particle is about 27 nm by negatively stained electron microscopy, and consists of sixty copies each of the virion proteins VP1, VP2, VP3 and VP4 arranged with icosahedral symmetry about a single strand of infectious positive sense RNA.

The genomic RNA is about 7500 nucleotides in length terminating at the 3′ end in a polyadenylate tail of 70–100 residues, and at the 5′ end in a small protein, VPG which is covalently bound to the RNA. The functional layout of the genome is shown in Fig. 1. A 5′ non coding region of about 740 residues precedes a single large open reading frame in which the structural proteins are encoded in the 5′ portion. The remainder of the proteins are non structural and involved in RNA synthesis or processing of proteins. The genome concludes with a short 3′ non coding region of 70 bases.

Molecular basis of the attenuation and eversion of the Sabin vaccine strains of poliovirus

A feature of the use of live attenuated poliovirus vaccines is that although the incidence of disease can be reduced to very low levels, there remains a low number of cases which does not appear to fall further. While it was suspected that at least some of the residual cases were due to the live vaccine itself [1] it was technically difficult to prove because of the general similarity of virus strains of the same serotype both antigenically and biologically which made strain differentiation difficult. It was not until the use of molecular approaches that it was conclusively shown that the isolates obtained from some cases were derived from the vaccine strains [10,18]. Three doses of vaccine are given in a course of immunisation, and the incidence of vaccine associated cases of poliomyelitis is about one per 530 000 for first time vaccinees and about one per 2×10^6 in vaccinees overall [14]. The incidence of vaccine associated disease is therefore low.

The Sabin vaccine strains of poliovirus were obtained by passage of wild type isolates under a variety of conditions, which differed for each of the serotypes but gave rise to viruses of very low virulence and high phenotypic

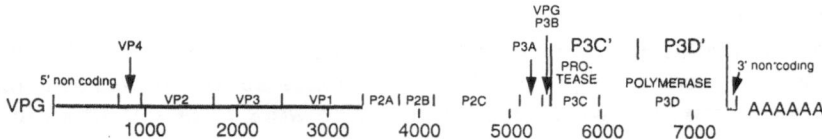

Fig. 1. Organisation of the genome of poliovirus. The protease (3C) and polymerase (P3D) can undergo an alternative cleavage within 3D mediated by P2A to give P3C^l and PD^l

stability when tested in animals [19]. For both the type 1 and type 3 strains the initial virus was highly virulent while the progenitor type 2 strain was isolated from a healthy child.

It was shown early on that full length cDNA copies of the poliovirus genome were infectious [17] and that RNA produced from such cloned copies under the control of a suitable promoter was of similar infectivity to the genomic RNA itself [27]. It is therefore in principle possible to define differences affecting virulence by producing viruses which are recombinants between closely related virulent and avirulent strains. The strategy has been to identify regions and mutations in the vaccine strains which will attenuate the virulent virus, and then to remove the mutations identified from the vaccine strain by site directed mutagenesis, to generate a virus of comparable virulence to the parental or revertant strain. This strategy was first taken to its conclusion for the type 3 Sabin strain, which is designated P3/Leon 12a, b and its virulent precursor, originally isolated from a fatal case of poliomyelitis in 1937 and designated P3/Leon/USA/1937 [29].

The Sabin type 3 vaccine strain and P3/Leon/USA/1937 differ by ten base differences, summarised in Fig. 2 [22]. By use of convenient restriction sites and site-directed mutagenesis a number of recombinant viruses were produced and tested in accordance with the neurovirulence test procedures used for poliovaccines [31] except that small numbers of animals were used.

Two base differences between Leon and the Sabin type 3 vaccine strain contributed most to differences in their neurovirulence. They were at base 2034, which produced a change in residue 91 of the capsid protein VP3 from a serine in the virulent virus to a phenylalanine in the vaccine strain, and in the 5' non coding region at base 472. Subsequently a third mutation which is believed to have an attenuating effect was identified by other workers at residue 2493, leading to an amino acid change from isoleucine to threonine at residue 6 of capsid protein VP1 [26,28]. This difference was not observed in the original clones probably because the virus from which they were derived had been plaque purified and passaged twice in HEp2C cells, and it has been shown that the mutation is rapidly lost on passage.

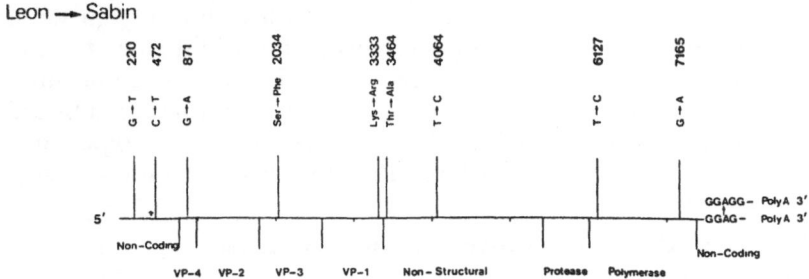

Fig. 2. Comparison of the sequences of cDNA clones of the type 3 vaccine strain P3Leon 12a, b and its virulent precursor strain P3Leon/USA/1937, vertical lines indicate locations of point mutational differences. The base or amino acid found in P3Leon 12a, b is indicated at the tip of the arrows

The type 2 vaccine strain was derived from an isolate from a healthy child. The virus was passaged in a chimpanzee and then found suitable for use as regards stability and attenuation. While the progenitor strain from the child is not generally available, and in any case is probably attenuated, the type 2 vaccine strain is capable of reversion and causing vaccine associated poliomyelitis, especially in contacts of vaccinees. Virus isolated from one such case is designated P2/117 and differs from the type 2 vaccine strain at 23 bases [16]. The generation of recombinant viruses has identified bases 481 and 2903 as significant mutations. The difference at 2903 results in an amino acid change from isoleucine valine at residue 143 of capsid protein VP1. A possible third mutation in or just before the sequence encoding VP4 may also have a slight effect. (Ren et al 191, Macadam et al, in preparation).

Work by Nomoto and co-workers suggested that the attenuation of the type 1 strain was more complex. A major attenuating mutation in the 5' non coding region at residue 480 was identified by generation of recombinants between the Sabin 1 strain and its virulent precursor Mahoney. However many other mutations with an attenuating effect were scattered throughout the rest of the genome [15]. This has been proposed as an explanation for the higher degree of safety of the type 1 Sabin vaccine strain of poliovirus compared to the type 2 and type 3 strains. However Christodoulou and co-workers reported that the Sabin type 1 strain could apparently revert to virulence by the introduction of only a few mutations. Vaccine virus was passaged at successively higher temperatures and became more virulent. On sequencing the isolates it was found that almost total reversion to the virulence of Mahoney was accomplished by two or three mutations, one at residue 525 in the 5' non coding region, one at base 6203 which introduces a change from histidine to tyrosine in residue 73 of the polymerase protein 3D, and one at the extreme 3' end of the genome [3]. It is therefore possible that the attenuated phenotype of the type 1 Sabin vaccine strain of poliovirus is attributable to as few mutations as that of type 2 and type 3, although this awaits confirmation by the construction of a Sabin type 1 vaccine strain in which bases 480 and 6203 are reverted. It is striking that all three live attenuated poliovirus vaccine strains prepared by Sabin by different passage routes have attenuating mutations in the 5' non coding region within the same short sequence; this presumable reflects the consistency of the criteria by which suitable vaccine strains were chosen. It is believed that mutations in this region alter the efficiency of initiation of protein translation by affecting RNA secondary structure [9,21,23,24,25]. The mutation in type 3 in capsid protein VP3 renders virus growth sensitive to elevated temperatures [13].

In vivo growth of the Sabin vaccine strains of poliovirus

Mutations known to attenuate the vaccine strains for primates can be shown to be reverted or suppressed in isolates made from vaccine associated cases of poliomyelitis, consistent with their having an attenuating effect in humans.

However early work also indicated some degree of change in isolates obtained from healthy vaccinees. In particular isolates of type 3 tended to lose their temperature sensitive growth phenotype and increase somewhat in neuro-virulence, and isolates of type 1 changed slightly in their antigenic properties [30]. It was therefore of interest to study the virus excreted by healthy vaccinees following routine immunisation, using molecular biological methods. One child was studied in particular detail [11] although the results obtained have now been confirmed in at least forty.

Immunisation was with trivalent vaccine containing each of the three Sabin vaccine strains given orally, and the first dose was given at 4 months of age. Stool samples were collected and attempts to isolate virus made; type 1 virus was isolated for only a few days, while type 2 and type 3 virus could be isolated for 73 days. Poliovirus excretion appears to have finally ceased when there was evidence of infection with an adenovirus. The duration of the period of excretion of virus was rather longer than expected; while 1% of vaccinees are reported to excrete virus for 10 weeks as in this case, 50% no longer excrete virus after five or six weeks [11].

The nature of the type 3 virus excreted was surprising, however, as virus from the fourth specimen, collected 47 hours post immunisation, had lost the attenuating mutation at base 472, and could be shown to be of increased although not wild type neurovirulence [5]. It was later shown that eleven days after immunisation, the virus excreted was no longer temperature sensitive in its growth properties. This was effected by a second site mutation in capsid protein VP2 at residue 18, a position involved in the interactions

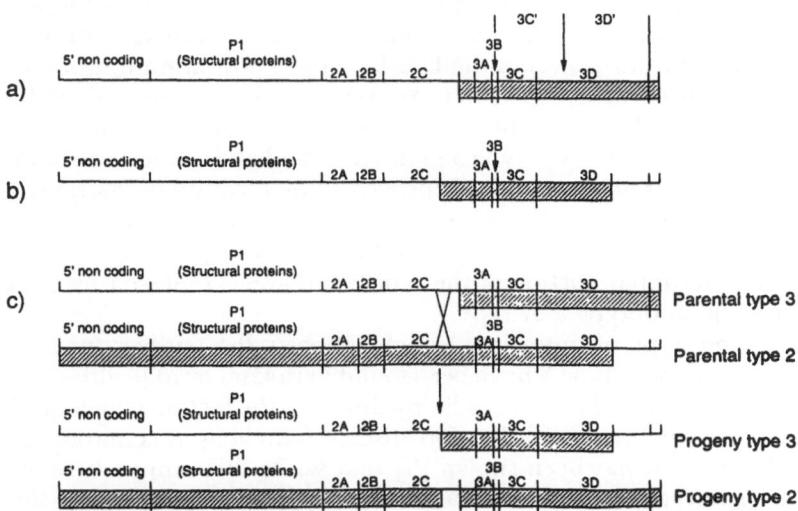

Fig. 3. Structure of recombinants found in vaccinee. **a** recombinant type 3 virus (A) excreted from day 11 to day 42; **b** recombinant type 3 virus (B) excreted from day 42 to day 73; **c** Possible mechanism for generating recombinant B. Type 2 and type 3 sequences are shown in different intensities

between pentamers in the virus structure [6,8]. However, more dramatic genomic rearrangements also occurred; oligonucleotide mapping and sequence analysis showed that the isolate made at day eleven was a recombinant in which the structural proteins and some of the non structural proteins derived from type 3 while the non structural proteins from the middle of P2C to the extreme 3' end derived from the type 2 vaccine strain as shown in figure 3a. Virus with this genomic structure was excreted until day 42, when a second recombination event occurred, resulting in the regaining of a portion of the 3' end of the type 3 genome (Fig. 3b), so giving a virus in which the central portion of the region of the genome coding for the non structural proteins derived from type 2, while the remainder derived from type 3. The 5' recombination site of the second recombinant was nearer the 5' end of the genome than the first, suggesting that it may have arisen by recombination between the first type 3 recombinant and a type 2–type 3 recombinant, as illustrated diagrammatically in figure 3c. No evidence for a suitable partner type 2–type 3 recombinant could be found in the type 2 isolates made from the child or as yet from other healthy vaccinees [24] but such recombinant type 2 strains have been identified in isolates from vaccine associated cases [7] (A Macadam, unpublished). The second recombination event was also associated with the loss of the 3' terminal guanosine residue preceding the polyadenylate tract [2]. In addition to these changes, point mutations in antigenic sites defined by monoclonal antibodies and sequencing also occurred in the recombinant strains.

The extensive rearrangement and mutation of the genome is surprising in view of the proven high degree of safety of the vaccines, but studies in other vaccinees have shown that a very similar course of events is followed. In particular with one exception the 5' non coding mutation at 472 is lost in all isolates studied from babies in the UK by 5 days post immunisation, [4,13]. The temperature sensitive growth phenotype which is suppressed by the same mutations as the attenuated phenotype, is lost by 11 days post immunisation [9]. Moreover all isolates of type 3 poliovirus made from vaccinees later than eleven days after vaccination have proved to be recombinant strains, either between type 3 and type 1 or type 3 and type 2. Where a type 3–type 2 recombinant is isolated a complex recombinant is generated at a later stage, when the 3' terminal portion of the type 2 genome is replaced by the corresponding region of type 3 or type 1.

The same type of mutations which suppress the temperature sensitive growth phenotype of type 3 have been found in isolates from healthy vaccinees and vaccine associated cases of poliomyelitis and the factors which determine whether an individual will develop disease following vaccination are not clear. However, it has been shown [9] that while the temperature sensitive growth phenotype and the attenuated phenotype are suppressed by the same mutations, recombinant strains of virus from either healthy vaccinees or vaccine associated cases are of slightly lower virulence than equivalent non recombinant strains. The basis of this is not obvious, in that no mutation attenuating the type 2 srain has been identified in the region of the type 2

Table 1. Stability of attenuating mutations of the Sabin type 3 vaccine strain in vaccinees

Vaccinee	Loss of		
	472	2493	ts phenotype
DM	2 days	2 days	11 days
EM	3 days	4 days	11 days
AM	5 days	<2 days	12 days
CM	6 days	<2 days	12 days

genome incorporated into the recombinants. However, the revertant type 2 strains studied tend to be less virulent in animals than the type 3 strain, and this may be reflected in the attenuating effects of the exchanged segment. It is likely that the generation of the recombinants which is on the face of it undesirable, may contribute to the safety of the vaccine. It would be predicted that vaccination with monovalent rather than trivalent vaccine should be associated with a higher incidence of vaccine associated poliomyelitis.

The selective pressures leading to the emergence of the recombinant strains are not known although it would seem that some protein or combination of proteins encoded by the type 3 genome to the 3' side of protein P2C is selected against in the gut, and that the 3' terminal portion of the type 2 genome is also selected against in the gut.

Finally the mutation identified by Weeks-Levy et al in VP1 is apparently also lost rapidly, being undetectable in four vaccinees after four days (A Macadam, unpublished). The duration of attenuating mutations in excreted type 3 virus is summarised in Table 1 for four vaccines.

The type 1 and type 2 strains are also subject to selection and rapid modification in vaccinees. While no recombinant type 1 or type 2 strains have yet been identified as the principal strain in healthy vaccinees, they have been found in isolates of type 2 virus from vaccine associated cases [7] (A Macadam, personal communication). The attenuating mutation in the type 2 strain at residue 143 of VP1 is lost in a significant proportion of isolates from healthy vaccinees, and in all isolates from vaccine associated cases. The 5' non coding region mutation in type 2 is lost slightly later than that in type 3, at about seven days post immunisation, but in all vaccinees, while that in type 1 is lost at about seven days post immunisation in only half of vaccinees [4].

Safety of live attenuated polioviruses vaccines

The live attenuated poliovirus vaccines derived by Sabin are among the safest and most effective vaccines in use. In rare instances, estimated at one per 530 000 primary vaccinees, live poliovaccines are implicated in poliomyelitis. Their effect in eliminating polio from countries where they have been used, including developing countries in South America, clearly demonstrates their safety in general use. The high degree of molecular variation detected in vaccinees is therefore surprising but does not affect the epidemiological observation of safety and efficacy. However, most infections with wild type poliovirus are asymptomatic and the ease with which vaccine strains alter in the human gut implies that the wild virus occupies a well defined ecological niche, to which it is highly adapted.

References

1. Assaad F, Cockburn WC (1982) The relation between acute persisting spinal paralysis and poliomyelitis vaccine – results of a ten-year enquiry. Bull WHO 60: 231–242
2. Cammack N, Phillips A, Dunn G, Patel V, Minor PD (1989) Intertypic genomic rearrangements of poliovirus strains in vaccinees. Virology 167: 507–514
3. Christodoulou C, Colbere-Garapin F, Macadam A, Taffs LF, Marsden S, Minor PD, Horaud F (1990) Mapping of mutations associated with monkey neurovirulence of Sabin 1 poliovirus revertants selected at high temperature. J Virol 64: 4922–4929
4. Dunn G, Begg NT, Cammack N, Minor PD (1990) Virus excretion and mutation by infants following primary vaccination with live oral poliovaccine from two sources. J Med Virol 32: 92–95
5. Evans DMA, Dunn G, Minor PD, Schild GC, Cann AJ, Stanway G, Almond JW, Currey K, Maizel JV (1985) A single nucleotide change in the 5' noncoding region of the genome of the Sabin type 3 poliovaccine is associated with increased neurovirulence. Nature 314: 548–550
6. Filman DJ, Syed R, Chow M, Minor PD, Macadam AJ, Hogle J (1989) Structural factors that control conformational transitions and serotype specificity in type 3 poliovirus. EMBO Journal 8: 1567–1579
7. Lipskaya GY, Muzychenko AR, Kutitova OK, Maslova SV, Equestre M, Drozdov SG, Perez-Besoff R, Agol VI (1991) J Med Virol 35: 290–296
8. Macadam AJ, Arnold C, Howlett J, John A, Marsden S, Taffs F, Reeve P, Hamada N, Wareham K, Almond J, Cammack N, Minor PD (1989) Reversion of attenuated and temperature sensitive phenotypes of the Sabin type 3 strain of poliovirus in vaccinees. Virology 174: 408–414
9. Macadam AJ, Ferguson G, Burlison J, Stone D, Skuce R, Almond JW, Minor PD (1993) Correlation of RNA structure and attenuation of Sabin vaccine strains of poliovirus in tissue culture. Virology (in press)
10. Minor PD, Schild GC (1981) Identification of the origin of poliovirus isolates. Report of World Health Organization informal meetings 1978-1980. Lancet i: 968–969
11. Minor PD, John A, Ferguson M, Icenogle JP (1986) Antigenic and molecular evolution of the vaccine strain of type 3 poliovirus during the period of excretion by a primary vaccinee. J Gen Virol 67: 693–706
12. Minor PD, Dunn G (1988) The effect of sequences in the 5' non coding region on the replication of polioviruses in the human gut. J Gen Virol 69: 1091–1096
13. Minor PD, Dunn G, Evans DMA, Magrath DI, John A, Howlett J, Phillips A, Westrop G, Wareham K, Almond JW, Hogle J (1989) The temperature sensitivity of the Sabin type 3 vaccine strain of poliovirus: molecular and structural effects of a mutation in capsid protein VP3. J Gen Virol 70: 1117–1123
14. Nkowane BM, Wassilak SG, Oversteen WA, Bart KJ, Schonberger LB, Hinman AR, Kew OM (1987) Vaccine associated paralytic poliomyelitis United States. 1973 through 1984. J Am Med Ass 257: 1335–1340
15. Omata T, Kohara M, Kuge S, Komatsu T, Abe S, Semler BC, Kameda A, Itoh H, Arita M, Wimmer E, Nomoto A (1986) Genetic analysis of the attenuation phenotype of poliovirus type 1. J Virol 58: 348–358

16. Pollard SR, Dunn G, Cammack N, Minor PD, Almond JW (1989) Nucleotide sequence of a neurovirulent variant of the type 2 oral poliovirus vaccine. J Virol 63: 4949–4951
17. Racaniello VR, Baltimore D (1981) Cloned poliovirus complementary DNA is infectious in mammalian cells. Science 214: 916–919
18. Ricoh-Hesse R, Pallansch MA, Nottay BK, Kew OM (1987) Geographic distribution of wild poliovirus type 1 genotypes. Virology 160: 311–322
19. Sabin AB, Boulger LR (1973) History of Sabin attenuated poliovirus oral live vaccine strains. J Biol Stand 1: 115–118
20. Salk JE (1960) Persistence of immunitye after administration of formalin treated poliovirus vaccine. Lancet ii: 715–723
21. Skinner MA, Racaniello VR, Dunn G, Cooper J, Minor PD, Almond JW (1989) A new model for the secondary of the 5' non coding RNA of poliovirus is supported by biochemical and genetical data which also show that RNA secondary structure is important in neurovirulence. J Mol Biol 207: 379–392
22. Stanway G, Hughes PJ, Mountford RC, Reeve P, Minor PD, Schild GC, Almond JW (1984) Comparison of the complete nucleotide sequence of the genomes of the neurovirulent poliovirus P3/Leon/37 ant its attenuated Sabin vaccine derivative P3/Leon/12a,b. Proc Natl Acad Sci USA 81: 1539–1543
23. Svitkin YV, Maslova TV, Agol VI (1985) The genomes of attenuated and virulence poliovirus strains differ in their in vitro translation efficiencies. Virology 147: 243–252
24. Svitkin YV, Pestova TV, Maslova SV, Agol VI (1988) Point mutations modify the response of poliovirus RNA to a translation on initiation factor: a comparison of neuro-virulent and attenuated strains. Virology 166: 394–404
25. Svitkin Y, Cammack N, Minor PD, Almond JW (1990) Translation deficiency of the Sabin type 3 poliovirus genome: associations with an attenuating mutation C472-V Virology 175: 103–109
26. Tatem JM, Weeks-Levy C, Georgia A, DiMichele SJ, Gorgaez EJ, Racaniello VR, Cano FR, Mento SJ (1992) A mutation present in the amino terminus of Sabin 3 polio-virus VP1 proteins is attenuating. J Virol 66: 3194–3197
27. van der Werf S, Bradley J, Wimmer E, Studier FW, Dunn JJ (1986) Synthesis of infectious poliovirus RNA by purified T7 polymerase. Proc Natl Acad Sci USA 83: 2330–2334
28. Weeks-Levy C, Tatem JM, DiMichele SJ, Waterfield W, Georgia AF, Mento SJ (1991) Idenification and characterization of a new base substitution in the vaccine strain of Sabin 3 poliovirus. Virology 185: 934–937
29. Westrop GD, Evans DMA, Dunn G, Minor PD, Magrath DI, Taffs F, Marsden S, Wareham KA, Skinner M, Schild GC, Almond JW (1989) Genetic basis of attenuation of the Sabin type 3 oral poliovaccine. J Virol 63: 1338–1344
30. WHO (1969) Evidence on the safety and efficacy of live poliomyelitis vaccines currently in use, with special reference to type 3 poliovirus. Bull WHO 40: 925–945
31. WHO (1983) Requirements for poliomyelitis vaccine (oral) (Revised 1982). Technical Report Series 687: 107–174; World Health Organisation, Geneva

Subject Index